Waste Treatment and Disposal

Waste Treatment and Disposal

Second Edition

PAUL T. WILLIAMS

Professor of Environmental Engineering
The University of Leeds, UK

John Wiley & Sons, Ltd

Other Wiley Editorial Offices

John Wiley & Sons Inc., 111 River Street, Hoboken, NJ 07030, USA

Jossey-Bass, 989 Market Street, San Francisco, CA 94103-1741, USA

Wiley-VCH Verlag GmbH, Boschstr. 12, D-69469 Weinheim, Germany

John Wiley & Sons Australia Ltd, 33 Park Road, Milton, Queensland 4064, Australia

John Wiley & Sons (Asia) Pte Ltd, 2 Clementi Loop #02-01, Jin Xing Distripark, Singapore 129809

John Wiley & Sons Canada Ltd, 22 Worcester Road, Etobicoke, Ontario, Canada M9W 1L1

Wiley also publishes its books in a variety of electronic formats. Some content that appears
in print may not be available in electronic books.

Library of Congress Cataloging in Publication Data

Williams, Paul T.
  Waste treatment and disposal / Paul T. Williams. — 2nd ed.
    p. cm.
  ISBN 0-470-84912-6
  1. Refuse and refuse disposal—Great Britain.   2. Refuse and refuse disposal—Government policy—
Great Britain.   I. Title.
  TD789.G7W54 2005
  363.72′8′094—dc22

                            2004016990

British Library Cataloguing in Publication Data

A catalogue record for this book is available from the British Library

ISBN-10 0470-84913-4 (P/B)
ISBN-13 978-0470-84913-2 (P/B)
ISBN-10 0470-84912-6 (H/B)
ISBN-13 978-0470-84912-5 (H/B)

Typeset in 10/12pt Times by Integra Software Services Pvt. Ltd, Pondicherry, India
Printed and bound in Great Britain by Antony Rowe Ltd, Chippenham, Wiltshire
This book is printed on acid-free paper responsibly manufactured from sustainable forestry
in which at least two trees are planted for each one used for paper production.

This book is dedicated with lots of love to Lesley, Christopher, Simon and Nicola

Contents

Preface

This second edition arises from the 1998 first edition (published by John Wiley & Sons, Ltd 1998) which was largely based on the UK. This new book has been substantially revised and rewritten to cover waste treatment and disposal with particular emphasis on Europe. Increasingly in Europe the European Commission legislation has had a major influence on the management of solid waste and hence the need for a European focussed text. The book is aimed at undergraduate and postgraduate students undertaking courses in Environmental Science and Environmental, Civil, Chemical and Energy Engineering, with a component of waste treatment and disposal. It is also aimed at professional people in the waste management industry.

The first chapter is an historical introduction to waste treatment and disposal. The major legislative and regulatory measures emanating from the European Commission dealing with waste treatment and disposal are described.

Chapter 2 discusses the different definitions of waste. Estimates of waste arisings in Europe and the rest of the world are discussed as well as the methods used in their estimation. Various trends in waste generation and influences on them are also discussed. Several categories of waste are discussed in terms of arisings, and treatment and disposal options. The wastes described in detail are: municipal solid waste; hazardous waste; sewage sludge; clinical waste; agricultural waste; industrial and commercial waste. Other wastes described are: construction and demolition waste; mines and quarry waste; end-of-life vehicles and scrap tyres. The chapter ends with a discussion of the different types of waste containers, collection systems and waste transport.

Chapter 3 is concerned with waste reduction, re-use and recycling, with the emphasis on recycling. Municipal solid waste and industrial and commercial waste recycling are discussed in detail. Examples of recycling of particular types of waste, i.e., plastics, glass, paper, metals and tyres are discussed. Economic considerations of recycling are discussed.

Chapter 4 is concerned with waste landfill, the main waste disposal option in many countries throughout Europe. The EC Waste Landfill Directive is covered in detail. Landfill design and engineering, the various considerations for landfill design and operational practice, are described. The different main types of waste which are landfilled, i.e., hazardous, non-hazardous and inert wastes and the processes operating within and outside the landfill are discussed. The major different landfill design types are discussed in detail. The formation of landfill gas, landfill gas migration, management and monitoring of landfill gas are discussed, as is landfill leachate formation and leachate management and treatment. The final stages of landfilling of wastes, i.e., landfill capping, landfill site completion and restoration are described. The recovery of energy through landfill gas utilisation is discussed in detail. The problems of old landfill sites are highlighted.

Chapter 5 is concerned with incineration, the second major option for waste treatment and disposal in Europe. The EC Waste Incineration Directive is described in detail and the various incineration systems are discussed. Concentration is made on mass burn incineration of municipal solid waste, following the process through waste delivery, the bunker and feeding system, the furnace, and heat recovery systems. Emphasis on emissions formation and control is made with discussion of the formation and control of particulate matter, heavy metals, toxic and corrosive gases, products of incomplete combustion, such as polycyclic aromatic hydrocarbons (PAHs), dioxins and furans. The contaminated waste-water and contaminated bottom and flyash arising from waste incineration is discussed. Energy recovery via district heating and electricity generation are described. Other types of incineration including fluidised bed incinerators, starved air incinerators, rotary kiln incinerators, cement kilns, liquid and gaseous waste incinerators and the types of waste incinerated in each different type is discussed.

Chapter 6 discusses other options for waste treatment and disposal. Pyrolysis of waste, the types of product formed during pyrolysis and their utilisation as well as the different pyrolysis technologies, are discussed. Gasification of waste, gasification technologies and utilisation of the product gas, are described. Combined pyrolysis–gasification technologies are discussed. Composting of waste is described, including the composting process and the different types of composter. Anaerobic digestion of waste, the degradation process and operation and technology for anaerobic digestion are discussed. Examples of the different types of pyrolysis, gasification, combined pyrolysis–gasification, composting and anaerobic digestion systems are described throughout.

The concluding chapter discusses the integration of the various waste treatment and disposal options described in the previous chapters to introduce the concept of 'integrated waste management'. The different approaches to integrated waste management are described.

1

Introduction

Summary

This chapter is an historical introduction to waste treatment and disposal. The development of waste management in the European Union through the use of various policy, strategy and legislative measures are discussed. The adoption of sustainable development by the EU through the various Environment Action Programmes is presented. The main EU Directives, Decisions and Regulations in relation to waste management are described. The Waste Strategy of the EU is presented and the policy initiatives related to its implementation are discussed. The economics of waste management across Europe are discussed. The main treatment and disposal routes for wastes in the European Union are briefly described.

1.1 History of Waste Treatment and Disposal

The need for adequate treatment and disposal of waste by man, arose as populations moved away from disperse geographical areas to congregate together in communities. The higher populations of towns and cities resulted in a concentration of generated waste, such that it became a nuisance problem. Waste became such a problem for the citizens of Athens in Greece that, around 500 BC, a law was issued banning the throwing of rubbish into the streets. It was required that the waste be transported by scavengers to an open dump one mile outside of the city. The first records that waste was being burned as a disposal route appear in the early years of the first millennium in Palestine. The Valley of

Waste Treatment and Disposal, Second Edition   Paul T. Williams
© 2005 John Wiley & Sons, Ltd   ISBNs: 0-470-84912-6 (HB); 0-470-84913-4 (PB)

Gehenna outside Jerusalem contained a waste dump site at a place called Sheol where waste was regularly dumped and burned. The site became synonymous with hell.

Throughout the Middle Ages, waste disposal continued to be a nuisance problem for city populations. Waste was often thrown onto the streets causing smells and encouraging vermin and disease. For example, in 1297 a law was passed in England requiring house-holders to keep the front of their houses clear of rubbish. More than a 100 years later, in 1408, Henry IV ruled that waste should be kept inside houses until a 'raker' came to cart away the waste to pits outside the city (Project Integra 2002). In 1400 in Paris, the huge piles of waste outside the city walls began to interfere with the city defences.

In Europe, the industrial revolution between 1750 and 1850 led to a further move of the population from rural areas to the cities and a massive expansion of the population living in towns and cities, with a consequent further increase in the volume of waste aris-ing. The increase in production of domestic waste was matched by increases in industrial waste from the burgeoning new large-scale manufacturing processes. The waste gener-ated contained a range of materials such as broken glass, rusty metal, food residue and human waste. Such waste was dangerous to human health and, in addition, attracted flies, rats and other vermin which, in turn, posed potential threats through the transfer of disease. This led to an increasing awareness of the link between public health and the environment.

To deal with this potential threat to human health, legislation was introduced on a local and national basis in many countries. For example, in the UK, throughout the latter half of the nineteenth century, a series of Nuisance Removal and Disease Prevention Acts were introduced in the UK which empowered local authorities to set up teams of inspectors to deal with offensive trades and to control pollution within city limits. These Acts were reinforced by the Public Health Acts of 1875 and 1936, which covered a range of meas-ures some of which were associated with the management and disposal of waste. The 1875 Act placed a duty on local authorities to arrange for the removal and disposal of waste. The 1936 Act introduced regulation to control the disposal of waste into water and defined the statutory nuisance associated with any trade, business, manufacture or process which might lead to degradation of health or of the neighbourhood (British Medical Association 1991; Reeds 1994; Clapp 1994). In the US, early legislation included the 1795 Law introduced by the Corporation of Georgetown, Washington DC, which prohibited waste disposal on the streets and required individuals to remove waste them-selves or hire private contractors. By 1856, Washington had a city-wide waste collection system supported by taxes. In 1878, the Mayor of Memphis organised the collection of waste from homes and businesses and removal to sites outside the city. By 1915, 50% of all major US cities provided a waste collection system which had risen to 100% by 1930 (Neal and Schubel 1989; McBean et al 1995).

One of the main constituents in domestic dust bins in the late nineteenth century was cinders and ash from coal fires, which represented a useful source of energy. The waste also contained recyclable materials such as old crockery, paper, rags, glass, iron and brass and was often sorted by hand by private contractors or scavengers to remove the useful items. Much household waste would also be burnt in open fires in the living room and kitchen as a 'free fuel' supplement to the use of coal. The combustible content of the waste was recognised as a potential source of cheap energy for the community as a whole and the move away from private waste contractors to municipally organised waste collection, led to an increase in incineration. Purpose-built municipal waste incinerators were introduced in the UK in the late 1870s and, by 1912, there were over 300 waste incinerators in the UK, of which 76 had some form of power generation (Van Santen 1993). One of the first municipal incinerators introduced in the US was in 1885 in Allegheny, Pennsylvania (Neal and Schubel 1989). By 1914, there were about 300 waste incinerators in the US. However, many of the waste incinerators were small-scale, hand-fed plants which were poorly designed and controlled and their operation was not cost-effective.

However, the growth of incineration was secondary to the main route to disposal, which was dumping, either legally or illegally. The ease of waste disposal to land and the move to centralised waste management through town or city authorities meant that this route increasingly became the preferred waste disposal option. Particularly as incineration plants were difficult and expensive to maintain. As these incineration plants reached the end of their operational lifetime they tended to become scrapped in favour of landfill. The waste dumps themselves however, were poorly managed, open tips, infested with vermin and often on fire. The environmental implications of merely dumping the waste in such open sites was recognised, and increasingly waste began to be buried. Burying the waste had the advantages of reducing odours and discouraging rats and other vermin and consequently the sites became less dangerous to health. Through the first half of the 20th century some improvements in landfill sites were seen, with improved site planning and site management. However, this did not apply to all areas and many municipal sites still had the minimum of engineering design and the open dump was still very common. When such sites were full, they were covered with a thin layer of soil and there was minimum regard to the effects of contaminated water leachate or landfill gas emissions from the disused site (McBean et al 1995).

The First and Second World Wars and the inter-war periods saw a rise in waste reclamation and recycling, and waste regulation and the environment became a less important issue. Following the Second World War, waste treatment and disposal was not seen as a priority environmental issue by the general public and legislature, and little was done to regulate the disposal of waste. However, a series of incidents in the late 1960s and 70s, highlighted waste as a potential major source of environmental pollution. A series of toxic chemical waste dumping incidents led to increasing awareness of the importance of waste management and the need for a more stringent legislative control of waste. Amongst the most notorious incidents were the discovery, in 1972, of drums of toxic cyanide waste dumped indiscriminately on a site used as a children's playground near Nuneaton in the UK, the leaking of leachate and toxic vapours into a housing development at the Love Canal site, New York State in 1977, the dumping of 3000 tonnes of arsenic and cyanide waste into a lake in Germany in 1971, and the leak of polychlorinated biphenyls (PCBs) into rice oil in Japan in 1968, the 'Yusho' incident (Box 1.1, British Medical Association 1991).

The massive adverse publicity and public outcry led to pressure for the problem of waste disposal to be more strictly controlled by the legislature. In the UK, as a direct result of the Nuneaton cyanide dumping incident, emergency legislation was introduced in the form of The Deposit of Poisonous Waste Act, 1972. Further legislation on waste treatment and disposal followed in 1974 with the Control of Pollution Act, which controlled waste disposal on land through a new licensing and monitoring system for waste disposal facilities. The late 1980s and 1990s saw further development of waste management legislation in the UK and the increasing influence of European Community legislation. For

Box 1.1

Waste Disposal Incidents which Influenced Waste Management and Legislation

1. Love Canal, Niagara City, New York State, USA: 1977

Love Canal, Niagara City was an unfinished canal excavated for a projected hydro-electricity project. The abandoned site was used as a dump for toxic chemical waste and more than 20 000 tonnes of waste containing over 248 different identified chemicals were deposited in the site between 1930 and 1952. Following the sale of the plot in 1953, a housing estate and school were built on the site. In 1977 foul smelling liquids and sludge seeped into the basements of houses built on the site. The dump was found to be leaking and tests revealed that the air, soil and water around the site were contaminated with a wide range of toxic chemicals, including benzene, toluene, chloroform and trichloroethylene. Several hundred houses were evacuated and the site was declared a Federal Disaster Area. There were also later reports of ill health, low growth rates for children and birth defects amongst the residents. As the actual and projected clean-up costs of the site became known, legislation in the form of the Comprehensive Environmental Response, Compensation and Liabilities Act, 1980, was introduced by Congress. This legislation placed the responsibility and cost of clean-up of contaminated waste sites back to the producers of the waste.

Source: British Medical Association 1991.

2. Cyanide Dumping, Nuneaton, Coventry, Warwickshire, UK: 1972

A series of toxic waste dumping episodes occurred in the early months of 1972. The most serious of which was the dumping of 36 drums of sodium cyanide in a disused brickworks at Nuneaton, on the outskirts of Coventry. The site was in constant use as a play area by local children. The drums were heavily corroded and contained a total of one and a half tonnes of cyanide, enough, police reported, to wipe out millions of people. Over the following weeks and months further incidents of toxic waste dumping were reported extensively in the press. Drums of hazardous waste were found in numerous unauthorised sites including a woodland area and a disused caravan site. The episodes generated outrage in the population, and emergency legislation was rushed through Parliament in a matter of weeks in the form of The Deposit of Poisonous Waste Act, 1972. The new Act introduced penalties of five years imprisonment and unlimited fines for the illegal dumping of waste, in solid or liquid form, which is poisonous, noxious or polluting. The basis of the legislation was the placing of responsibility for the disposal of waste on industry. Further legislation on waste treatment and disposal followed in 1974 with the Control of Pollution Act.

Source: The Times 1972.

example; the 1990 Environmental Protection Act; the 1995 Environment Act; the 1994 Waste Management Licensing Regulations; 1994 Transfrontier Shipment of Waste Regulations; the 1996 Special Waste Regulations; the 2000 Pollution Prevention and Control Regulations and the Landfill Regulations 2002, which all contain measures in direct response to EC Directives.

In the US, in response to the increasing concerns of indiscriminate waste disposal, landmark legislation covering waste disposal was developed with the Resource, Conservation and Recovery Act (RCRA) 1976, which initiated the separation and defining of hazardous and non-hazardous waste and the separate requirements for their disposal. The RCRA was an amendment to the 1965 Solid Waste Disposal Act which was the first Federal statutory measure to improve solid waste disposal activities. However, it was the RCRA which embodied the US approach to waste treatment and disposal, establishing a framework for national programs to achieve environmentally sound management of both hazardous and non-hazardous wastes. The Act has been amended several times since 1976, by such as the Hazardous and Solid Waste Amendments of 1984, the Federal Facilities Compliance Act of 1992 and the Land Disposal Program Flexibility Act of 1996.

1.2   European Union Waste Management Policy

The European Union had its origins in the European Economic Community (EEC) which was established by the Treaty of Rome in 1958. Since then a series of Acts and Treaties, including the Single European Act (1987), the Maastricht Treaty (1993) and the Treaty of Amsterdam (1997) have resulted in the development of the organisation and governance of the European Union (Box 1.2). Included in these Acts and Treaties are the general

---

Box 1.2

European Governance

There are a number of bodies which are involved in the process of implementing, monitoring and further developing the legal system of the European Union. EU law is composed of three interdependent types of legislation. Primary legislation includes the major Treaties and Acts agreed by direct negotiation between the governments of the Member States, for example, the Single European Act (1987), the Maastricht Treaty (1992) and the Treaty of Amsterdam (1997). These agreements are then ratified by the national parliaments of each country. Secondary legislation is based on the Treaties and Acts and takes the form of Directives, Regulations and Decisions. The third type of legislation is Case Law based on judgements from the European Court of Justice.
   There are four institutions that serve to govern the European Union.

1. The European Commission – The European Commission initiates all legislative proposals and ensures their implementation in all Member States. The Commission has a President and nineteen commissioners who are each responsible for one or more policy areas. The European Commission also has the important responsibility of administration of the EU budget. The Commission is divided into 25 Directorates-General which cover specific areas such as sustainable development, natural resources and environment and health.
2. The Council of the European Union – Laws initiated by the European Commission are put before the Council of the European Union, also known as the Council of

Continued on page 6

Continued from page 5

Ministers, for adoption or rejection. The Council is therefore the main legislative body of the EU and is also responsible for major EU policy decisions. The Council is made up of one Minister from each Member State who is empowered to make decisions on behalf of their Government. Each Member State of the EU acts as President of the Council for a period of six months in rotation. The Council of the European Union, which comprises representatives at ministerial level, should not be confused with the European Council which brings together the heads of governments of each Member State.

3. The European Parliament – The European Parliament is made up of directly elected members and, since 1987, has acquired a legislative power via co-decisions with the Council of Ministers. Co-decisions cover a limited number of areas such as research and technology, environment, consumer affairs and education, but in areas such as tax the Parliament may only give an opinion. The European Parliament is involved in the formulation of Directives, Regulations and Decisions (see Box 1.4) by giving its opinion and proposing amendments to proposals brought forward by the European Commission. The European Council of Ministers and the European Commission are democratically accountable to the European Parliament.

4. The European Court of Justice – The European Court of Justice is made up of fifteen judges appointed from the Member States whose responsibility is to ensure that the European Treaties are implemented in accordance with EU law. The judgements of the Court overrule those of national courts.

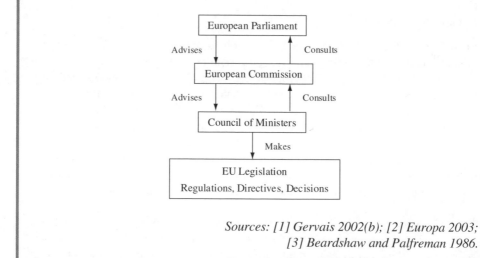

Sources: [1] Gervais 2002(b); [2] Europa 2003; [3] Beardshaw and Palfreman 1986.

objectives of protecting and improving the quality of the environment. Additionally, more detailed policy statements in relation to the environment are included in Environmental Action Programmes. These Action Programmes include EU policy development in relation to waste treatment and disposal. There have been six Environmental Action Programmes since 1973. The approach and strategy in terms of waste in the successive Environmental Action Programmes has been from one of pollution control to pollution prevention and latterly to a sustainable development approach (Gervais 2002(b)). The First Environmental Action Programme (1973–76) regarded waste as a remedial problem requiring control at Community level. The Second (1977–81) and Third (1982 86) Environmental Action Programmes emphasised the need for waste prevention, recycling, re-use and final disposal, via environmentally safe means. The need for action in regard to waste minimisation at the production process through the use of clean technologies was the policy of the Fourth Environmental Action Programme (1987–92). The Fourth Programme also emphasised the hierarchical approach to waste management of the first three Programmes. During the period of the Fourth Environmental Action Programme, a Community Strategy for Waste Management was drawn up by the EU which set out the hierarchical structure of waste management as a long-term strategy for the EU (Gervais 2002(b)). The Fifth (1993–2000) and Sixth (2001–2010) Environmental Action Programmes incorporate into the policies and strategies of the EU, the concepts of 'sustainable development' and the integration of environmental decision-making and policy formulation into all major policy areas of the EU. One of the main objectives of the Sixth Environment Action Programme focuses on the sustainable management of natural resources and waste. The Programme identifies the reduction of waste as a specific objective and sets a target of reducing the quantity of waste going to final disposal by 20% by 2010 and by 50% by 2050. The actions required to achieve these targets include:

- the development of a strategy for the sustainable management of natural resources by laying down priorities and reducing consumption;
- the taxation of natural resource use;
- establishing a strategy for the recycling of waste;
- the improvement of existing waste management schemes;
- investment into waste prevention and integration of waste prevention into other EU policies and strategies.

The concept of 'sustainable development' has developed from the 1992 United Nations Rio Conference on Environment and Development, through to the Johannesburg World Summit on Sustainable Development (2002). The concept requires that society takes decisions with proper regard to their environmental impacts. The concept tries to strike a balance between two objectives, the continued economic development and achievement of higher standards of living both for today's society and for future generations, but also to protect and enhance the environment. The economic development of society clearly has an impact on the environment since natural resources are used and by-product pollution and waste are produced in many processes. However, *sustainable* development promotes development by encouraging environmentally friendly economic activity and by discouraging environmentally damaging activities. Such activities include energy efficiency measures, improved technology and techniques of management, better product design and marketing, environmentally friendly farming practices, making better use of land and buildings and improved transport efficiency and waste minimisation (Sustainable Development 1994; This Common Inheritance 1996).

The Gothenburg European Council of 2001 resulted in the European Union Heads of Government adopting a Sustainable Development Strategy. The strategy is based on the principle that the economic, social and environmental effects of all policies should be examined in a co-ordinated way and taken into account in decision-making (Sustainable

Development 2001). This includes the proposal that all major policy proposals should include a sustainability impact assessment.

The treatment and disposal of waste is one of the central themes of sustainable development. The approach of the European Union and its member states for the management of waste has developed via a series of Directives and Programmes into a strategy concerning the treatment of waste which has the key objectives of minimising the amount of waste that is produced and to minimise any risk of pollution of the environment.

1.3   Waste Strategy of the European Union

1.3.1   Community Strategy for Waste Management

The waste management policy of the European Union set out in the various Environment Action Programmes is implemented through the Waste Management Strategy and subsequent legislative measures such as Directives, Regulations and Decisions of the European Union on specific waste management issues.

The initial EU strategy document on waste, the Community Strategy for Waste Management (SEC (89) 934 Final), was drawn up in 1989 as part of the Fourth Environmental Action Programme (1987–92). It was presented as a 'communication' to the European Commission and to the Council of the European Parliament. The Strategy set out the principles of the hierarchy of waste management through the prevention of waste by clean and improved technologies, the re-use and recycling of waste, and optimisation of the final disposal. The proximity principle, whereby waste should be dealt with as near as possible to its source and also the goal of self-sufficiency in waste treatment and disposal, were emphasised. In 1996 the European Environment Ministers adopted by Resolution a revised Waste Strategy (COM (96) 399 Final 1996). This was a review of the 1989 document and re-emphasised the need for sustainable approaches to waste management in the EU with a high level of environmental protection. Waste prevention was therefore seen as the priority and waste recovery via re-use, recycling, composting and energy from waste were hierarchical objectives. The difficulties in harmonisation of the various terms used in defining 'wastes' across the EU and, consequently, implementation of EU-wide legislation was recognised as a problem. Therefore the strategy called for a review of the waste definitions and catalogue of wastes. The measure of the successful implementation of waste legislation, with the aim of moving waste treatment processes up the hierarchy of waste management, depends on accurate and reliable statistical data. The common use of waste terminology and the reliable collection of accurate data via the European Environment Agency, was therefore stressed. The use of quantitative targets to reduce waste production and to increase re-use recycling and the recovery of waste, were recommended for the Member States. The Strategy also called for the need for specific emission standards in the area of waste incineration and the control of waste landfills. Specific Directives were subsequently implemented via the 1999 Landfill of Waste Directive and the 2000 Incineration of Waste Directive. The strategy also set out recommendations in the area of transfrontier shipments of waste, waste management planning at local and regional level and encouraged the use of a broad range of instruments, including economic instruments, to achieve the policy objectives of the Strategy.

Through the measures of the Waste Framework Directive (1975) as amended in 1991 and 1996, the member states of the European Union are required to have a National Waste Strategy that sets out their policies in relation to the recovery and disposal of waste. In particular, the Strategy must identify the type, quantity and origin of waste to be recovered or disposed of, as well as the general technical requirements and any special arrangements for particular waste and suitable disposal sites or installations.

The objectives of the National Waste Strategy of Member States (Environment Act 1995; Lane and Peto 1995; Gervais 2002(b)) include:

- ensuring that waste is recovered or disposed of without endangering human health and without using processes or methods which could harm the environment;
- establishing an integrated and adequate network of waste disposal installations, taking account of the best available technology but not involving excessive costs;
- ensuring self-sufficiency in waste disposal;
- encouraging the prevention or reduction of waste production and its harmfulness by the development of clean technologies;
- encouraging the recovery of waste by means of recycling, re-use or reclamation, and the use of waste as a source of energy.

Underlying National Waste Strategies, is the 'self-sufficiency principle' which states that Member States shall take appropriate measures to establish an integrated and adequate network of disposal installations which enable the Union as a whole to become self-sufficient in waste disposal. A move towards individual member-state self-sufficiency, is also recommended. The development of a waste strategy should also reflect the 'proximity principle' under which waste should be disposed of (or otherwise managed) close to the point at which it is generated. This creates a more responsible approach to the generation of wastes, and also limits pollution from transport. It is therefore expected that each region should provide sufficient waste treatment and disposal facilities to treat or dispose of all the waste it produces.

The EU strategy on waste has developed into the concept of the 'hierarchy of waste management' (Sustainable Development 1994; Making Waste Work 1995; Waste Not Want Not 2002). The hierarchy was originally developed through the aims of the original 1975 Waste Framework Directive which encouraged, waste reduction, re-use and recovery with disposal as the least desirable option. The hierarchy was formally adopted in the 1989 EU Community Strategy for waste Management (Gervais 2002(b)). A more detailed version of the hierarchy has also been proposed (Figure 1.1, Waste Not Want Not 2002).

1. *Waste reduction.* Uppermost in the hierarchy is the strategy that waste production from industrial manufacturing processes should be reduced. Reduction of waste at source should be achieved by developing clean technologies and processes that require less material in the end products and produce less waste in their manufacture. This may involve the development of new technologies or adaptations of existing processes. Other methods include the development and manufacture of longer lasting products and products which are likely to result in less waste when they are used. The manufacturing process should also avoid producing wastes which are hazardous, or reduce the

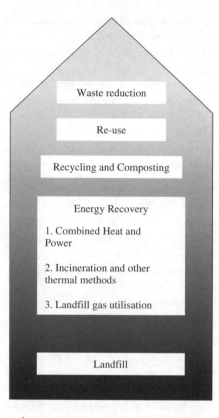

Figure 1.1   *The hierarchy of waste management. Source: Waste Not Want Not, 2002.*

toxicity of such wastes. Waste reduction has the incentive of making significant savings in raw materials, energy use and production and waste disposal costs.

2. *Re-use.* The collection and re-use of materials, for example, the re-use of glass bottles, involves the collection, cleaning and re-use of the same glass bottle. Tyre re-treading would also come into this category, where many truck tyres may be re-treaded many times throughout their lifetime. Re-use may also include new uses for the item once they have served their original purpose. For example, the use of used tyres for boat fenders and as silage covers. Re-use can be commercially attractive in some circumstances. However, re-use may not be desirable in all cases since the environmental and economic cost of re-use in terms of energy use, cleaning, recovery, transportation etc., may outweigh its benefits.

3. *Recycling and composting*

   (i) Materials Recycling. The recovery of materials from waste and processing them to produce a marketable product, for example, the recycling of glass and aluminium cans is well established, with a net saving in energy costs of the recycled material compared with virgin production. The potential to recycle material from waste is high, but it may not be appropriate in all cases, for example, where the abundance of the raw material, energy consumption during collection and re-processing, or the emission of pollutants has a greater impact on the environment or is not cost-effective. Materials recycling

also implies that there is a market for the recycled materials. The collection of materials from waste, where there is no end market for them, merely results in large surpluses of unwanted materials and also wastes additional energy with no overall environmental gain.

(ii) Composting. Decomposition of the organic, biodegradable fraction of waste to produce a stable product such as soil conditioners and growing material for plants. Composting of garden and food waste has been encouraged for home owners as a direct way of recycling. It has been extended to the larger scale for green waste from parks and gardens and also to municipal solid waste and to sewage sludge. The quality of compost produced from waste, compared with non-waste sources, has been an issue for waste composting, particularly in the area of contamination.

4. *Energy recovery.* Recovery of energy from waste incineration or the combustion of landfill gas. Many wastes, including municipal solid waste, sewage sludge and scrap tyres, contain an organic fraction which can be burnt in an incinerator. The energy is recovered via a boiler to provide hot water for district heating of buildings or high-temperature steam for electricity generation. The incinerator installation represents a high initial capital cost and sophisticated emissions control measures are required to clean-up the flue gases. Producing energy by combined heat and power (CHP) enables the maximum recovery of energy from waste by producing both electricity and district heating. The waste is again incinerated, but CHP systems would use a different type of steam turbine which would generate a lower amount of electricity, then the steam effluent from the turbine would be at a higher temperature, enabling district heating also to be incorporated. The production of landfill gas from the biodegradation of the organic fraction of wastes such as domestic waste and sewage sludge in a landfill site, produces a gas consisting mainly of methane which can be collected in a controlled, engineered way and burnt. Again the derived energy is used for either district heating or power generation. Additionally, there are newer technologies such as pyrolysis and gasification which can recover energy in the form of gas or liquid fuels. These can then be exported to power stations or used to generate energy on site.

5. *Landfill.* Under the hierarchy, landfill is seen as the least desirable option. Biological processes within the landfill ensure that, over a period of time, any biodegradable waste is degraded, neutralised and stabilised to form an essentially inert material. However, methane and carbon dioxide which are 'greenhouse gases' are generated throughout the degradation period. The European Union, through the Waste Landfill Directive (Council Directive 1999/31/EC 1999) has set targets for the reduction of biodegradable waste going to landfill, to encourage more recycling and to reduce emissions of the greenhouse gases. Where disposal to landfill occurs, the process is controlled, ensuring that human health is not endangered or harm to the environment does not occur. Landfill sites are often used mineral workings, which are required to be infilled after use and consequently, the disposal of certain types of waste such as treated and inert wastes into landfill, can be beneficial and eventually result in recovered land. A further major consideration for landfill disposal is the leachate, the potentially toxic liquid residue from the site, which may enter the water course.

The EU waste management strategy, encompassing sustainable development, requires that waste management practices move up the hierarchy such that waste is not merely disposed of, but should where possible be, recovered, reused or minimised. However, this may not be achievable in all cases and in some cases may not be desirable. For example, some wastes are best landfilled or incinerated since the environmental and economic cost of trying to sort and decontaminate the waste to produce a useable product outweighs the benefits. Consequently, the principle of Best Practicable Environmental Option (BPEO) has been developed (Box 1.3).

Box 1.3
Best Practicable Environmental Option (BPEO)

Best Practicable Environmental Option (BPEO) has been defined in the UK as:
   'The outcome of a systematic consultative and decision-making procedure which emphasises the protection and conservation of the environment across land, air and water. The BPEO procedure establishes, for a given set of objectives, the option that provides the most benefits or least damage to the environment as a whole, at acceptable cost, in the long term as well as in the short term'.

   The principle was introduced in the UK to take account of the total pollution from a process and the technical possibilities for dealing with it. BPEO is an integrated multi-media approach which applies to polluting discharges to air, water or land and should take into account the risk of transferring pollutants from one medium to another. The option chosen requires an assessment of the costs and benefits of the appropriate measures, but does not imply that the best techniques should be applied irrespective of cost. The concept is also applied in a wider context to policy and strategy planning for waste disposal and to the management of particular waste streams. The concept implies that different alternative options have been investigated before the preferred option is chosen which gives the best environmental outcome, in terms of emissions to land, air and water, at an acceptable cost. All feasible options which are both practicable and environmentally acceptable should be identified, and the advantages and disadvantages to the environment analysed. Whilst the selection of the preferred option is subjective, the decision makers should be able to demonstrate that the preferred option does not involve unacceptable consequences for the environment. The strategy of sustainable waste management has re-emphasised the need for BPEO to be applied in a wider context such that BPEO should not be restricted to the disposal of a particular waste stream without also examining the production process to determine whether the waste can be minimised, recovered or recycled. The use of the term 'practicable' involves a number of parameters including that the option chosen must be in accordance with current technical knowledge and must not have disproportionate financial implications for the operator. However, the best practicable option may not necessarily be the cheapest.

   Although a UK term, the link between the environmental benefits and economic effects have been discussed in terms of waste treatment in the EU Community Strategy for Waste Management (1997). There the choice of option in regard to waste recovery operations should be the best environmental option. However, the choice made should have regard both to environmental and to economic effects.

Sources: Royal Commission on Environmental Pollution 1988;
Council Resolution 97/C, 76/01 1997.

1.4   Policy Instruments

The aims of the EU Strategy on sustainable waste management and the objective of moving waste management options up the waste hierarchy may be achieved by a range of policy instruments. These include, the use of regulatory measures, market-based instruments, waste management planning and statistical data policy instruments, all of which are available to the EU or Member States of the EU. For example:

1. The regulatory policy is based on the extensive EU legislative and regulatory provisions covering the management of waste. A number of key European Community Directives, Regulations and Decisions influence the management of waste across the EU including:

   - Waste Framework Directive (75/442/EEC 1975);
   - Transfrontier Shipments of Waste (Council Regulation 259/93/EEC);
   - European Waste Catalogue (Commission Decision 2000/532/EC 2000);
   - Assessment of the Effects of Certain Public and Private Projects on the Environment Directive (85/337/EEC 1985);
   - Integrated Pollution Prevention and Control Directive (Council Directive 96/61/EC 1996);
   - Waste Incineration Directive (Council Directive 2000/76/EC 2000);
   - Waste Landfill Directive (Council Directive 1999/31/EC 1999);
   - End-of-Life Vehicle (ELV) Directive (2000/53/EC 2000);
   - Packaging and Packaging Waste Directive (Council Directive 94/62/EC 1994);
   - Waste Management Statistics Regulation (COM (99) 31 Final 1999);
   - Electrical and Electronic Equipment Waste Directive (Proposal COM (2000-0158) 2000).

2. The EU emphasises that waste management should reflect, as far as practicable, the costs of any environmental damage, whilst being carried out on a commercial and competitive basis. The costs of the various waste management options should fall, as far as possible, on those responsible for the creation of the waste. The fifth Environmental Action Programme included, as one of its key priorities, the broadening of the range of environmental policy instruments (Europa 2003). Environmental taxes and charges in the area of waste management can be a way of implementing the 'polluter pays' principle, by encouraging the use of more sustainable waste treatment and disposal options. However, there are no EU-wide economic measures, but each Member State of the EU is encouraged to develop such economic instruments to influence the choice of the waste management option. Amongst the economic instruments introduced by Member States are landfill taxes, incineration taxes, direct waste charging schemes and tradeable waste allowances.

3. The use of planning may be used as a policy measure to control and plan the location of waste management facilities. In addition, it ensures that there is adequate provision of waste management facilities, such as recycling, recovery, landfill, composting and incineration, leading to an integrated waste management structure. The 'proximity principle', whereby the treatment and disposal of waste should be carried out close to the point of waste production, confers more responsibility on the communities which produce the waste. In addition, regional self-sufficiency in waste management should

be a guiding principle of the planning authority. The Waste Framework Directive of 1975 was the main enabling legislation regarding waste management introduced by the EU. The Directive contained the requirement that Member States draw up a waste management plan identifying the appropriate locations and installations for waste treatment plants. For major waste management projects an environmental assessment is required, ensuring that the planned site is the most suitable location with minimum impact on the environment. Thereby, the planning authority ensures that the waste management plan or strategy of the region is implemented.

4. The statistical data policy required to meet the aims of the EU and Member State Waste Management policy, is based on the key role of statistically accurate data in waste management. To enable suitable waste strategies to be determined and industry and waste targets to be set, information on the sources, types and volumes of waste are produced, but also the proportions re-used, recovered or disposed are required. A key role in this context are the appropriate 'competent authorities' of each Member State of the EU, such as the UK Environment Agency, the German Federal Environment Agency (Umweltbundesamt), the Danish Environmental Protection Agency, etc. These Member State agencies are supported by the European Environment Agency. The Agency's objective is to 'provide the Community and the Member States with information which is objective, reliable and comparable at European level and which will enable them to take the measures required to protect the environment, evaluate the implementation of the measures and ensure that the public is properly informed on the state of the environment' (Gervais 2002(b)). The European Environment Agency carries out the following functions (Europa 2003):

- recording, collecting, assessing and transmitting data on the state of the environment;
- providing the European Community and the member States with the objective information that they require to draw up and implement appropriate and effective environment policies;
- helping to monitor environmental measures;
- working on the comparability of data at European level;
- promoting the development and application of environmental forecasting techniques;
- ensuring that reliable information on the environment is widely circulated.

The European Environment Agency was established in 1990 as a consultative body with the aims of supporting sustainable development and helping to achieve a significant and measurable improvement in Europe's environment. This is achieved through the provision of targetted and reliable information which is made available to policy-making agents in the European institutions and in the Member States. To this end, the Agency aims to provide a Europe-wide environmental data gathering and processing network. All the statistical data provided by the Member States is transmitted to Eurostat, the Statistical Office of the European Union. The Agency is therefore also able to evaluate the effectiveness of legislation already passed (European Environment Agency 1999). A major section of the European Environment Agency concentrates on the Theme of Waste. It has been recognised for some time that, across Europe, the data in relation to waste generation statistics, treatment and disposal routes, is inconsistent and incomplete. Consequently, in the Waste Theme, the Agency provides an EU-wide data gathering system, specifically on waste. In addition, detailed reports based on trends of waste generation, the implementation of waste legislation and emissions data, are produced. These data feed into the various policy making and legislative bodies of the EU.

1.5 EU Waste Management Legislation

The strategy of the European Union regarding the management of waste throughout the Union has developed from the various Policy and Strategy documents of the EU. However, its direct applicability to the Member States of the Union is through the various Directives, Regulations and Decisions of the legislature. These include a number of key measures which apply to various waste sectors, waste streams and waste treatment and disposal processes (Box 1.4). The main EU legislation in the area of waste management is described below.

Box 1.4
European Waste Legislation

European measures do not usually operate directly in member states of the European Union but set out standards and procedures which are then implemented by the Member States via their own legislative systems. The exception are 'Regulations', which are directly applicable and binding in all the Member States. However, most European Community (EC) law is set down mainly in 'Framework Directives' a term which is commonly shortened to 'Directives', which set general standards and objectives. Directives may contain differing requirements which take into account the different environmental and economic conditions in each Member State. Directives are implemented into national legislation by each Member State parliament. More detailed, subsidiary 'daughter Directives' deal with specific subjects within the Framework Directive. 'Decisions' are usually very specific in nature and are individual legislative acts which are binding on the sectors involved. Enforcement of EC law is devolved to Member States, but each state is answerable to the Community as a whole for the implementation of that law.

The main legislation introduced by the EC in relation to waste are:

Council Directive/Regulation	Area Covered
Council Directive 75/442/EC (1975)	Establishment of waste disposal authorities, proper waste control regimes and the requirement for a waste plan, amended by Council Directive 91/156/EEC (1991)
Council Directive 78/319/EEC (1978)	Toxic and dangerous waste, amended by Council Directive 91/689/EEC (1991)
Council Directive 80/68/EEC (1980)	Protection of groundwater against pollution from certain dangerous substances
Council Directive 84/360/EEC (1984)	Control of air pollution from industrial plant requires authorisation and the uses of BATNEEC for specified incinerators

Continued on page 16

Continued from page 15

Council Directive 84/631/EEC (1984)	Supervision and control of transfrontier shipments of hazardous waste within the EC
Council Directive 85/337/EC (1985)	Requirement of an environmental assessment for certain prescribed developments, e.g., incinerators and landfill sites
Council Directive 86/278/EEC (1986)	Control of sewage sludge to land
Council Directive 91/156/EEC (1991)	Amends the 75/442/EEC Directive and introduces the polluter pays principle and encourages recycling
Council Directive 91/689/EEC (1991)	Control of hazardous waste
Council Directive 91/271/EEC (1991)	Restriction of sewage sludge disposal to sea
Council Regulation 259/93/EEC (1993)	Supervision and control of shipments of all wastes, within, into and out of the EC
Council Directive 94/62/EC (1994)	Recycling and recovery of packaging and packaging waste
Council Directive 94/67/EC (1994)	Describes operational standards and emission limits for new and existing hazardous waste incinerators
Council Directive 96/61/EC (1996)	Introduced integrated pollution prevention and control (IPPC)
Council Directive 1999/31/EC (1999)	Landfill of waste, setting targets for reduction in biodegradable waste going to landfill, banning co-disposal
Council Directive 2000/76/EC (2000)	Incineration and co-incineration of all wastes
Council Directive 2000/53/EC (2000)	Recycling of end-of-life vehicles

Sources: Garbutt 1995; Gervais 2002(b); Europa 2003.

1.5.1   Waste Framework Directive

The most important EU Directive concerning waste was the main controlling Waste Framework Directive introduced in 1975 (75/442/EEC 1975), which established the general rules for waste management. The Directive has been subsequently amended several times, including 1991 (Council Directives 91/156/EEC and 91/692/EEC) and 1996 (Commission Decision 96/350/EC and Council Directive 96/59/EC). The 1975 Directive set out the key objective that waste should be recovered or disposed of without endangering human health and without using processes or methods which could harm the environment. In particular, without risk to water, air, soil, plants or animals, without causing nuisance through noise or odours and without adversely affecting the countryside or places of special interest (Murley 1999). Member States of the EU were required to take the necessary measures to prohibit the abandonment, dumping or uncontrolled disposal of waste. Waste Management Plans were required to be drawn up by each of the Member States to set out how the objectives of the Directive could be met. These plans included the types and quantities of waste, general technical requirements and the identification of suitable disposal sites. These plans developed into the National Waste Strategies of each Member State. The Framework Directive set out the principle of the hierarchy of waste management, first to develop clean technologies which minimised the use of natural resources and minimised the production of waste, and second to recover secondary materials from waste by means of recycling, re-use or reclamation and the use of waste as a source of energy.

The Directive also defined what was meant by 'waste' but only in general terms as 'any substance or object listed in the Directive, which the holder discards or intends or is required to discard'. The Directive also states that 'the uncontrolled discarding, discharge and disposal of waste is prohibited'. Member States are required to promote the prevention, recycling and re-use of wastes and to use waste as a source of energy. Each of the Member States of the EU was also required to establish competent authorities to control waste management processes through a system of permits and authorisations (Murley 1999). The competent authorities would be the Environment Agencies or their equivalent in each Member State, such as the UK Environment Agency, the German Federal Environment Agency (Umweltbundesamt), the Danish Environmental Protection Agency, etc. The 'polluter pays' principle was also stressed, in that the producer of the waste should bear the cost of disposal. The Directive set out the need for each Member State to produce a waste management plan or strategy to implement the measures outlined in the Directive. In particular, the types, quantities and origins of the wastes to be treated, the general technical requirements and the appropriate locations and installations for waste treatment and disposal.

The Waste Framework Directive set out a list of categories of waste, and a list of waste disposal and–waste recovery operations which were covered by the provisions of the Directive (Tables 1.1–1.3). The limited category of wastes was later superseded

Table 1.1 *Categories of waste as set out in the 1975 Waste Framework Directive*

1. Production or consumption residues not otherwise specified below.
2. Off-specification products.
3. Products whose date for appropriate use has expired.
4. Materials spilled, lost or having undergone other mishap including any materials, equipment, etc., contaminated as a result of the mishap.
5. Materials contaminated or soiled as a result of planned actions (e.g., residues from cleaning operations, packing, materials, containers, etc.).
6. Unusable parts (e.g., reject batteries, exhausted catalysts, etc.).
7. Substances which no longer perform satisfactorily (e.g., contaminated acids, contaminated solvents, exhausted tempering salts, etc.).
8. Residues of industrial processes (e.g., slags, still bottoms, etc.).
9. Residues from pollution abatement processes (e.g., scrubber sludges, baghouse dusts, spent filters, etc.).
10. Machining or finishing residues (e.g., lathe turnings, mill scales etc.).
11. Residues from raw materials extraction and processing (e.g., mining residues, oil field slops, etc.).
12. Adulterated materials (e.g., oils contaminated with PCB's etc.).
13. Any materials, substances or products whose use has been banned by law.
14. Products for which the holder has no further use (e.g., agricultural, household, office, commercial and shop discards, etc.).
15. Contaminated materials, substances or products resulting from remedial action with respect to land.
16. Any materials, substances or products which are not contained in the above categories.

Source: Waste Framework Directive 1975.

Table 1.2   *Disposal operations covered by the 1975 Waste Framework Directive*

1. Deposit into or onto land (e.g., landfill).
2. Land treatment (e.g., biodegradation of liquids or sludge wastes in soils, etc.).
3. Deep injection (e.g., injection of pumpable wastes into wells, salt domes, etc.).
4. Surface impoundment (e.g., placement of liquids or sludge wastes into pits, ponds or lagoons, etc.).
5. Specially engineered landfill (e.g., placement of waste into lined discrete cells which are capped and isolated from one another and the environment, etc.).
6. Release into a water body except seas/oceans.
7. Release into seas/oceans including sea bed insertion.
8. Biological treatment not specified elsewhere in the list which results in materials which are discarded using the disposal operations in the list.
9. Physico-chemical treatment not specified elsewhere in the list which results in materials which are discarded using the disposal operations in the list.
10. Incineration on land.
11. Incineration at sea.
12. Permanent storage (e.g., emplacement of containers in a mine, etc.).
13. Blending or mixing prior to disposal operations.
14. Repackaging prior to disposal operations.
15. Storage pending disposal operations (excluding temporary storage, pending collection, on the site where it is produced.

Source: Waste Framework Directive 1975.

Table 1.3   *Recovery operations covered by the 1975 Waste Framework Directive*

1. Waste used principally as a fuel or other means to generate energy.
2. Solvent reclamation/regeneration.
3. Recycling/reclamation of organic substances which are not used as solvents (including composting and other transformation processes).
4. Recycling/reclamation of metals and metal compounds.
5. Recycling/reclamation of other inorganic materials.
6. Regeneration of acids or bases.
7. Recovery of components used for pollution abatement.
8. Recovery of components from catalysts.
9. Oil re-refining or other re-uses of oil.
10. Land treatment resulting in benefit to agriculture or ecological improvement.
11. Use of wastes obtained from the processes listed above.
12. Exchange of wastes for submission for the processes listed above.
13. Storage of wastes pending operations listed above (excluding temporary storage, pending collection, on the site where it is produced).

Source: Waste Framework Directive 1975.

by the European Waste Catalogue. Similarly, the regulations relating to individual waste disposal and recycling operations were covered in more detail by later Directives, daughter Directives, Regulations and Decisions of the EU.

1.5.2   Transfrontier Shipment of Waste Directive

The first Directive on the Transfrontier Shipment of Waste (Council Directive 84/631/EEC 1984) in 1984 concerned hazardous waste only and was later amended in 1986 (Council

Directive 86/279/EEC 1986). The Directive stipulated that pre-notification of transfrontier shipments of hazardous waste was required and the wastes could only be sent to designated facilities which could dispose of the waste without endangerment to the environment and human health. The earlier Directives were replaced in 1993 by Council Regulation 259/93/EEC on the Supervision and Control of Transfrontier Shipments of Waste (1993). Because the measure was a Regulation rather than a Directive, it was directly applicable to all Member States of the EU (Box 1.4). The 1993 Regulation covered all movements of waste whether hazardous or not, as defined in the general definition of waste outlined in the 1975 Waste Framework Directive. The movement of waste within and into or out of the EU was covered and also the movement of waste between countries outside the EU but which might pass through one of the EU Member States en route (Europa 2003). The Regulation includes detailed procedures for a compulsory pre-notification scheme, waste description, authorisation and consignment note system. The final treatment and disposal of the waste must be guaranteed by a certificate from the receiver of the waste that disposal has been safely dealt with in an environmentally acceptable way. This must be received within 180 days of shipment. A common compulsory notification and a standard consignment note system has been introduced across the EU. If the waste has not been treated or dealt with properly, the waste must be taken back to the originator of the waste. The waste must be covered by a financial guarantee in case it has to be returned or shipped elsewhere for treatment. The regulations draw the distinction between wastes for disposal, for example, via landfill or incineration, and wastes for recycling. Waste for recovery or recycling is categorised into three types designated as red, amber and green, according to how hazardous it is. Red is the most hazardous including, for example, toxic wastes such as asbestos, dioxins and polychlorinated biphenyls (PCBs). Amber wastes include waste oils and gasoline sludges, and green wastes include non-hazardous wastes such as paper, plastics and glass, wastes from mining operations, textiles and rubber (Murley 1999). Underling the transfrontier shipment of waste is the EU principle of self-sufficiency in waste disposal, where waste should be dealt with closest to the point of generation. However, this may not always be economically viable, hence the need for regulation of the movement of waste.

The import and export of waste into and out of the EU is strictly regulated. Pre-authorisation of the appropriate authority in the EU Member State and the third-party state is required. Export of waste to many countries such as those in Africa, the Caribbean and the Pacific is banned completely.

1.5.3 European Waste Catalogue

The 1975 Waste Framework Directive (Council Directive 75/442/EEC 1975) defined waste only in general terms as 'any substance or object which the holder disposes of or is required to dispose of' and led to each Member State of the EU defining waste differently (Laurence 1999). The 1991 amendment, 91/156/EEC, listed certain broad categories of wastes such as residues from industrial process, spilled materials, residues from pollutant abatement processes, machinery processes, contaminated materials, etc. To ensure that all categories were covered, a 'catch all' phrase of 'any materials, substances or products which are not contained in the categories listed' was included. The 1991 amendment 91/156/EEC also required the European Commission to draw up a list of wastes belonging to each of the categories listed. This the Commission did in 1994 as a Council Decision by publishing the European Waste Catalogue (94/3/EC 1994). Also in 1991, the 1975 Waste Framework Directive was amended by Council Directive 91/689/EEC which dealt specifically with hazardous waste. As was the case for non-hazardous wastes, the amendment listed various categories of hazardous waste such as pharmaceuticals, wood preservatives, inks, dyes, resins, tarry materials, mineral oils etc. The amendment required that a list of the wastes in each category of hazardous waste be drawn up and the list consisting of over 200 different types of hazardous waste was produced in Council Decision 94/904/ EC (1994). The 1991 amendment (91/689/EEC) also identified the properties of hazardous waste which make it hazardous such as, explosive, oxidising, flammable, irritant, harmful, toxic, carcinogenic, corrosive, infectious, etc.

Commission Decision 2000/532/EC in 2000 (Commission Decision 2000) replaced earlier lists of wastes and hazardous waste in one unified document. The list of wastes is divided into twenty different categories known as 'chapters', each with a two-digit code, which are listed in Table 1.4. Within each chapter are between 1 and 13 sub-chapters, which

Table 1.4   *European Waste Catalogue: chapters and sub-chapters of the list*

01   Wastes resulting from exploration, mining, dressing and further treatment of minerals and quarry
    01 01   Wastes from mineral excavation
    01 02   Wastes from mineral dressing
    01 03   Wastes from further physical and chemical processing of metalliferous minerals
    01 04   Wastes from further physical and chemical processing of non-metalliferous minerals
    01 05   Drilling muds and other drilling wastes

02   Wastes from agricultural, horticultural, hunting, fishing and aquacultural primary production, food preparation and processing
    02 01   Wastes from mineral excavation
    02 02   Wastes from mineral dressing
    02 03   Wastes from further physical and chemical processing on metalliferous minerals
    02 04   Wastes from further physical and chemical processing on non-metalliferous minerals
    02 05   Drilling muds and other drilling wastes

03   Wastes from wood processing and the production of paper, cardboard, pulp, panels and furniture
    03 01   Wastes from wood processing and the production of panels and furniture
    03 02   Wood preservation wastes
    03 03   Wastes from pulp, paper and cardboard production and processing

04   Wastes from the leather, fur and textile industries
    04 01   Wastes from the leather and fur industry
    04 02   Wastes from the textile industry

05   Wastes from petroleum refining, natural gas purification and pyrolytic treatment of coal
    05 01   Oily sludges and solid wastes
    05 02   Non-oily sludges and solid wastes
    05 04   Spent clay filters
    05 05   Oil desulphurisation wastes
    05 06   Wastes from the pyrolytic treatment of coal
    05 07   Wastes from natural gas purification
    05 08   Wastes from oil regeneration

06   Wastes from inorganic chemical processes
       06 01   Waste acidic solutions
       06 02   Waste alkaline solutions
       06 03   Waste salts and their solutions
       06 04   Metal-containing wastes
       06 05   Sludges from on-site effluent treatment
       06 06   Wastes from sulphur chemical processes
       06 07   Wastes from halogen chemical processes
       06 08   Waste from production of silicon and silicon derivatives
       06 09   Wastes from the phosphorus chemical processes
       06 10   Waste from nitrogen chemical processes and fertiliser manufacture
       06 11   Waste from the manufacture of inorganic pigments and opacifiers
       06 12   —
       06 13   Wastes from other inorganic chemical processes

07   Wastes from organic chemical processes
       07 01   Wastes from Manufacture, Formulation, Supply and Use (MFSU) basic organic
                  chemicals
       07 02   Wastes from the MFSU of plastics, synthetic rubber and man made fibres
       07 03   Wastes from the MFSU of organic dyes and pigments
       07 04   Wastes from the MFSU of organic pesticides
       07 05   Wastes from the MFSU of pharmaceuticals
       07 06   Wastes from the MFSU of fats, grease, soaps, detergents, disinfectants and
                  cosmetics
       07 07   Wastes from the MFSU of fine chemicals and chemical products not otherwise
                  specified

08   Wastes from the Manufacture, Formulation, Supply and Use (MFSU) of coatings
       (paints, varnishes and vitreous enamels), adhesives, sealants and printing inks
       08 01   Wastes from MFSU and removal of paint and varnish
       08 02   Wastes from MFSU of other coatings (including ceramic materials)
       08 03   Wastes from MFSU of printing inks
       08 04   Wastes from MFSU of adhesives and sealants
       08 05   Wastes not otherwise specified

09   Wastes from the photographic industry
       09 01   Wastes from the photographic industry

10   Inorganic wastes from thermal processes
       10 01   Wastes from power stations and other combustion plants (except 19)
       10 02   Wastes from the iron and steel industry
       10 03   Wastes from aluminium thermal metallurgy
       10 04   Wastes from lead thermal metallurgy
       10 05   Wastes from zinc thermal metallurgy
       10 06   Wastes from copper thermal metallurgy
       10 07   Wastes from silver, gold and platinum thermal metallurgy
       10 08   Wastes from other non-ferrous thermal metallurgy
       10 09   Wastes from casting of ferrous pieces
       10 10   Wastes from casting of non-ferrous pieces
       10 11   Wastes from manufacture of glass and glass products
       10 12   Wastes from manufacture of ceramic goods, bricks, tiles and construction
                  products
       10 13   Wastes from manufacture of cement, lime, plaster, articles and products made
                  from them

Table 1.4   *Continued*

11   Inorganic metal-containing wastes from metal treatment and coating of metals, and
      non-ferrous hydrometallurgy
      11 01   Liquid wastes and sludges from metal treatment and coating of metals
      11 02   Wastes and sludges from non-ferrous hydrometallurgical processes
      11 03   Sludges and solids from tempering processes
      11 04   Other inorganic metal-containing wastes not otherwise specified

12   Wastes from shaping and surface treatment of metals and plastics
      12 01   Wastes from shaping (e.g., forging, welding pressing, cutting, etc.)
      12 02   Wastes from mechanical surface treatment processes (e.g., blasting, grinding
               polishing, etc.)
      12 03   Wastes from water and steam degreasing processes (except 11)

13   Oil wastes (except edible oils, 05 and 12)
      13 01   Waste hydraulic oils and brake fluids
      13 02   Waste engine, gear and lubricating oils
      13 03   Waste insulating and heat transmission oils and other liquids
      13 04   Bilge oils
      13 05   Oil/water separator contents
      13 06   Oil waste not otherwise specified

14   Wastes from organic substances used as solvents (except 07 and 08)
      14 01   Wastes from metal degreasing and machinery maintenance
      14 02   Wastes from textile cleaning and degreasing of natural products
      14 03   Wastes from the electronic industry
      14 04   Wastes from coolants, foam/aerosol propellants
      14 05   Wastes from solvent and coolant recovery

15   Waste packaging; absorbents, wiping cloths, filter materials and protective clothing
      not otherwise specified
      15 01   Packaging
      15 02   Absorbents, filter materials, wiping cloths and protective clothing

16   Wastes not otherwise specified in the list
      16 01   End-of-life vehicles and their components
      16 02   Discarded equipment and its components
      16 03   Off-specification batches
      16 04   Waste explosives
      16 05   Chemicals and gases in containers
      16 06   Batteries and accumulators
      16 07   Wastes from transport and storage tank cleaning (except 05 and 12)
      16 08   Spent catalysts

17   Construction and demolition wastes (including road construction)
      17 01   Concrete, bricks, tiles, ceramics, and gypsum based materials
      17 02   Wood, glass and plastic
      17 03   Asphalt, tar and tarred products
      17 04   Metals (including their alloys)
      17 05   Soil and dredging spoil
      17 06   Insulation materials
      17 07   Mixed construction and demolition waste

18   Wastes from human or animal health care and/or related research (except kitchen and
      restaurant wastes not arising from immediate health care)
      18 01   Wastes from natal care, diagnosis, treatment or prevention of disease in humans
      18 02   Wastes from research, diagnosis, treatment or prevention of disease involving
               animals

19 Wastes from waste treatment facilities, off-site wastewater treatment plants and the water industry
 19 01 Wastes from incineration or pyrolysis of waste
 19 02 Wastes from specific physico/chemical treatments of industrial waste
 19 03 Stabilised/solidified wastes
 19 04 Vitrified waste and wastes from vitrification
 19 05 Wastes from aerobic treatment of solid wastes
 19 06 Wastes from anaerobic treatment of waste
 19 07 Landfill leachate
 19 08 Wastes from wastewater treatment plants not otherwise specified
 19 09 Wastes from the preparation of drinking water or water for industrial use
 19 10 Wastes from shredding of metal-containing waste

20 Municipal wastes and similar commercial, industrial and institutional wastes including separately collected fractions
 20 01 Separately collected fractions
 20 02 Garden and park wastes (including cemetery waste)
 20 03 Other municipal wastes

Source: Commission Decision 2000/532/EC 2000.

are also shown in Table 1.4. Within each sub-chapter are the specific waste categories. Consequently, each category of waste has a specific six-digit code, examples of which are shown in Table 1.5. Within each sub-chapter there is usually a six-digit designation code entitled 'wastes not otherwise specified' to ensure that all wastes arising in a particular

Table 1.5 *European Waste Catalogue: examples of specific waste categories on the Waste Catalogue list*

01 Wastes resulting from exploration, mining, dressing and further treatment of minerals and quarries
 01 01 Wastes from mineral excavation
   01 01 01 Waste from metalliferous excavation
   01 01 02 Waste from non-metalliferous excavation
 01 05 Drilling muds and other drilling wastes
   01 05 01 Oil containing drilling muds and wastes
   01 05 99 Wastes not otherwise specified

16 Wastes not otherwise specified in the list
 16 01 End-of-life vehicles and their components
   16 01 03 End-of-life tyres
   16 01 04 Discarded vehicles
   16 01 06 End-of-life vehicles, drained of liquids and other hazardous components
   16 01 99 Wastes not otherwise specified

18 Wastes from human or animal health care and/or related research (except kitchen and restaurant wastes not arising from immediate health care)
 18 01 Wastes from natal care, diagnosis, treatment or prevention of disease in humans
   18 01 01 Sharps
   18 01 02 Body parts and organs including blood bags and blood preserves
   18 01 08* Cytotoxic and cytostatic medicines
   18 01 10* Amalgam waste from dental care

Table 1.5   *Continued*

19	Wastes from waste treatment facilities, off-site wastewater treatment plants and the water industry	

19 01   Wastes from incineration or pyrolysis of waste
        19 01 02   Ferrous materials removed from bottom ash
        19 01 06*   Aqueous liquid waste from gas treatment and other aqueous liquid waste
        19 01 07*   Solid waste from gas treatment
        19 01 08*   Spent activated carbon from flue gas treatment
        19 01 11*   Bottom ash and slag containing dangerous substances
        19 01 11   Bottom ash and slag not containing dangerous substances
        19 01 13*   Flyash containing dangerous substances
        19 01 99   Wastes not otherwise specified

19 06   Wastes from anaerobic treatment of waste
        19 06 01   Anaerobic treatment sludges of municipal and similar wastes
        19 06 02   Anaerobic treatment sludges of animal and vegetal wastes
        19 06 99   Wastes not otherwise specified

19 07   Landfill leachate
        19 07 01   Landfill leachate

20   Municipal wastes and similar commercial, industrial and institutional wastes including separately collected fractions

20 01   Separately collected fractions
        20 01 01   Paper and cardboard
        20 01 02   Glass
        20 01 03   Organic kitchen waste
        20 01 21   Fluorescent tubes and other mercury containing waste
        20 01 23   Discarded equipment containing fluorocarbons

20 03   Other municipal wastes
        20 03 01   Mixed municipal waste
        20 03 02   Waste from markets

---

* = hazardous.

Source: Commission Decision 2000/532/EC 2000.

industrial sector are listed. There are more than 650 waste categories on the list but it is not regarded as complete and exhaustive and other waste categories with their individual six-digit codes could be added at a later date. The European Commission will periodically review and revise the list (Commission Decision 2000). Some categories of waste are asterisked to denote that they are considered as hazardous.

1.5.4   Environmental Assessment Directive

The need for an Environmental Assessment of the impact of large-scale projects in the European Union was introduced as a requirement for member States of the European Union in 1985. The EC Directive concerned was the Assessment of the Effects of Certain Public and Private Projects on the Environment, 85/337/EEC. The assessment was required for those projects which are likely to have a significant effect on the environment, due to their nature, size or location. Included in the Directive were, for example, crude-oil refineries, power stations, iron and steel works, chemical installations and motorways. Also included were large-scale energy from municipal waste incinerators, clinical and hazardous waste incinerators and large-scale (greater than 75 000 tonnes/year) landfill sites (Environmental Assessment Guidance 1989). Not all waste treatment and disposal projects require an assessment, only those which may have a significant impact on the environment. What constitutes a 'significant' impact might be where the project is of large-scale or destined for a site of special scientific interest or perhaps whether the project is likely to give rise to significant pollution.

For a large-scale waste treatment and disposal project, such as a municipal waste incinerator or landfill site, the environmental assessment would include assessments of a wide range of criteria (Energy from Waste 1996 and Petts and Eduljee 1994). Such criteria may include:

- **Visual Impact.** The visual impact of a large and prominent industrial plant or landfill site upon the existing landscape and visual amenity.
- **Air Emissions**
  - Incinerators. The existing air quality, concentration, volume and dispersion characteristics of pollutant gases, ground level concentrations, considerations of local topography and meteorology, comparison with legislative and guidance limits.
  - Landfill Sites. Fugitive emissions of landfill gas from the site, dispersion of the gases, odour problems, increase in ambient concentration, explosion risk, comparison with legislative and guidance limits.
- **Water Discharges**
  - Incinerators. The treatment and disposal options for scrubber liquor and cooling water.
  - Landfill Sites. The treatment and disposal options for leachate, effects on downstream treatment works and water resources.
- **Ash Discharges**
  - Incinerators. The treatment and disposal options for bottom ash and flyash.
- **Human Health.** The impacts and pathways of exposure to the pollutant emissions, ingestion via the food chain and water and inhalation. Estimation of hazard and risk.
- **Fauna and Flora.** The impact of emissions on local fauna and flora and loss of habitat, particularly for sites of special scientific interest.
- **Site Operations.** The management controls and analysis of risks associated with the plant operation and consequence of operational failure. Impact of plant operation noise.
- **Traffic.** The number of heavy goods vehicles and other vehicle movements, impacts on existing road network and traffic flows, noise from increased traffic, accident statistics and routing considerations.
- **Socio-economic Impacts.** The effects of the project on adjoining residents and the existing industry, including economic benefits such as employment and investment.
- **Land-use and Cultural Heritage.** Compatibility of the project with existing and proposed adjacent land-use and conformity with local development plans.

The environmental assessment must identify, describe and assess the direct and indirect effects of the project on human beings, fauna, flora, soil, water, air, climate and landscape, material assets and the cultural heritage. The assessment is carried out by a project team of experts for the developer and would include a description of the project, comprising information on the site, design, size and scale and the main characteristics of the process. In addition, information about likely environmental effects, including expected residues and emissions, is required and also a description of the proposed measures to prevent, reduce or offset any adverse effects on the environment. The assessment would include an appraisal of alternative sites and processes considered, baseline surveys of the site and surrounding area, a review of the options and proposed methods to prevent or minimise any environmental effects, prediction and evaluation of the environmental impacts. The assessment procedure involves full disclosure of information and consultation with the public. Clearly the environmental assessment can be a complex, difficult, time-consuming and expensive task. It has been estimated that the assessment can cost up to 5% of the total capital costs of the project, although more typically it is around 1% and a timescale of one year is likely (Energy from Waste 1996).

Figures 1.2 and 1.3 show the sources of impacts and effects on the environment for an incinerator and a landfill site (Petts and Eduljee 1994).

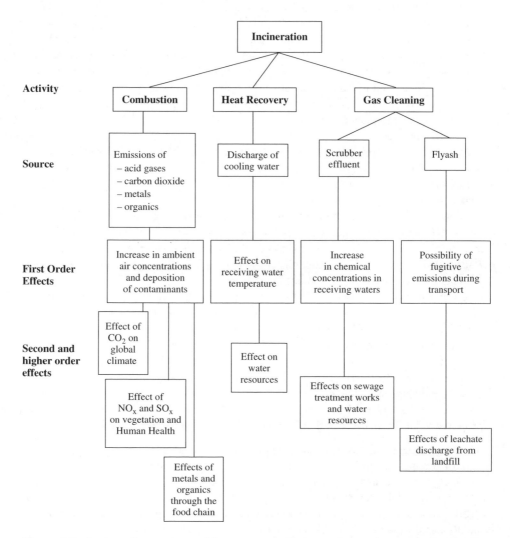

Figure 1.2 *Incineration sources of impacts and effects on the environment. Source: Petts and Eduljee 1994. Copyright © 1994 John Wiley & Sons Ltd. Reproduced with permission.*

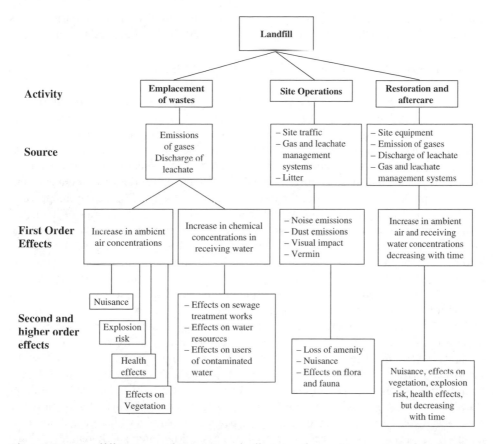

Figure 1.3 *Landfill sources of impacts and effects on the environment. Source: Petts and Eduljee 1994. Copyright © 1994 John Wiley & Sons Ltd. Reproduced with permission.*

The 1985 Directive is under review and a proposal for a new Directive has been presented by the European Commission (COM (2000) 839 Final 2001). One of the main objectives of the proposed new Directive is to increase the effectiveness of public participation in environmental decision-making. The proposal makes provision for a public participation procedure in relation to projects involving a range of waste management activities. Member States are required to ensure that the public are informed, together with any relevant information, about proposals for such projects; that they are entitled to express comments and opinions before decisions on the plans and programmes are made; and that in making those decisions, due account is required to be taken of the results of the public participation. These measures are required before any permit or licence for the project, such as a waste incinerator or waste landfill site is granted. The grounds for the final decision and the considerations taken into account in reaching that decision should be made available to the public.

There is also provision in the proposal for participation of the public in decision-making for large environmental projects which have significant effects on territories other than where the project is being carried out. To this end, trans-boundary public participation in the decision-making process can be made between Member States of the European

Union. Information on the environmental impact of the project on neighbouring states should be supplied to those states and any objections should be taken into account.

1.5.5   Integrated Pollution Prevention and Control Directive

The objective of the 1996, EU Directive concerning integrated pollution prevention and control (Council Directive 96/61/EC 1996) is to prevent or minimise air, water and soil pollution by emissions from industrial installations in the Community, with a view to achieving a high level of environmental protection. The Directive is aimed at particular processes in viewing the pollution impact of the process in a wider sense with the aim of greater environmental sustainability. Through the Directive, a series of Reference Documents are produced which set out the best environmental way of operating the process. More specifically, it provides for a permitting system for certain categories of industrial installations requiring both operators and regulators to take an integrated, overall look at the polluting and consuming potential of the installation (BREF Waste Incineration 2003). Through this approach, operators of industrial installations should take all appropriate preventative measures against producing pollution. The environmental performance of the installation is improved by the application of the best available techniques to prevent or control pollution. Integrated pollution prevention and control concerns a range of highly polluting industrial processes, including the energy industries, the production and processing of metals and the mineral and chemical industries, as well as waste management. The waste management category includes, for example, waste incineration, waste landfill and hazardous waste management facilities. The basic obligations required by the plant or process operator are set out as a list of measures for preventing the pollution of water, air and soil. The Directive covers both new and existing industrial activities. The process is regulated by the issuing of operating licences or permits which are authorised by the designated authority. For each industrial process, a Management Guideline is issued by the designated authority which sets out in detail the minimum requirements to be included in any authorising permit. These require the process to operate to the emission limits set for pollutants, give the monitoring requirements and require the minimisation of long-distance or trans-boundary pollution. The emission-limit values for each industrial activity is set by the Council or via the relevant Directive, such as the Waste Incineration Directive. The Member States of the EU, through the various State authorities, are responsible for issuing permits or licences, for inspecting the industrial installations and for ensuring that the regulations are complied with.

The term 'best available techniques' is defined by the EU as 'the most effective and advanced stage in the development of activities and their methods of operation which indicate the practical suitability of particular techniques for providing in principle the basis for emission-limit values designed to prevent and, where that is not practicable, generally to reduce emissions and the impact on the environment as a whole' (BREF Waste Incineration 2003). In the context of Integrated Pollution Prevention and Control, the term 'best' means the most effective in achieving a high general level of protection of the environment as a whole. Techniques which are defined as 'available' are those developed on a scale which allows implementation in the relevant industrial sector, under economically and technically viable conditions, taking into consideration the costs and advantages. The techniques should be reasonably accessible to the operator, whether or not they are used or produced in the

Table 1.6   *Examples of industrial installations covered by Reference Documents on Best Available Techniques Reference Document (BREF) for Integrated Pollution Prevention and Control*

General principles of monitoring
Glass manufacturing industry
Pulp and paper industry
Production of iron and steel
Cement and lime manufacturing industries
Industrial cooling systems
Chlor-alkali manufacturing industry
Ferrous metals processing industry
Large volume organic chemical industry
Wastewater and waste gas treatment/management systems in the chemical sector
Large combustion plants
Management of tailings and waste rock in mining activities
Waste treatment industries
Waste incineration
Waste landfill sites (except inert landfills)
Manufacture of large-volume inorganic chemicals (ammonia, acids, fertilisers)

Source: BREF Waste Incineration 2003.

operators Member State. The term 'techniques' includes both the technology used and the way in which the installation is designed, built, maintained, operated and decommissioned.

The European Commission has set up Technical Working Groups comprising experts in the various industrial sectors covered by the Directive on Integrated Pollution Prevention and Control. The Technical Working Groups are a requirement of the EU Directive and are set up to provide a European-wide reference document for the best available techniques (BAT) to prevent and control pollution for each industrial process. These BAT reference documents (BREF) are very detailed describing the various types of processes in a particular industrial sector and the options and costs for the control of emissions to air, water and land. The European wide Integrated Pollution Prevention and Control (IPPC) of pollutants from industrial activities is co-ordinated by the European IPPC Bureau. Reference documents describing the processes and the best available techniques for the prevention and control of pollutants are listed in Table 1.6. Because the best available techniques for preventing and controlling pollutants from a particular industrial installation develop continuously, the competent authority in each Member State (e.g., the UK Environment Agency, the German Federal Environment Agency (Umweltbundesamt), the Danish Environmental Protection Agency, etc.) is required to monitor developments in the area and update the Reference Documents as required.

The application for a permit or licence to operate a waste treatment plant or any IPPC controlled plant should include a description of:

- the installation and its activities;
- the raw materials and the energy used in or generated by the installation;
- the sources of emissions from the installation;
- the conditions of the site of the installation;
- the nature and quantities of emissions from the installation into air, water and on to land, as well as identification of significant effects of the emissions on the environment;

- the proposed technology and other techniques for preventing or, where this not possible, reducing emissions from the installation where necessary and measures for the prevention and recovery of waste generated by the installation;
- measures planned to monitor emissions into the environment.

The permit or licence is issued by the competent authority in each of the Member States of the EU.

1.5.6 Waste Incineration Directive

The Waste Incineration Directive (Council Directive 2000/76/EC) introduced in 2000 represents a single text on the incineration of waste and repeals earlier Directives on incineration of waste (Council Directives 89/369/EEC and 89/429/EEC) and incineration of hazardous waste (Council Directive 94/67/EC). Whilst Directive 2000/76/EC covers emissions to the atmosphere, the Directive also cites other Directives applicable to the discharge of wastewater from the incinerator (Council Directive 91/271/EEC amended by 98/15/EC and 76/464/EEC) and also the landfilling of waste (Council Directive 1999/31/EC). The Directive applies to existing plants from December 2005 and for new plants from December 2002. The distinction between hazardous waste and non-hazardous waste incinerators in previous EU Directives was not deemed to be correct, since the same emission limits should apply, no matter what type of waste was being incinerated. However, differences in the techniques and conditions of incineration may be appropriate. The revision of the emission limits to air for incinerators from the 1989 levels to the 2000 levels resulted in the imposition of more stringent emission-limit values as shown in Table 1.7.

Table 1.7 *Comparison of emission-limit values to air for large-scale municipal waste incineration for the 1989 and 2000 EU Waste Incineration Directives*

Pollutant	Emission limits (mg/m^3)	
	1989	2000
Total dust	30	10
Total organic carbon (TOC)	20	10
HCl	50	10
HF	2	1
CO	100	50
SO_2	300	50
NO_x (expressed as NO_2)	—	200
Metals		
Total class I	0.2	
Total class II	1.0	
Total class III	5.0	
Cd and Tl		0.05
Hg		0.05
Sb, As, Pb, Cr, Co, Cu, Mn, Ni, V (Total)		0.5
Dioxins and furans (TEQ) ng/m^3		0.1

Sources: Council Directive 2000/76/EC 2000; Council Directive 89/369/EEC 1989; Council Directive 89/429/EEC 1989.

There is a great deal of concern from the public and environmental groups regarding the emissions from the incineration of waste. Therefore the EU, through the Waste Incineration Directive (2000), seeks to attain a high level of environmental protection and human health protection through the setting of stringent operational conditions, technical requirements and emission-limit values for waste incinerators.

The main objective of the 2000 Waste Incineration Directive is to prevent or reduce, as far as possible, air, water and soil pollution caused by the incineration or co-incineration of waste, as well as the resulting risk to human health. The Directive covers the incineration and co-incineration of waste. A co-incineration plant is any plant whose main purpose is the generation of energy or production of material products and which uses waste as a regular or additional fuel or in which waste is thermally treated for the purpose of disposal.

Also specified in the Waste Incineration Directive are the emission-limit values for discharges of wastewater from the cleaning of exhaust gases. It might be suggested that the pollutants in the wastewater could be reduced by merely adding clean water to dilute the levels found. However, dilution of the wastewater to comply with emission-limit values is not permitted.

The solid residues resulting from the incineration or co-incineration plant are also covered by the EU Waste Incineration Directive. The residues should be minimised in their amount and harmfulness and should be recycled, where appropriate, directly in the plant or outside in accordance with relevant Community legislation. Transport of dry residues should be in closed containers. Prior to determining the routes for disposal or recycling, appropriate tests should be carried out to establish the physical and chemical characteristics and polluting potential of the different incineration residues. The analysis includes the total soluble fraction and heavy metals soluble fraction.

The Directive concerns the incineration of all types of waste, including municipal solid waste, hazardous waste, sewage sludge, tyres, clinical waste, waste oils and solvents, etc. As any Integrated Pollution Prevention and Control (IPPC) process, the incinerator requires a permit which is issued and monitored by the environment agencies or competent authorities of the Member States. The permits set out the categories and quantities of waste which can be incinerated, the plant capacity and the sampling and measurement procedures which are to be used (Europa 2003). If hazardous waste is to be incinerated, full administrative information is required of the process source of the waste, the physical and chemical composition of the waste, and the characteristics of the waste which make it hazardous. Because incineration of waste comes under IPPC regulations, a Best Available Techniques (BAT) reference document is produced, which details the range of incinerator systems available and the operational, gas cleaning, wastewater cleaning, sampling, monitoring and reporting requirements, etc., for the incineration of waste. A waste incinerator is a major project with the potential to have a significant effect on the environment. Therefore, before a permit is granted, an environmental impact assessment is required. This would be required under the EC Directive on the Assessment of the Effects of Certain Public and Private Projects on the Environment, 85/337/EEC.

The Directive covers the complete process operation of the incinerator, including the requirements to be included in the authorisation permit, waste handling, the operating conditions of the incinerator and the detailed emission-limit values to air and wastewater from the gas cleaning process. The process control and flue gas and wastewater monitoring requirements are detailed. Public participation in the waste incineration process is encouraged through access to information prior to the granting of the permit. The functioning parameters of the incinerator and the measured emissions produced from the incinerator must be available to the public. The procedures to be undertaken if an unavoidable stoppage, disturbance or failure occurs, are also stipulated.

The process of incineration of waste to ensure efficient combustion is controlled through the Directive in that the temperature of combustion is also specified in the Directive. The Directive states that the gases derived from incineration are to be raised to a temperature of 850 °C for 2 s, if chlorinated hazardous waste is used with a chlorine content of over 1% then the temperature has to be raised to 1100 °C. The operating conditions also set out that waste incinerators should achieve a level of incineration such that the slag and bottom ashes shall have a total organic carbon (TOC) content of less than 3%. The TOC represents the degree of complete burnout of the waste organic materials.

The Directive sets out the limit values for emission to air for a whole range of toxic gases, including heavy metals such as mercury, cadmium, chromium and lead, dioxins and furans, carbon monoxide, dust, hydrogen chloride, hydrogen fluoride, sulphur dioxide, nitrogen oxides and gaseous organic compounds (expressed as TOC). The emission-limit values for discharges of wastewater from the cleaning of exhaust gases, concentrates on the total suspended solids in the wastewater and the levels of heavy metals including mercury, cadmium, arsenic, lead and chromium. The permitted maximum concentration of dioxins and furans in the wastewater are also stipulated.

1.5.7 Waste Landfill Directive

Council Directive 1999/31/EC (1999) on the Landfill of Waste has the main objective to prevent or reduce as far as possible the negative effects on the environment from the landfilling of waste, by introducing stringent operational and technical requirements for waste and landfills. The Directive provides for measures, procedures and guidance to prevent or reduce pollution of surface water, soil and air, and also to the global environment, including global warming effects, as well as the risk to human health from the practice of landfilling of waste.

Throughout the EU, each landfill should be categorised into different classes of landfill site defined as:

- landfills for hazardous waste;
- landfills for non-hazardous waste;
- landfills for inert waste.

Each type of designated landfill can only accept the particular waste for which it is designated. Consequently, only hazardous wastes are permitted in hazardous waste landfill sites, non-hazardous landfills can accept municipal solid waste and other non-hazardous wastes and the inert landfill can only accept inert waste. Some types of waste are not permitted to go to landfill at all. These include liquid waste, flammable waste, explosive or oxidising wastes, infectious clinical or hospital waste and used tyres.

As landfilling of waste is included in the EU Directive on Integrated Pollution Prevention and Control (IPPC), it is a designated IPPC process and consequently, is covered by the IPPC permitting process. The permits must contain a description of the types and total quantity of waste to be deposited, the capacity of the site, a description of the site including its hydrogeological and geological characteristics and the proposed methods of pollution prevention and control. In addition, the permit must state the proposed operation, monitoring and control plan as well as the plan for closure of the site and aftercare procedures. Because landfilling of waste comes under IPPC regulations, a Best Available Techniques (BAT) reference document will be produced describing in detail the best techniques for the reception of the waste, operational procedures, liner systems required, gas and leachate sampling and monitoring systems, etc.

Waste landfills represent a long-term process, including the long timescales relating to waste stabilisation in the landfill and the long timescales over which emission of landfill gases and leachate can occur together with the related aftercare monitoring and control. Consequently, the applicant for a permit to operate a landfill should demonstrate financial security to ensure that the commitments to safeguarding the environment are in place for the future. Since a landfill site is a major project with the potential to have a significant effect on the environment, a full-scale environmental impact assessment would be required by the EC Directive on the Assessment of the Effects of Certain Public and Private Projects on the Environment, 85/337/EEC. The operator of the site is required to report each year to the relevant Member State Environment Agency or Environmental Protection Agency, on the types and quantities of waste accepted and the results of the environmental monitoring programme.

The Directive also states that waste must be treated before it is landfilled. Treatment is defined in the Directive as the physical, thermal, chemical or biological processes, including sorting, that change the characteristics of the waste in order to reduce its volume or hazardous nature, facilitate its handling or enhance recovery.

The basis of the Directive is to harmonise standards throughout the EU for waste landfill facilities on the basis of a high level of environmental protection. It is clear that the European Commission regards landfilling of waste as the least favourable option due to the fact that 'landfilling does not make use of waste as a resource' and may result in 'substantial negative impacts on the environment'. The most important of which have been highlighted by the Commission and include 'emissions of hazardous substances to soil and groundwater, emissions of methane into the atmosphere, dust, noise, explosion risks and deterioration of land' (Commission of the European Communities, COM (97) 105 Final 1997).

Biodegradable waste is biologically degraded in the landfill to produce gas and liquid leachate. Biodegradable waste is therefore the source of landfill gas which is composed of mainly methane and carbon dioxide which are greenhouse gases, contributing to global warming. The reduction in biodegradable wastes going to landfill has a central aim of reducing emissions of methane as a major greenhouse gas in line with a further European Community Strategy in relation to climate change and reduction of methane emissions. To comply with this Strategy, reducing the contribution of methane to the atmosphere by reducing the biodegradable waste going to landfill is seen as the most effective and cheapest option. Consequently, the Waste Landfill Directive sets targets to reduce the amount of biodegradable waste sent to landfill to:

- 75% of 1995 levels by 2006;
- 50% of 1995 levels by 2009;
- 35% of 1995 levels by 2016.

Countries that landfilled more than 80% of their municipal solid waste in 1995 can extend the deadlines shown by four years to 2010, 2013 and 2020 respectively. Within the principles of subsidiarity, Member States are allowed to decide how the reduction targets for biodegradable waste are to be met. Several European countries, including Germany, Austria, Finland, France and the Netherlands have already introduced limits or guidelines for biodegradable wastes going to landfill (Commission of the European Communities, COM (97) 105 Final 1997). The Directive also states that the diversion of biodegradable waste away from landfill should aim to encourage the separate collection of biodegradable waste and increased recovery and recycling.

Very detailed documentation on the Best Available Techniques to be adopted to operate the landfill will be provided via a Reference Document by the IPPC Technical Working Groups of the European IPPC Bureau. However, the Directive itself contains significant operational detail required to be implemented by the EU Member States, regarding the landfilling of wastes. For example, the acceptance of waste at the landfill site involves the checking of waste characteristics documentation, visual inspection of the waste and maintaining a register of the types, origin, producer and collector of the waste. In the case of hazardous waste, the precise location of the place where the waste was deposited on the site should be documented. The minimum permeability and thickness of the landfill liner system, which contains the waste and acts as a barrier to the surrounding environment, is stipulated for the three different types of landfill. In addition to the barrier system a leachate collection and bottom-sealing system should be incorporated and the collected leachate must be treated to meet the appropriate environmental standard. The landfill gas resulting from the biodegradation of waste must be collected via a system of porous pipework within the landfill site and the gas treated and used. If the gas collected cannot be used to produce energy it should be flared. Procedures for the sampling and analysis of leachate and gases (methane and carbon dioxide) are required to be set up, including monitoring outside the landfill site to assess any impact on the surrounding environment. The monitoring of the site is carried out during operation but also after the site is closed, for example, at six-monthly intervals, until the regulating authority agrees that stabilisation of the site has occurred and the landfill poses no hazard to the environment.

1.5.8   End-of-Life Vehicle Directive

The EU End-of-Life Vehicle (ELV) Directive (Council Directive 2000/53/EC 2000) was introduced to address the problem of the estimated 14 million motor vehicles scrapped in the Europe Union each year. Seventy-five percent of the typical vehicle is composed of ferrous and non-ferrous metals which are readily recycled back into the metals industries. However, the remaining 25%, which is composed of plastics, rubber and other components, is currently disposed of to waste landfill. The European Commission with its increasing emphasis on the re-use, recycling and recovery of waste, views this as a waste of a resource and potentially harmful to the environment. The approach of the EU to the problem of vehicle waste is through minimising waste through reduced use of the raw materials of the car and improved product design, coupled with increased recycling and re-use of waste. The End of Life Vehicle (ELV) Directive has therefore set stringent targets for the recovery of scrap vehicles (cars and vans). The Directive requires Member States of the EU to re-use and recover 85 wt% of the average vehicle weight by 2006, increasing to 95 wt% by

2015. The Commission makes the distinction in the Directive between targets for re-use and recycling which must make up 80 wt% of the 2006 target and 85 wt% of the 2015 target. The remaining percentage is a recovery target, which is mainly achieved through incineration with energy recovery.

Previous legislation enacted by the EU concerned particular waste streams associated with scrap motor vehicles, including the disposal of waste oils (Council Directive 75/439/EEC amended by Council Directives 87/101/EEC and 91/692/EEC) and waste batteries (91/157/EEC, amended by Council Directives 93/86/EEC and 98/101/EC).

The End-of-Life Vehicle Directive applies specifically to cars and vans and stipulates (Europa 2003) that vehicle manufacturers and material and equipment manufacturers must:

- endeavour to reduce the use of hazardous substances;
- design and produce vehicles which facilitate the dismantling, re-use, recovery and recycling of end-of-life vehicles;
- increase the use of recycled materials in vehicle manufacture;
- ensure that vehicle components do not contain mercury, chromium(VI), cadmium or lead (batteries are exempt).

Member States of the EU must also set up collection systems for scrap vehicles and for waste used parts and then transfer these to authorised treatment facilities. A 'free-take-back' system enables the vehicle owner to take the car or van to the authorised treatment facility. The principle of 'producer responsibility' applies to the end-of-life vehicle, in that the automotive manufacturer must meet all or a significant part of the free-take-back costs. The treatment facility must dismantle the vehicles before treatment to recover all the environmentally hazardous components such as the brake fluids, engine oil and battery acids. Priority must be given to recycling of vehicle components such as the batteries, tyres and engine oil. To aid the dismantling procedure, the vehicle manufacturers and component manufacturers must use material coding standards to aid the identification of the material of components. In addition, the vehicle manufacturers should provide information or manuals for the dismantling of vehicles to enable the recovery and recycling of components.

The Directive also sets out the requirements for the vehicle treatment or dismantling facility. A suitable storage area on site is required to store vehicles safely before dismantling. The site should also have a range of suitable containers for storage of the dismantled spare parts, storage tanks for hazardous liquids and storage for used tyres and glass. The dismantling process is divided into a de-pollution operation and a component recycling operation. De-pollution involves the removal of batteries, liquefied gas tanks, removal of potentially explosive air bags and removal of components which contain mercury. Removal and separate collection of vehicle fluids including fuel, engine oil, transmission oil, hydraulic oil, antifreeze, brake fluids and air conditioning fluids, is required. Regarding component recycling, this involves the removal of catalysts, tyres, glass, large plastic components such as bumpers and components containing copper, aluminium and magnesium.

1.5.9 Packaging and Packaging Waste Directive

The Packaging and Packaging Waste Directive (Council Directive 94/62/EC 1994) was introduced in 1994 as a means of harmonising across the European Union, the various measures taken within the Member States concerning the management of packaging and packaging waste. A priority in terms of management was the prevention of packaging waste by, for example, reductions in the amount of packaging used, an increased level of re-use of packaging and recycling of the packaging waste. Using these measures, the overall level of waste packaging would be reduced. Through the Directive, targets for recovery and recycling of packaging waste for each Member State have been set by the EU, consisting of between 50 and 65% by weight for recovery and, within that general target, between 25 and 45% by weight for recycling, for the total packaging waste in each Member State. A minimum of 15% by weight for each packaging material was also set. Recycling is defined as the reprocessing in a production process to produce new packaging materials or materials for other uses. Whereas 'recovery' is a defined process as set out in the Waste Framework Directive (Council Directive 75/442/EEC) of 1975, processes shown in Table 1.3. As such the *recycling* target cannot include incineration of the packaging waste for energy recovery.

Packaging covers materials such as plastic and glass bottles, paper, plastic and cardboard wrapping, cartons, aluminium and steel food and drink cans. Industrial and commercial packaging includes metal and plastic drums, wooden pallets and cardboard and plastic crates and containers. All types of packaging are covered by the EU Directive. The Directive requires EU Member States to ensure that systems are set up for the return and collection of used packaging and packaging waste from the consumer or from the waste stream, in order to channel it towards the most appropriate waste management process for re-use, recycling or recovery.

Packaging containing trace levels of heavy metals is controlled by a gradual reduction in the concentration of lead, cadmium, mercury and chromium(VI) over a five-year period. Member States are also required to set up databases, detailing the amount and characteristics of packaging waste and the levels of recycling, recovery and disposal. A proposal for an identification system for packaging was also outlined in the Directive using a numbering system for plastics, paper and cardboard, metal, wood, textiles and glass.

1.5.10   Waste Electrical and Electronic Equipment Directive

Total electrical and electronic equipment waste generated in Europe is estimated at up to 6 million tonnes annually (COM (2000-0158) 2000). The waste stream has been identified as one of the fastest growing waste streams in the EU, constituting 4% of municipal waste and increasing at a rate of between 16 and 28% every five years (Europa 2003). It is also one of the largest known sources of heavy metals and organic pollutants in municipal solid waste. Across the EU, the majority of the waste, more than 90%, is disposed of mainly to landfill and incineration (Waste Management World 2002).

The proposed Waste Electrical and Electronic Equipment Directive (COM (2000-0158) 2000) aims to promote the re-use, recycling and other forms of recovery of electrical and electronic waste. Thereby the Directive seeks reduce the amount of such waste and bring about an improvement in the environmental performance of the industry (Europa 2003). Coupled with the proposed Directive is a further proposed EU Council Directive that aims to restrict the use of hazardous substances in electrical and electronic equipment (COM (2000-0159) 2000). The restriction on the use of hazardous substances in electrical and electronic equipment, such as heavy metals and brominated flame retardants, outlined in the second Directive will improve the re-use, recycling and recovery of the equipment required in the first Directive. The manufacture of electrical and electronic equipment also requires the intensive use of natural resources. Consequently, increased levels of recycling will lead to significant savings of natural resources. The prevention of pollution and a reduction in the wasteful use of natural resources was a key objective of the fifth European Environmental Action Programme (Europa 2003).

Included in the category of waste electrical and electronic equipment are:

- Household appliances, e.g., refridgerators, washing machines, microwave ovens, irons, toasters, hair dryers, etc.
- Information technology and telecommunications equipment, e.g., personal computers, printers, telephones, mobile phones, calculators, mainframe computers, etc.
- Consumer equipment, e.g., radios, television sets, video recorders, musical instruments, etc.
- Lighting equipment, e.g., fluorescent lamps, sodium lamps, metal halide lamps etc.
- Electrical and electronic tools, e.g., electrical drills, electrical saws, sewing machines, etc.
- Toys, e.g., electric trains, car racing sets, video games, etc.
- Medical equipment systems, e.g., radiotherapy equipment, dialysis equipment, analysers, freezers, etc.
- Monitoring and control instruments, e.g., smoke detectors, heating regulators, thermostats, etc.
- Automatic dispensers, e.g., hot drink dispensers, cold drink dispensers, automatic dispensers for solid products, etc.

The Directive requires Member States to set up collection systems where the consumer will be able to return the equipment free of charge. The Directive therefore requires the separate collection of electrical and electronic waste as a separate waste stream. In line with the polluter responsibility principle, producers of the equipment are required to organise and finance the treatment, recovery and disposal of the waste. Underlying the principle of polluter or producer responsibility gives the manufacturer the incentive to develop approaches to design and manufacture of their products to ensure the longest possible product life and, in the event that it is scrapped, the best methods for recovery and disposal. The collected waste should be transferred to an authorised treatment facility where any potentially hazardous components and materials such as fluids, asbestos waste, batteries, mercury-containing components, polychlorinated biphenyl-containing capacitors, cathode ray tubes, liquid crystal displays, etc. are removed. The separate components would be stored separately in separate containers.

The associated proposed Directive on the restriction of the use of hazardous substances in electrical and electronic equipment provides for the reduction of certain hazardous substances used in the manufacture of the equipment. Significant levels of lead, mercury, cadmium, chromium(VI), polybrominated biphenyls and polybrominated diphenylethers are currently found in electrical and electronic waste. For example, mercury is found in fluorescent lamps and laboratory equipment, lead in the glass of cathode ray tubes, cadmium in photocells, chromium(VI) is used as an anti-corrosion additive in certain refridgerators. The brominated biphenyls and diphenylethers are used as fire retardants in a range of electrical and electronic equipment. By restricting the use of such hazardous materials in the manufacture of new equipment and diverting the waste away from landfill and incineration, through increased levels of recycling and recovery, the environmental problems associated with these hazardous chemicals could be significantly reduced. Consequently, the manufacturers are required to phase out the use of the hazardous materials and substitute other materials.

1.5.11   Waste Management Statistics Regulation

An essential requirement for the effective monitoring of the various policy initiatives of the EU regarding waste management requires detailed statistical knowledge of the production of waste, recycling rates, re-use and disposal of waste in each Member State of the EU. Therefore, the European Commission has introduced a draft Regulation (Box 1.4, COM (99) 31 Final 1999). The Regulation aims to harmonise the collection of waste management statistics across the EU in order to have comparability of results between Member States of the EU. Statistics are to be prepared and made readily available in the following areas:

- Waste production and recycling in relation to economic activity, for example, mining and quarrying, manufacture of food products, manufacture of chemicals, construction industry, etc.
- Household and similar waste collection by businesses and municipal collection schemes including household waste, separately collected wastes, such as paper and cardboard, glass, plastics, end-of-life vehicles, textiles, etc. For each category of waste, the quantity recovered, incinerated and landfilled is required.
- Waste incineration, composting and disposal by businesses and municipal authorities. The number and capacity of waste treatment facilities, quantity of waste treated, recycling facilities, imported waste treated, energy production, etc., are required.

The different categories of waste used in compiling the waste management statistics are based on those of the European Waste Catalogue (Commission Decision 2000/532/EC 2000, Section 1.2.4). All the statistical data provided by the Member States is transmitted to Eurostat, the Statistical Office of the European Union.

1.6   The Economics of Waste Management

1.6.1   Economic Policy Instruments

The key instrumentation of EU policy in regard to waste management is the use of the legislature through the implementation of the various waste management Directives, Regulations and Decisions of the European Commission. However, other policy instruments are available to Member States, including economic instruments such as taxes, charges and trading tariffs. These have the impact of delivering the objectives of the waste management strategy using the incentive of the price mechanism (Gervais 2002(b)). There are no European Union-wide waste management taxes or charges. However, the EU encourages the use of such environmental taxes introduced by the individual Member States (Gervais 2002(b)).

For the manufacturer, increasing the costs of waste disposal through taxation or charges, or the setting of regulatory recycling targets, encourages the development of cleaner, lower waste-producing processes and is an incentive to recycle. Underlying this approach is the 'polluter pays' principle, which was introduced in the Fifth Environmental Action Programme (1993). The producer of waste thereby has the incentive to develop production processes that minimise waste or increase the levels of recycling in order to minimise their costs. The obligation on the producer to recycle or recover waste is most easily achieved where there is a clearly identified specific waste stream, such as packaging waste or waste electrical and electronic equipment.

On a wider scale, the introduction of taxes such as a landfill tax or an incineration tax has the impact that businesses and local authorities who bear the increased cost are encouraged to increased levels of recycling and a reduction in waste production. There are a number of countries throughout Europe that have introduced a landfill tax. Table 1.8 shows examples of landfill tax and incineration tax introduced in Europe (Waste Not Want Not 2002). Landfill tax has resulted in the costs of landfilling better reflecting the environmental impact of landfilling of waste. In addition, it has encouraged industry and consumers to produce less waste and provided an incentive to re-use or recycle waste (ENDS Report 265, 1997). Alongside the introduction of a landfill tax would be the impact of legislation, such as the EU Waste Landfill Directive and mandatory recycling targets such as those in the EU Packaging and Packaging Waste Directive. In such cases for Member States of the EU they would use the implementation of the landfill tax as a means to divert waste away from landfill to more recycling by increasing the costs of waste disposal to landfill for businesses and local authorities. Underlying the introduction of an incineration tax would be voluntary or mandatory recycling targets.

The use of direct charging of the public for waste management is an economic instrument to encourage more public participation in recycling. Direct charging involves the direct charging of households for the volume or weight of waste produced or the frequency of the collection. There would be no charge for the collection of the separated recyclable

Table 1.8 *Landfill tax and incineration taxes in selected EU Countries*

	Country	Tax Rate
Landfill Tax		
	Denmark	€50/tonne
	The Netherlands	€12.6–65.4/tonne depending on type of waste
	Ireland	€19/tonne
	Sweden	€31.3/tonne
	France	€9.1/tonne
	Austria	€43.6/tonne for treated waste only
	UK	€22.4/tonne
	Belgium	€52–55/tonne for thermally treated MSW
	Italy	€10–50/tonne depending on region
Incineration Tax		
	Belgium	€6–20/tonne
	Denmark	€38/tonne
	Norway	€9.4–28.3/tonne depending on plant efficiency

Source: Waste Not Want Not, 2002.

fraction. Direct charging of communities for waste production has been practised in the USA for more than 80 years, resulting in an average reduction of waste away from land-fill of 40% and in some communities up to 74% (Waste Not Want Not 2002). In some EU countries, direct charging for the amount of waste produced is widely practiced in, for example, Austria, Belgium, Luxembourg, Sweden, the Netherlands, Denmark and Germany. For example, in the Netherlands, charges are made to households on the basis of both volume and frequency of collection and volume and weight of waste collected, with differential tariffs for each option. Sweden has direct charging systems for waste based on the volume, frequency of collection, and the weight of waste collected. Germany has a charging system for waste collection based on the number of containers, size of container and frequency of collection. There are dangers that direct charging of households for waste production may result in an increase of 'fly tipping', the illegal dumping of waste. However, this has been less of a problem where recycling facilities are readily available to the public or door-step collection of recycled materials has been implemented (Waste Not Want Not 2002).

Other economic instruments suggested to control the various options for waste management are tradeable landfill permits or allowances, which are being introduced in the UK for biodegradable waste (Waste Not Want Not 2002). The system requires the development of a market for waste going to landfill between local authorities. Each local authority is allocated a certain tonnage of waste or an allowance. The authority will only be allowed to send waste to landfill up to the levels of the allowance which they hold. However, where a local authority diverts waste away from landfill by, for example, increased recycling or composting, the authority may sell their unused allowance to another local authority who have more waste to landfill than the allowance allocated to them permits. This trade in landfill allowances generates a market, bringing income to high recycling authorities, with low rates of waste going to landfill, but increased costs to a local authority with low recycling, and high levels of waste sent to landfill. This landfill allowance market acts as an incentive to encourage recycling and diversion of waste away from landfill, consequently controlling waste management.

1.6.2   The Costs of Waste Management

There is increasing awareness throughout the EU that waste treatment and disposal options should reflect their full environmental and economic costs, as outlined in the fifth European Action Programme (1993–2000). Consequently, to the capital and operational costs of the particular waste management option chosen, should also be added the external costs to the environment. In addition, implicit in the hierarchy of waste management is that disposal of waste by landfill is the least desirable option, followed by waste incinera-tion, recycling and re-use, with waste minimisation being at the top of the hierarchy. Therefore, the lowest cost option may not be the most environmentally acceptable option, or may not fulfil obligations required either by the EU legislation, the EU industry or waste sector targets. In general, the costs of waste management are dependent on the type of waste requiring treatment and the associated environmental hazard. For example, the disposal costs of inert waste into landfill are very low compared with very expensive, hazardous chemical waste incineration. The wastes such as chemical sludges, chlorinated solvents, scrap tyres, clinical waste, etc., are difficult, and consequently more expensive, to dispose of compared with domestic and commercial waste.

In many countries throughout the EU, landfilling wastes represent the cheapest option (Eunomia 2003). Incineration is capital intensive and also generally has higher operating costs than landfill. However, offset against this fact is the greater income from the sale of heat, power or steam. Capital costs for incinerators have increased in recent years to meet EC emissions limits and represent a major cost of the overall plant. It is often difficult to assess the cost in general terms of incinerators, particularly large-scale municipal incinerators, where the costs are site-dependent. For industrial waste incineration, costs will depend on the calorific value of the waste, the amount of gas clean-up, the throughput, the requirement for the energy generated, etc. Hazardous waste disposal costs, in particular, tend to be high due to the greater technical requirements of the plant. Landfill costs are also expected to increase as the implementation of the EC Waste Landfill Directive will mean more stringent gas and leachate collection, treatment and monitoring and aftercare requirements. Additionally, the air pollution control residue from incineration, which typically represents less than about 5% by weight of the original waste, is regarded as hazardous and also requires disposal by expensive special landfill. The bottom ash residue is not regarded as hazardous and may be sold as building aggregate, but in some cases where ready markets do not exist, may also have to be landfilled, an additional cost to the incineration process.

The influence of energy recovery on both landfill and incineration costs can significantly reduce the costs of disposal. Municipal solid waste has a significant energy content (calorific value) and the combustion of the waste in an incinerator can be used to generate electricity, or the heat can be used for district heating. Similarly, the generation of combustible methane generated from the anaerobic decomposition of the waste in a landfill site can be also be used to generate power or heat. Selling the energy produces income for the facility and reduces costs. Other wastes, such as clinical waste, dried sewage sludge and scrap tyres, have high calorific values and incinerators with energy recovery are available. A significant influence on disposal costs for landfill, in particular, is transport cost. Increasingly, suitable landfill sites close to the point of waste generation are becoming scarcer and in some cases waste is being transported over long distances to the landfill site. Waste management facilities such as incinerators, recycling and composting plants, which can be sited within city limits, clearly have a distinct advantage in such cases. Also, as the scale of operation is increased for either landfill or incineration, the gate fee becomes reduced. For example, a landfill site or incineration plant of 300 000 tonnes per annum throughput has an estimated gate fee (cost/tonne) of almost half that for a 100 000 tonne per annum facility (Royal Commission on Environmental Pollution 1993). To support such large facilities, with their associated large capital investments, a long-term guaranteed supply of waste is required and, where energy recovery is employed, a guaranteed long-term end market for the energy.

In regard to the costs of a particular waste treatment and disposal option, the difference between the 'cost' and the 'gate fee' should be determined. The cost usually given as cost/tonne, represents all the costs involved in the process, including capital and operating costs, repayment of debts, etc. The gate fee represents the price, usually as a price/tonne, that the operating company can get for the service, which depends on the market place.

Table 1.9 *Comparative costs per tonne collected of residual waste collection in selected different member States of the EU (€/tonne)*

Country	Costs			Collection frequency
	Low	High	Best Estimate	
Austria			70	Every 2 weeks, sometimes more frequently in summer
Belgium	58	92	75	Mostly every 2 weeks, sometimes weekly
Denmark			126	Weekly
Finland	15^U	32^R		Weekly, biweekly or monthly depending on route or area
France	54^U 63^R	65^U 74^R	60^U 70^R	Examples include 5 times per week urban areas and 2 times per week in rural areas
Germany	39^U 48^R	81^U 91^R	67^U 71^R	May be every 2 weeks, weekly in summer months
Greece	25^U 40^R	36^U 67^R	30^U 55^R	Ranging from daily for some urban areas, weekly for some rural
Ireland	60	70	65	Weekly
Italy	48	255	75	Weekly or twice weekly in cost optimised systems collecting food waste, may be 3 or 4 times daily in some areas with no food waste collection
Luxembourg	85	104	85	Every 2 weeks
The Netherlands	75	123	100	Weekly
Portugal			45^E	
Spain	19	91	60	Likely to be daily in urban areas
Sweden	59	80	65	Every 2 weeks in single family houses, weekly in urban areas with multi-occupancy buildings
UK	32^U 50^R	50^U 80^R	42^U 60^R	Usually weekly, a few local authorities alternate residual waste collection with collection of biowaste (2 weekly)

^{U} = Urban, ^{R} = Rural, ^{E} = Estimated.

Source: Eunomia 2003.

However, for municipal solid waste, the market for waste management tends to be poorly developed. For example, for most municipalities there is no choice of a range of landfills, with incinerators or recycling centres competing for the waste generated. Contracts tend to be long-term to enable the development of high-cost facilities, such as waste incinerators, or large-scale projects, such as landfills.

Collection Comparative costs of waste collection are shown in Table 1.9 (Eunomia 2003). The data refer to residual waste collection, which represents municipal solid waste after the potentially recyclable materials, such as dry recyclable materials and biowaste, such as garden waste or food waste, have been removed. The costs of collection of the recyclable materials are not included here. The cost comparisons in Table 1.9 are very heavily influenced by the local conditions which apply; For example (Eunomia 2003):

- Variations may occur in the number of collection points per unit time which are passed by the collection vehicle, which is influenced by population density and local traffic conditions.
- The nature of the collection containers such as the size of bins, number of bins, bags, costs of containers, etc., which all influence the costs of collection and time taken for collection.
- A variation in the amount of source-separated waste collected will influence the amount of residual waste collected, which should reduce costs.
- The size of the vehicle, since larger vehicles can carry more waste and consequently larger rounds can be programmed.
- Labour costs, which vary with the number of operatives, the nature of the area from which the waste is collected, such as density of housing, and the unit labour costs in each country. The labour costs often represent a substantial fraction of the total collection costs.
- The frequency of collection, which is related to the type of housing stock, the collection mechanism, the climate and the presence of food waste. Where malodorous biowaste, such as kitchen waste, is removed from the waste stream for composting, this allows longer times between collections to be used at lower costs, since the residual waste does not therefore represent a nuisance.
- The sophistication of the collection vehicle.
- The transport costs to the final destination. For example, where the final destination is a landfill site, this may involve additional significant transport costs.

From Table 1.9, there are no clear patterns of collection costs between countries. The labour costs involved in collection are a major proportion, typically 50%, of the total costs and therefore differences between countries may be reflected in differences in labour costs.

Reporting of collection costs alone may also be misleading in that the costs do not include the approach used to recycle, whether a bring or collect system is used. Higher recycling rates are achieved with a collect system, but are more costly. Also, whether the sorting is undertaken at the vehicle, which increases the collection costs, or at a materials recycling facility, where collection costs are not included. For collection of recyclable materials, the influences on cost include the size of the recycling installation, the quality of the incoming and outgoing flows of recyclable material, the proportion of non-treatable material, the degree of automation and the utilisation rate of the recycling installation (Hannequart and Radermaker 2003).

Treatment   The costs of the main treatment options for municipal solid waste in the EU, namely incineration and waste landfill, are shown in Tables 1.10 and 1.11. The variations shown for each treatment option are variable, due to a range of site-specific characteristics, the technologies used, the costs of land, the scale, etc. (Eunomia 2003). For example, specific influences on the costs of incineration (Table 1.10) include:

- Costs of land acquisition.
- The scale of the incineration plant, since economies of scale lead to much higher costs for the smaller-scale plant of less than approximately 50 000 tonnes per year throughput.
- The utilisation rate of the plant.

Table 1.10   *Comparative costs of municipal solid waste incineration in selected different Member States of the EU*

Country	Costs (net of income)	Energy Supply Income (€/kwh)	Costs of Ash Treatment (€/tonne)
Austria	326 @ 60 ktpa 159 @ 150 ktpa 97 @ 300 ktpa	Electricity 0.036 Heat 0.018	Bottom ash 63/t Flue gas residues 363/t
Belgium	71–75 @ 150 ktpa (not including tax)	Electricity 0.025	
Denmark	30–45/tonne (not including tax)	Electricity 0.05	Bottom ash 34/t Flue gas residues 134/t
Finland	None	For gasification; Electricity 0.034 Heat 0.17	
France	118–129 @ 18.7 ktpa 86–101 @ 37.5 ktpa 80–90 @ 75 ktpa 67–80 @ 150 ktpa	Electricity 0.023	13–18/t input
Germany	250 @ <50 ktpa 105 @ 200 ktpa 65 @ 600 ktpa	Electricity 0.046	Bottom ash 28.1/t Flyash and flue gas control residues 255.6/t
Greece	None		
Ireland	46 @ 200 ktpa[E]		
Italy	41.3–93 @ 350 ktpa	Electricity 0.14 (old) 0.04 (market) 0.05 (green certificate)	Bottom ash 75/t Flyash and flue gas control residues 129/t
Luxembourg	97 @ 120 ktpa	Electricity 0.025[E]	Bottom ash 16/t input waste Flue gas residues 8/t input waste
The Netherlands		Electricity 0.05[E]	
Portugal	46–76[E]		
Spain	34–56	Electricity 0.036	
Sweden	21–53	Electricity 0.03 Heat 0.02	
UK	69 @ 100 ktpa 46 @ 200 ktpa	Electricity 0.032	Bottom ash recycled Flyash 90/t[E]

ktpa = kilotonnes per annum, [E] = estimated.
Source: Eunomia 2003.

- The requirements of flue gas treatment required by each Member State. Although the EU Waste Incineration Directive sets out minimum requirements in regard to flue gas clean-up of pollutant emission, Member States can impose stricter requirements which will influence capital and operational costs.
- The treatment and disposal process for the ash residues. In some Member States a large proportion of the bottom ash is sold for secondary building aggregate, whilst in other countries it is landfilled with consequent added cost. The flyash is a hazardous waste and requires special treatment, such as stabilisation or landfilling in a high-cost hazardous landfill site. The required treatment for flyash will vary from country to country.

Table 1.11 *Comparative costs of municipal solid waste landfill in selected different Member States of the EU*

Country	Costs (excluding tax) (€/tonne)	Gate fee (excluding tax) (€/tonne)	Tax (€/tonne)	Total costs (€/tonne)
Austria	67		43	110
Belgium (Flanders)		47.5	52–55	100
Denmark		44	50	94
Finland		37–46	15	52–61
France	31–85		9	40–94
Germany	20 @ 300 ktpa 51 @ 50 ktpa	35–220		30–51
Greece				9–30
Ireland		35–78	19	60–95^E
Italy	52@1.25×10^6 m^3 site		Variable tax based on region and pre-treatment	70–75
Luxembourg	123–147			123–147
The Netherlands		43–100	64	107–164
Portugal		6–15^E		6–15
Spain	25–35^E	6–40		25–35
Sweden		20–60	30.6	50.6–90.6
UK	28 @ 175 ktpa @1.75×10^6 m^3 site	8–35	19.2	40–48

ktpa = kilotonnes per annum, ^{E} = Estimated.

Source: Eunomia 2003.

- The efficiency of energy recovery of the incineration plant and the income received for the electricity or heat supplied.
- The recovery of metals from the bottom ash and the income from their sales will influence the overall costs of incineration.
- In some countries such as Denmark and Belgium (Flanders), there is a separate incineration tax, which will be added to the costs.

For landfill sites (Table 1.11, Eunomia 2003), the costs are influenced by:

- Site acquisition costs. These will vary from country to country, from region to region and even from landowner to landowner. Additionally, the land may be leased from the owner or bought outright or may involve a royalty payment to the owner.
- Site technical requirements, which include the requirements for the landfill liner system, the geology and hydrogeology, etc. The Waste Landfill Directive (Council Directive 1999/31/EC) requires a high level of environmental protection and monitoring for landfilling of waste through the use of a landfill liner system which contains the waste and involves leachate and gas collection systems, which can be expensive. Sampling and analysis of landfill gas and leachate are required, which also add to the costs.
- The scale of the landfill, which influences the rate of filling of the void space with waste and the overall total tonnage of waste than can be landfilled. Together, these factors determine the period over which the waste can be accepted at the site and hence the

time period over which capital can be depreciated. The longer timescale over which depreciation of capital occurs results in lower annual costs. Longer timescales also allow a fund to be built up to pay for the aftercare costs of the site.

- In some cases, landfill gases are collected and combusted to produce energy, this will generate income from energy or heat sales.
- Aftercare costs. These include site restoration and gas and leachate sampling and analysis. Additionally, because the landfill is a long-term process, the waste Landfill Directive (Council Directive 1999/31/EC) requires that financial provision or a fund is made available for the full aftercare lifetime of the landfill. This fund represents a cost to the project.
- Landfill tax. Different Member States of the EU have different rates of landfill tax, which influences the total costs of waste landfill.

Underlying the costs associated with waste treatment and disposal are the systems in place to manage the waste and the funds allocated to waste management (ISWA 2002). In fully developed countries, such as Western Europe and North America, the allocation of resources for waste management is assured by the established structures in place and by the public's willingness and capacity to pay for those services. In contrast, many developing countries have a public who are generally not concerned about waste and therefore the allocated resources for waste management are low (ISWA 2002). Table 1.12 shows a comparison of the total expenditure in a range of city urban areas and the proportion of expenditure allocated to waste management for each city. The data are in US dollars/capita in relation to the year the data was collected (ISWA 2002). Also shown are the Gross National Product (GNP)/capita and the percentage of the GDP/capita spent on waste management for each country. There are wide variations in the proportion of each city's income that is spent on waste. The data are only indicative, since each city will have unique waste management services and citys will have different accounting systems.

Table 1.12   *Comparative expenditure on urban waste for selected world cities*

City	Year	Total urban expenditure US$/capita	Solid waste expenditure US$/capita	GNP/capita	% GNP spent on waste
Accra	1994	2.76	0.66	390	0.17
Ahmedabad	1995	24.27	1.61	350	0.46
Bogata	1994	—	7.75	1620	0.48
Bucharest	1995	94.75	2.37	1450	0.16
Budapest	1995	310	13.8	4130	0.33
Chennai	1995	14.75	1.77	350	0.51
Dhaka	1995	8.31	1.46	270	0.54
Hanoi	1994	—	2.0	250	0.80
Mumbai	1995	63.65	3.92	350	1.12
New York	1992	5804	97	23 240	0.42
Riga	1995	153	6.0	2420	0.25
Strasbourg	1995	1600	63	24 990	0.25
Sydney	1995	—	38	18 720	0.20
Toronto	1994	2043	48	19 510	0.25

Source: ISWA 2002.

However, broad comparisons can be made, for example, since the fully developed cities of New York, Sydney and Toronto are in countries which have high GNPs and consequently the total expenditure per head in each city will be high, leading to a greater amount of funds available for waste management. Conversely, the public in countries with low GNPs will have less to spend on the total urban expenditure, leading to a lower amount of money spent on waste management for the cities in those developing countries, for example, Mumbai, Chennai, and Ahmedabad.

1.6.3 Project Financing

Large-scale waste treatment and disposal projects, whether mass-burn municipal solid waste incineration, waste landfill, or hazardous waste treatment facilities, require a large-scale investment. Local municipalities often do not have the large financial inputs required which may represent tens of millions of pounds investment. Increasingly, such projects are developed as joint ventures between the municipality and the private sector, although private sector involvement differs between Member States of the EU (Eunomia 2003). Table 1.13 shows examples of the different involvement of municipalities and the extent of private sector involvement in waste management in the EU (Eunomia 2003). Private sector finance, particularly for the larger projects, would require bank loans, secured against on-going company profits and assets. The risks for such a loan lie with the borrowing company (Chappell 1995).

An alternative approach is 'project financing', where the risk to the investor is based on the success of the project. Repayment of the loan comes from the financial success of the project. Where the project fails there is no recourse to recover the debt from the company.

Table 1.13 *Examples of the different involvement of municipalities and private sector involvement in waste management in the EU*

Country	Municipal Involvement	Private Sector Involvement
Austria	Municipalities Waste Associations	Residual waste collection – 50% Recycling – 80% Composting – 50% Increasing in residual waste management
Belgium	Municipalities Inter-municipal waste associations Net Brussels	Collection and treatment often contracted out to private sector Brussels – collection and treatment is contracted out to Net Brussels
Denmark	Municipalities Inter-municipal waste associations	Collection – 80% Landfills and most incinerators are in public hands
Finland	Municipalities Co-operative municipal waste management companies	Municipalities dominate collection Private sector owns co-incineration and some other treatment plants
France	Communes Departements	Treatment – 28% (operating and capital) Collection – 50% (operating) Treatment – 54% (operating)

Table 1.13   *Continued*

Country	Municipal Involvement	Private Sector Involvement
Germany	Municipalities	Landfill – Few involvement of private sector Incinerators – Some private involvement Contracts through DUALES system
Greece	Municipalities	Limited private role in collection and transport
	Associations of municipalities	No involvement in recycling
Ireland	Municipalities	Movement from public to private sector in collection and treatment Involvement in most recycling schemes Collection – 40% involvement
Italy	Municipality	Collection – 46% involvement
	Ambito Territoriale Ottimale	Treatment mostly in public hands
Luxembourg	Municipalities	Incineration – some private sector involvement
	Inter-communal Syndicates	Composting – some private sector involvement Collection – some private sector involvement
The Netherlands	Municipalities	Collection – 33%
	Independent publicly owned companies	Residual waste treatment – little involvement Biological treatment – some involvement
Portugal	Municipalities	Involvement in treatment
Spain	Municipalities	Private sector involvement in recycling collection
	Autonomous regions	
	Public companies	
Sweden	Municipality	Collection – 60% private sector involvement
UK	Municipalities – divided into:	Collection – 50% private involvement
	Waste Collection Authorities	Treatment – almost all private sector, some landfills still owned by municipalities
	Waste Disposal Authorities	Some community level composting

Source: Eunomia 2003.

Contracting under such financial arrangements is complex, since each investor requires clearly defined risks and returns from their investment. Such projects are financed from various sources of funding. For example, a typical capital structure for a financing project is shown in Figure 1.4 (Chappell 1995). The senior debt is usually the major portion of the total cost and consists of a bank loan. Subordinate debt represent additional funds which bridge the gap between the large bank loan and the direct investing participants. The equity part of the finance would come from shareholders investments from the direct equity participants in the project, such as the fuel or waste supplier, i.e., the local authority, the energy purchaser, equipment supplier and project developer. The size of the bank loan will be determined by the risk involved in the project. The energy from waste schemes, such as incineration or landfill gas projects with guaranteed sources of waste and markets for the generated energy would be liable for such loans, whereas untried technologies would find financing difficult (Chappell 1995).

Each member of the project financing team will have different objectives for the project, for example, the local authority requires a waste treatment plant to dispose of its

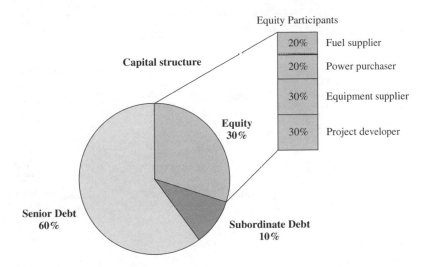

Figure 1.4   *Typical capital structure for project financing. Source: Chappell 1995.*

waste, the equipment supplier wishes to develop its business, the banks would merely require a return on investment. Therefore, detailed contract arrangements are required to apportion suitable risk and consequent return on the investment.

1.7   Options for Waste Treatment and Disposal

Waste management is a major issue in every European country. The generation of waste reflects a loss of materials and energy and results in economic and environmental costs on society for its collection, treatment and disposal (European Environment Agency 2003). Waste forms an increasing part of the total flow of materials through the European economy (Figure 1.5). It has been estimated that 3000 million tonnes of waste of all categories are generated throughout Europe each year. The hierarchy of waste management suggests that, over time, the management of waste will result in an increase in waste minimisation, re-use and recycling and a decrease in landfill. However, the current situation across Europe shows that landfilling of waste is the dominant disposal method for all categories of waste. Landfill involves the controlled and managed disposal of waste such as untreated or treated municipal solid waste, construction and demolition waste, industrial waste, etc., into a hole in the ground. The landfill in most cases would be lined to contain any leachate production. Landfill gas, which is composed of mainly methane and carbon dioxide, is derived from the biodegradation of the organic fraction of the waste. The amount of landfill gas produced would depend on the type of waste placed in the landfill, biodegradable wastes such as sewage sludge and the biodegradabale fraction of municipal solid waste, generate high concentrations of landfill gas. Construction and demolition waste, composed mainly of inert non-biodegradable materials, would produce low amounts of landfill gas and in some cases none, depending on the fraction of biodegradable waste, such as wood. Where a sufficient amount is generated, it would be collected and flared or used to

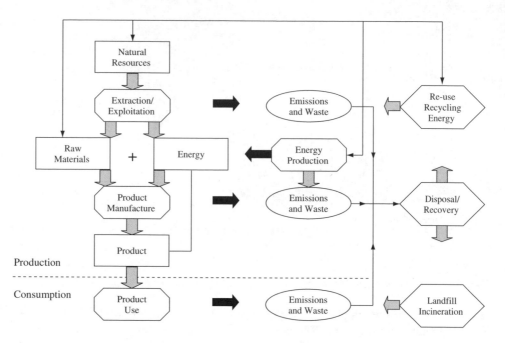

Figure 1.5 *Waste flows in relation to production and consumption of energy and raw materials. Source: European Environment Agency 2003.*

produce energy. Industrial and commercial wastes are also largely disposed of to landfill and, in many countries, often co-disposed with municipal solid waste. Although regarded as the lowest option in the hierarchy of waste management, there is an increasing emphasis on minimising the environmental impact of waste landfills by increasing the legislative control and monitoring of leachate and landfill gas.

Incineration of waste is suitable as a waste treatment option, depending on the category of waste and mainly applies to waste that is combustible such as municipal solid waste, some hazardous waste, clinical waste and some industrial waste. Incineration is the most widespread option for the treatment of solid wastes, after landfill. Incineration involves little pre-treatment prior to combustion. Municipal solid waste is incinerated in large-scale, mass-burn incinerators, which typically have throughputs of 50 000–400 000 tonnes of municipal solid waste per year. Investment in such large-scale incinerators is high and requires long contracts with guaranteed supplies of waste for the operational lifetime of the incinerator, which would be of the order of 25 years. Incineration of industrial waste, sewage sludge and clinical waste are typically smaller operations where throughputs are in the range of 10 000–20 000 tonnes per year. However, there is great public opposition to waste incineration. This is linked to the emissions which it generates. Of particular concern are the emissions of dioxins and furans, heavy metals and acid gases. However, the incineration industry, particularly in Western Europe, is highly regulated and subject to stringent emissions control.

Recycling is increasing in Western Europe, while Central and Eastern Europe and the countries of Eastern Europe, the Caucasus and Central Asia have relatively low recycling rates (European Environment Agency 2003). Western Europe comprises Austria, Belgium, Denmark, Finland, France, Germany, Greece, Ireland, Italy, Luxembourg, the Netherlands, Portugal, Spain, Sweden, United Kingdom, Iceland, Liechtenstein, Norway, Switzerland, Andorra, Monaco and San Marino. Central and Eastern Europe comprises Bulgaria, Czech Republic, Estonia, Hungary, Latvia, Lithuania, Poland, Romania, Slovak Republic, Slovenia, Cyprus, Malta, Turkey, Albania, Bosnia-Herzegovina, Croatia, Macedonia, Serbia, Montenegro. Eastern Europe, the Caucasus and Central Asia comprises Armenia, Azerbaijan, Belarus, Georgia, Republic of Moldova, Russian Federation, Ukraine, Kazakhstan, Kyrgyzstan, Tajikistan, Turkmenistan, Uzbekistan. The options for recycling would be dependent on the waste type. For example, construction and demolition waste has the potential to be recycled as secondary aggregate and in some countries the recycling of such waste is over 30%. In some countries such as Germany, Denmark and the Netherlands, recycling of construction and demolition waste has reached more than 90% (European Environment Agency 2003). Recycling of automobiles is also high at more than 75% due mainly to the recycling of the metal components. For certain wastes, such as hazardous waste, recycling is not an option and depending on the type of hazardous waste, incineration or biological, physical or chemical treatment is the best practicable environmental option. Recycling of municipal solid waste is dependent on the recycling of the components of the waste such as paper, plastics, metals and glass, where recycling rates differ for different categories, for example, across the OECD countries rates for recycling of more than 80% for metals, 35–40% for glass and 40–55% for paper and cardboard are reported (OECD 2004). Recycling rates vary between countries, for example, Ireland has a paper and cardboard recycling rate of 10% whilst in Germany it is 70% (OECD 2004).

By far the main route for the management of municipal solid waste in Western Europe and North America is landfill, followed by incineration, either with or without energy recovery, and then recycling and composting. For example, in 2000 the OECD countries, on average, landfilled 58% of municipal solid waste, incinerated 20%, of which 11% also included incineration with energy recovery, recycled 16% and composted 6% (OECD 2004). Composting is an aerobic treatment process which makes use of micro-organisms to breakdown the segregated biodegradable organic fraction of waste to carbon dioxide and water and to produce a residue suitable as a soil conditioner in agriculture, land reclamation or horticulture. Other treatment options for municipal solid waste include anaerobic digestion, which is carried out on the segregated biodegradable fraction of municipal solid waste under anaerobic conditions in a sealed vessel. The process produces a gas composed of methane and carbon dioxide and a residue suitable for agricultural and horticultural use, similar to compost. Mechanical and biological treatment (MBT) of MSW is a pre-treatment option prior to landfill. The raw MSW is treated by a set of mechanical or biological processes such as shredding, sieving, composting, anaerobic digestion, etc., which reduces the bulk of the waste and the proportion of biodegradable waste. Thermal treatment via pyrolysis and/or gasification techniques is receiving increased attention as an option for MSW.

1.7.1 The Choice of Waste Management Option

The EU strategy of sustainable waste management is being developed by encouraging upward movement through the hierarchy of waste management options. To attain that objective, a number of strategies have been used, including the strategy of setting targets on waste minimisation and recycling. Consequently, the economic case for each waste management option may not be the only criteria for the local authority decision makers. The concepts of sustainable waste management and also the principles of best practicable environmental option, will override the minimum cost option. However, the least attractive sustainable option of landfilling may be the best practicable environmental option in some cases, such as the restoration of former quarry workings by the disposal of residual, inert waste. Other guiding principles which should be taken into account in the choice of waste management option are: the proximity principle – that waste should be disposed of or dealt with close to the place where it arises; the self-sufficiency principle – that regions (and nations) accept responsibility for the wastes arising within them; and the 'polluter pays' principle – that the generator of the waste should pay for its disposal.

In selecting a waste disposal option for a particular waste the considerations which must be taken into account include the capital investment costs of the facility, the operating costs, decommissioning and aftercare, throughput of waste and environmental impact. These considerations are encompassed in the Best Practicable Environmental Option alongside the principle of sustainable development. There are a number of other factors involved in the choice of the waste management option chosen. The factors are summarised in Table 1.14 (Wilson et al 2001).

Part of the EU Waste Framework Directive requires The National Waste Strategy of the Member States to include the identification of suitable disposal sites or installations. However, the siting of waste treatment and disposal facilities has, in many cases, generated intense opposition in recent years, resulting in what has become known as NIMBY (Not In My Back Yard) syndrome. The public perception in relation to waste treatment and disposal facilities has been discussed in detail by Petts and Eduljee (1994). This opposition to local siting of such facilities has implications for the implementation of waste management strategies which encompass the proximity principle, whereby waste generated in a local area is the responsibility of that area.

Although for most countries the amount of municipal solid waste generated compared to the total waste arisings are small, that category of waste generates the most concern and attention. Increasingly, the public are concerned with the environmental impact of the various options for municipal solid waste management. This is consequently reflected in the legislature, which may seek to direct or encourage a particular waste option. The main environmental impacts of the main waste management options for municipal solid waste have shown that all waste management processes will have an environmental impact of some form or another (European Commission 2001). All waste management options involve vehicles transporting waste to and from the treatment plant, resulting in traffic congestion and emissions of carbon dioxide and other pollutants, noise and odour.

The environmental impact of landfilling of municipal solid waste arises through the emissions of landfill gas as methane and carbon dioxide, which contribute to global warming, and local impacts including odour and hazards, such as the risks of fires and explosions. The leachate formed from the degradation of the waste is hazardous, containing a range of chemicals and carries the potential risk of water-course pollution.

Incineration of municipal solid waste, impacts on the environment through the emissions of pollutants such as dioxins and furans, particulate matter, heavy metals, acid gases, such

Table 1.14 *Factors influencing the choice of waste management option*

Factors	Examples
Policy, management and institutional structures	• Local and regional politics and planning, strategy and stability • Societal support: public participation in the decision-making process • Political support • Form of government • Institutional and administrative structures for the management of waste • Managerial capacity and stability of personnel • Regulations and site specifications
Operational demands/ constraints	• Scale of operation in relation to waste generated • Infrastructure and security of waste supply and disposal • Existing contracts and obligations • Location and demography • Waste stream composition • Available technology versus proven technology – linked to cost
Economic and financial factors	• Available funding and subsidies • Costs of current system and other options • Best available technology • Local and regional budget issues • Economic tools used to influence waste management costs • Pricing system for waste services • Markets for secondary recycled materials
Legislation	• Prescriptive or enabling legislation, i.e., mandated targets for waste • International, national, regional and local legislation
Social considerations	• Public opinion and support of waste management in the region • Public participation in the decision-making process • Public ability to participate in an integrated waste management system (composting, recycling, etc. • Noise pollution, local pollution, increased vehicle and road traffic • Public resistance – not in my backyard (NIMBY)

Source: Wilson et al 2001.

as hydrogen chloride and sulphur dioxide and nitrogen oxides. The flyash and residues from the air pollution control system, require stabilisation and disposal as a hazardous waste. However, positive environmental impacts of MSW incineration are that the generation of energy from waste displaces the use of fossil fuels required to generate that energy, and bottom ash may be used as a secondary aggregate from which metals may be recovered for recycling.

Other municipal waste treatment options will have an environmental impact (European Commission 2001). Recycling municipal solid waste, generally results in overall energy savings since, in most cases, the manufacture of products from recycled materials requires less energy and hence lower overall emissions. Recycling prolongs the use of natural resources and reduces the environmental impact of producing virgin resources. Composting of the biodegradable fraction of MSW avoids the production of methane in landfills and produces a soil improver to replace fertilisers and peat. However, composting produces bio-aerosols and odour which require control. Mechanical and biological treatment (MBT) prior to final disposal into landfill reduces methane and leachate production, since the biodegradable fraction of the waste is degraded before landfilling. Materials may be recovered in the process for recycling and/or energy recovery.

Boxes 1.5–1.7 outline the management of waste in Japan, the USA and the UK.

Box 1.5
Waste Management in Japan

Japan has a population of 127 million people (2002) and because of the highly mountainous and volcanic nature of the country, only 10% of the land is suitable for residential purposes. The shortage of land in accessible areas limits the availability of suitable landfill sites and is the driving force behind Japan's waste management policy. Policies are based on waste reduction and recycling to minimise the amount of material that ultimately is destined for landfill, and the main route for waste disposal is incineration either with or without energy recovery. The main statutory control for waste management is set out in the waste Disposal Law 1970, which has had numerous subsequent amendments. Under the Law, waste is classified as either 'industrial' waste or 'general' (municipal) waste. The management of industrial waste, generated by business and industry, is the responsibility of the producer and is controlled by the 47 'Prefectures' which are the first tier of local government. The Prefectures authorise proposals for the construction or modification of industrial waste treatment and disposal facilities. Industrial waste production is estimated at 400 million tonnes per year (2000) and is dominated by waste from the iron and steel industry, animal waste, sludge, mining and construction industries. Approximately 50% of the industrial waste undergoes intermediate treatment, recycling is also a major route for the generated waste and the remainder is landfilled. Whilst the responsibility for treatment and disposal of industrial waste is the responsibility of the producer, increasingly the waste is treated by the municipalities.

The management of general waste which consists mainly of domestic waste and some commercial waste is the responsibility of the second tier of local government, the 3245 municipalities, although contracting out of collection and disposal is common. Local government is very strong in Japan and completely controls domestic waste collection and disposal, and byelaws enable them to control and raise local taxes to obtain finance for waste systems. Financial aid in the form of grants, subsidies and loans are available for the construction of waste treatment facilities. Some 52 million tonnes of municipal waste is generated each year (2000) in Japan, 77.4% of which is incinerated, 5.9% landfilled and 16.7% recycled. In addition, the ash generated from incineration at 7.4 million tonnes is landfilled each year, which is 14.3% of the municipal solid waste generated. Recycling of waste in Japan has been boosted by recent recycling laws including the Container and Packaging Recycling Law, effective since 2000, and the Home Electronics Recycling Law, effective since 2001.

Continued on page 55

Continued from page 54

Source separation of waste by households is well established with separation into either combustible or non-combustible material or recyclable materials such as glass, metal cans, newspapers etc. There is also a significant amount of composting at centralised facilities throughout Japan. A further 2.7 million tonnes of domestic waste is recycled privately by residents before collection.

Waste is collected from individual households in some municipalities, but in most cases the waste is left at collection sites or stations each typically serving 10–40 residential properties. For example, in Tokyo, there are 240 000 such collection points. The collection vehicles are small, of only 1–2 tonnes capacity to negotiate the narrow streets of the densely populated towns and cities. Collections of the waste may be up to three or four times per week. There are more than 1800 municipal waste incinerators in Japan and incineration technology is mainly of the grate, fluidised bed or rotary kiln combustor design. The ash residue from waste incineration is taken to landfill. There are approximately 2300 municipal landfill facilities operating in Japan and a further 1500 in the private sector. Operational standards for waste treatment and disposal are similar to European practices.

Sources: Waste Management in Japan 1995; Bonomo and Higginson 1988; Statistical Handbook of Japan 2003.

Box 1.6
Waste Management in the USA

The quantity of municipal solid waste (MSW) generated in the US is estimated to be approximately 230 million tonnes per year and industrial solid waste generation, 400 million tonnes per year. The primary sources of solid waste legislation at the central government level is covered by the Resource, Conservation and Recovery Act (RCRA) 1976, which was an amendment of the Solid Waste Disposal Act of 1965. The RCRA has subsequently been amended by several further amendments which mainly deal with landfilling of wastes. Municipal solid waste incineration is regulated under the Federal Clean Air Acts. The laws are administered by the Environmental Protection Agency (EPA) but where State enforcement programs are equal to or more stringent then the Federally established regulations they may be authorised by the EPA. No Federal regulations exist for materials recovery facilities, recycling systems, compost plants or transfer stations. Regulation of such facilities is at the State level.

The generation of solid waste in recent years has more closely followed the growth in the economy rather than the population increase. The majority of MSW, about 56% is landfilled, 30% is recovered/recycled/composted and 15% is incinerated. There are approximately 1800 landfill sites in the US (2001) and because of more stringent regulations requiring high standards of site lining, monitoring of gas and leachate and post closure liabilities, this has led to increased costs. Consequently, the recent trend has been toward fewer, larger landfills, often located further from the

Continued on page 56

Continued from page 55

source of the waste production. The majority of landfill sites are owned and operated by local government, but a significant number are privately owned. A further issue concerning landfill is that the Environmental Protection Agency (EPA) is required to enforce a law that bans hazardous waste disposal on land unless it can be proven to be less of a risk than other available technology. The reliance on landfill as the main route for waste disposal has led to capacity shortages in some areas of the US. The EPA policy is that MSW should be managed according to the hierarchy of: source reduction, re-use, recycling, incineration with energy recovery and landfilling. Consequently, many states have source waste reduction programs and recycled material market development programs and have set recycling goals of up to 50%. Across the different states of the US it is estimated that there are over 500 different legislative bills dealing with recycling. Approximately 9000 residential curbside recycling programs are in operation throughout the country with over 12 000 bring recycling centres. To process these recycled materials there are 480 materials recycling facilities across the US. The major markets for recycled products are aluminium, ferrous metal and waste paper. Incineration of MSW is carried out in 97 waste to energy plants (2001).

A further 400 million tonnes of industrial solid waste (not including mining or agricultural waste) is produced. Industrial chemicals, agro-chemicals, waste from the iron and steel industry, electric power station waste, and plastics and resins manufacturing constitute the majority of the waste, with other sources arising from, for example, the paper, food and textile industries. The majority of the industrial wastes are managed in on-site facilities, and, where the waste is transported off-site, landfill is the preferred disposal route. Apart from the industrial solid waste, it is estimated that more than 7000 million tonnes of industrial liquid wastewater is produced each year. More than 135 million tonnes of construction and demolition waste are generated in the US each year, the vast majority going to landfill; either MSW landfills or landfills specifically for construction and demolition waste.

Sources: Raymond Communications 1995; Hickman 1995; Steuteville 1996; EPA 2002; McCarthy and Tiemann 2002; ISWA 2002; US EPA Municipal Solid Waste in the United States 2003; US EPA 2004.

Box 1.7

Waste Management in the UK (England and Wales)

There have been several UK Government waste *strategy* documents in the last decade, including, Making Waste Work (1995), Waste Strategy 2000 and Waste Not Want Not (2002). However, the main enabling Parliamentary Act encompassing waste is the Environmental Protection Act of 1990. Part I of the Act deals with what are termed 'prescribed processes', which are large industrial processes including waste incineration, and Part II deals with disposal of waste on land, including landfill. The 1990 Act defined 'controlled waste' which is the main category of waste covered by the Act. Controlled waste refers to household, industrial and commercial waste.

Continued on page 57

Continued from page 56

The Environment Agency was established under the 1995 Environment Act and is responsible for the licensing for all waste disposal, storage, transfer and treatment plants and landfill sites and the responsibilities for 'prescribed processes' such as incinerators. Incinerators are covered by 'Guidance Notes' which set out the standards of operation of the process, including the substances released, emission levels, release levels etc. Smaller incinerator operations, such as clinical, small industrial and sewage sludge incinerators are regulated by Local Authorities which control only the emissions to air. Landfill operations have been viewed as not being in the same category as an industrial process. Therefore, the disposal of waste on land is covered by separate regulations and requires a 'Waste Management Licence' issued and regulated by the Environment Agency. Operational and environmental monitoring requirements are part of all licences and include both the operational lifetime and the post-closure phase of the facility.

The 1990 Environmental Protection Act also defined the duties of the Waste Collection Authorities and Waste Disposal Authorities. The Waste Disposal Authority is responsible for arranging for the disposal of waste collected by the Waste Collection Authority. Waste Collection Authorities provide receptacles for the waste for households, collect the waste in its area and deliver it to the place of disposal. The Collection Authority is also responsible for the development of a Recycling Plan dealing with the kinds and quantities of recyclable waste.

Within the UK there is also the concept of the 'duty of care'. The duty seeks to ensure the safe storage, handling and transport of waste by authorised people and to authorised sites for commercial and industrial waste.

The Environment Agency is required to prepare Waste Disposal Plans for the County or Metropolitan area. The Waste Disposal Plan includes such provisions as the arrangements needed for the treatment and disposal of household, industrial and commercial waste. The Waste Disposal Plans are not concerned with '*planning permission*' for the use of land, but only that suitable provision be made to deal with all the waste arising in a region adequately. Control over siting of facilities is exercised through the planning systems of the local authorities. Structure and Local Plans produced by the County and District Councils or the Unitary Development Plans produced at the Metropolitan Councils form the core of the planning procedures regarding land use in the UK. Both the Non-Metropolitan areas and the Metropolitan areas are required also to produce a Waste Local Plan dealing specifically with the treatment and disposal of waste in relation to the planning of land use. The plan gives detailed consideration of the preferred location for waste treatment and disposal facilities. However, they will be linked with the Waste Disposal Plan drawn up by the Environment Agency since that Plan will set out the types and quantities of waste arising in the area and the waste management facilities required to deal with that waste.

In England and Wales a total of 470 million tonnes of waste produced in 1998–99. Of this tonnage only 28 million tonnes was municipal solid waste. The vast majority of MSW is disposed of in landfill, at 83%. Incineration comprises 8%, and recycling and composting represents 9% of the management of MSW. This is reflected in the

Continued on page 58

Continued from page 57

fact that there are 1430 landfill sites and less than 20 MSW incinerators. Around 72.5 million tonnes of construction and demolition waste is produced each year (1998–99) with a high proportion recycled at 35%, some 42% is used in landfill site construction and 24% landfilled. Industrial and commercial waste production in England and Wales is approximately 75 million tonnes per year, 50 million tonnes from industry and 25 million tonnes from the commercial sector. More than 85% of the general industrial waste stream went to landfill. The largest category of waste is mines and quarry waste at 118 million tonnes, which is mostly disposed of in stable, above-ground tips at the production site, which is then landscaped and restored. Annual tonnage of agricultural waste is 96 million tonnes, mainly slurry and manure by-products from livestock, and mostly applied to land. Sewage sludge production is 26 million tonnes per year (wet sludge arisings, estimated on a 4% solids content) which is mostly landspread for agricultural use and also incinerated and landfilled. Dredgings comprise 57 million tonnes per year and represent the dredged material from harbours etc., which is largely dumped at sea.

Sources: Making Waste Work 1995; Waste Strategy 2000; Waste Not Want Not 2002; Williams 1998; Waste Statistics for England and Wales 2000; Gervais 2002(b); Environmental Services Association 2003.

References

Beardshaw J. and Palfreman D. 1986. The Organisation in its Environment. Pitman Publishing, London.

Bonomo L. and Higginson A.E. 1988. International Overview on Solid Waste Management. Academic Press, London.

BREF Waste Incineration. 2003. Draft Reference Document on Best Available Techniques for Waste Incineration, European Commission, Directorate-General JRC, Joint Research Centre, Technologies for Sustainable Development, European IPPC Bureau, Draft May.

British Medical Association. 1991. Hazardous Waste and Human Health. Oxford University Press, Oxford.

Chappell P. 1995. Drawing Board to Reality. Energy Technology Support Unit (ETSU), Harwell, Oxfordshire.

Clapp B.W. 1994. An Environmental History of Britain since the Industrial Revolution. Longman, London.

COM (96) 399 Final. 1996. Communication from the Commission on the review of the community strategy on waste management. Official Journal of the European Communities, Brussels, Belgium.

COM (97) 105 Final. 1997. Commission of the European Communities. Official Journal of the European Communities.

COM (99) 31 Final. 1999. C87, 29.03.1999. Draft Council Regulation (EC) on Waste Management Statistics. Official Journal of the European Communities.

COM (2000) 839 Final. 2001. Proposal for a Directive of the European Parliament and of the Council, Providing for Public Participation in Respect of the Drawing up of Certain Plans and Programmes Relating to the Environment and Amending Council Directives 85/337/EEC and 96/61/EC, Brussels 18.01.2001, 2000/031 (COD).

COM (2000-0159) 347. 2000. Final, 2000/0159 (COD), Proposal for a Directive of the European Parliament and of the Council on the restriction of the use of certain hazardous substances in electrical and electronic equipment. Official Journal of the European Communities, C 365, 19.12.2000, Brussels, Belgium.

COM (2000-0158) 347. 2000. Final, 2000/0158 (COD), Proposal for a Directive of the European Parliament and of the Council on waste electrical and electronic equipment. Official Journal of the European Communities, C 365, 19.12.2000, Brussels, Belgium.

Council Decision 94/904/EC. 1994. Establishing a list of hazardous waste. Official Journal of the European Communities, L226, Brussels, Belgium.

Council Directive 94/62/EC. 1994. Packaging and Packaging Waste Directive, L 365, 31.12.1994. Official Journal of the European Communities, Brussels, Belgium.

Council Directive 94/67/EC. 1994. Incineration of hazardous waste. Official Journal of the European Communities, L365, Brussels, Belgium.

Council Directive 96/61/EC. 1996. Integrated Pollution Prevention and Control. Official Journal, L257, Brussels.

Council Resolution 97/C, 76/01. 1997. Community Strategy for Waste Management (Council Resolution 97/C, 76/01), Brussels.

Council Directive 1999/31/EC. 1999. L 182/1, 26.4.1999. Landfill of Waste. Official Journal of the European Communities, Brussels, Belgium.

Council Directive 2000/76/EC. 2000. Incineration of waste. Official Journal of the European Communities, L332, Brussels, Belgium.

Council Directive 2000/53/EC. 2000. End-of-life Vehicles. Official Journal, L269, Brussels.

Commission Decision. 2000. 2000/532/EC. Commission Decision of 3 May 2000. Official Journal of the European Communities, L226/1, 6.9.2000, Brussels, Belgium.

Council Directive 75/442/EEC. 1975. Waste Framework Directive. Official Journal, L194, Brussels.

Council Directive 78/319/EEC. 1978. Toxic and dangerous waste. Official Journal of the European Communities, L84, Brussels, Belgium.

Council Directive 80/68/EEC. 1980. Protection of ground water against pollution from certain dangerous substances. Official Journal of the European Communities, L20, Brussels, Belgium.

Council Directive 84/360/EEC. 1984. Combating air pollution from industrial plant. Official Journal of the European Communities, L188, Brussels, Belgium.

Council Directive 84/631/EEC of 6 December 1984. On the Supervision and Control Within the European Community of the Transfrontier Shipment of Hazardous Waste. Official Journal L 326 Brussels.

Council Directive 85/337/EEC. 1985. The Assessment of the Effects of Certain Public and Private Projects on the Environment. Official Journal, L175, Brussels.

Council Directive 86/278/EEC. 1986. Use of sewage sludge in agriculture. Official Journal of the European Communities, L181, Brussels, Belgium.

Council Directive 91/156/EEC. 1991. Waste. Official Journal of the European Communities, L78, Brussels, Belgium.

Council Directive 91/271/EEC. 1991. Concerning urban waste water treatment. Official Journal of the European Communities, L135, Brussels, Belgium.

Council Directive 86/279/EEC. 1986. Transfrontier shipment of hazardous waste. Official Journal of the European Communities, L181, Brussels, Belgium.

Council Directive 91/689/EEC. 1991. Hazardous waste. Official Journal of the European Communities, L377, Brussels, Belgium.

Council Regulation 259/93/EEC. 1993. On the Supervision and Control of Transfrontier Shipments of Waste. Official Journal, L30, Brussels.

Council Directive 89/369/EEC. 1989. Reduction of Air Pollution from New Municipal Waste Incinerators. Official Journal of the European Communities, Brussels.

Council Directive 89/429/EEC. 1989. Reduction of Air Pollution from Existing Municipal Waste Incinerators. Official Journal of the European Communities, Brussels.

Council Directive 2000/76/EC. 2000. Incineration of Waste, L 332/91, 4.12.2000. Official Journal of the European Communities, Brussels, Belgium.

ENDS Report 265, 18–22. 1997. Environmental Data Services Ltd, Bowling Green Lane, London.

Energy from Waste, 1996. Best Practice Guide. Department of the Environment, HMSO, London.

Environment Act. 1995. HMSO, London.

Environmental Assessment Guidance Booklet. 1989. Department of the Environment, HMSO, London.

Environmental Services Association. 2003. London.

EPA. 2002. Municipal Solid Waste in the United States: 2000 Facts and Figures, Office of Solid Waste (OSW), US, EPA.

Eunomia. 2003. Eunomia Research and Consulting, Costs for Municipal Waste Management in the EU, Final Report to Directorate General Environment. European Commission, Brussels, Belgium.

Europa. 2003. The European Union On-line. European Union WEB Information, Brussels, Belgium.

European Commission. 2001. Waste Management Options and Climate Change. Office for Official Publications of the EC, European Commission, Luxembourg.

European Environment Agency. 1999. Information for Improving Europe's Environment. European Environment Agency, Copenhagen.

European Environment Agency. 2003. Europe's Environment: The Third Assessment. European Environment Agency, Copenhagen.

Garbutt J. 1995. Waste Management Law. A Practical Handbook, John Wiley & Sons, Chichester.

Gervais C. 2002(a). An overview of UK Waste and Resource Management Policy. Forum for the Future, London.

Gervais C. 2002(b). An overview of European Waste and Resource Management Policy. Forum for the Future, London.

Hannequart J.P. and Radermaker F. 2003. Methods costs and financing of waste collection in Europe: General review and comparison of various national policies. Association of Cities and Regions for Recycling (ACRR) in Europe, Brussels, Belgium.

Hickman H.L. 1995. MSW management, state-of-practice in the US. Presented at the APPA Regulatory Compliance Seminar, Arlington Virginia, 6–7 March.

ISWA. 2002. Waste Management. International Solid Waste Association, ISWA, Copenhagen.

Lane P. and Peto M. 1995. Blackstone's Guide to The Environment Act 1995. Blackstone Press Ltd, London.

Laurence D. 1999. Waste Regulation Law. Butterworths, London.

McBean E.A., Rovers F.A. and Farquar G.J. 1995. Solid Waste Landfill Engineering and Design. Prentice Hall, New Jersey.

Making Waste Work. 1995. Department of the Environment and Welsh Office, HMSO, London.

McCarthy J.E. and Tiemann M. 2002. Solid Waste Disposal Act/RCRA. Summaries of Environmental Law, Environment and Natural Resources Policy Division, US EPA.

Murley L. 1999. Pollution Handbook. National Society for Clean Air, Brighton, UK.

Neal H.A. and Schubel J.R. 1989. Solid Waste Management and the Environment: The Mounting Garbage and Trash Crisis. Prentice-Hall Inc., Englewood Cliffs, New Jersey.

OECD. 2004. Addressing the Economics of Waste. Organisation of Economic Cooperation and Development.

Petts J. and Eduljee G. 1994. Environmental Impact Assessment for Waste Treatment and Disposal Facilities. John Wiley & Sons Ltd, Chichester.

Project Integra. 2002. A History of Waste. Hampshire County Council.

Raymond Communications Inc. 1995. State Recycling Laws Update, 1995. Riverdale, MD, USA.

Reeds J. 1994. Controlling the Tips. The Taste Manager, September 18.

Royal Commission on Environmental Pollution. 1988. Best Practicable Environmental Option, 12th Report. HMSO, London.

Royal Commission on Environmental Pollution. 1993. 17th Report, Incineration of Waste. HMSO, London.

SEC (89) 934 Final. 1989. Community strategy for waste management. Official Journal of the European Communities, Brussels, Belgium.

Statistical Handbook of Japan. 2003. Statistics Bureau, Ministry of Public Management. Home Affairs, Posts and Telecommunications, Japan Statistical Association, Japan.

Steuteville R. 1996. The State of Garbage in America. Biocycle, April, 54–61.

Sustainable Development. 1994. The UK Strategy. HMSO, London.

Sustainable Development. 2001. Achieving a Better Quality of Life. Government Annual Report, HMSO, London.

The Environment Act. 1995. HMSO, London.

The Times. 1972. Times Newspapers.

This Common Inheritance, UK Annual Report. 1996. Reporting the UK's Sustainable strategy of 1994. HMSO, London.

US EPA. 2004. Municipal Solid Waste. Environmental Protection Agency, United States, Washington.

US EPA. 2003. Municipal Solid Waste in the United States: 2001 Facts and Figures. Office of Solid Waste, US EPA.

Van Santen A. 1993. Wastes Management Proceedings. 18 July.

Waste Framework Directive. 1975. Council Directive 75/442/EEC, OJ L 194, 25.7.75.

Waste Management in Japan. 1995. The Institute of Wastes Management, Department of Trade and Industry, Overseas Science and Technology Expert Mission Visit Report, IWM, Northampton.

Waste Management World. 2002. November/December, 20–30.

Waste Not Want Not. 2002. UK Government Strategy Unit. HMSO.

Waste Statistics for England and Wales 1998–99. 2000. HMSO, London.

Waste Strategy. 2000. Department of Transport and the Regions, HMSO, London.

Williams P.T. 1998. Waste Treatment and Disposal, 1st Edition. John Wiley & Sons, Chichester, UK.

Wilson E.J., McDougall F.R. and Willmore J. 2001. Euro-trash: Searching Europe for a more Sustainable Approach to Waste Management. Resources, Conservation and Recycling, 31, 327–346.

2

Waste

Summary

This chapter discusses the different definitions of waste. Estimates of waste arisings in Europe and the rest of the world are discussed as well as the methods used in their estimation. Several categories of waste are discussed in terms of their arisings, treatment and disposal options. The wastes described in detail are municipal solid waste, hazardous waste including clinical waste, household hazardous waste and sewage sludge. Other wastes described are agricultural waste, industrial waste, construction and demolition waste, mines and quarry waste, power station ash, scrap tyres and end-of-life vehicles. The chapter ends with a discussion of the different types of waste containers, collection systems and waste transport.

2.1 Definitions of Waste

The definition of waste can be very subjective, what represents waste to one person may represent a valuable resource to another. However, waste must have a strict legal definition to comply with the Law, because such strict definitions of waste have financial and legal implications for business, local authorities and Government.

In addition, for the requirement of a legal definition of waste, agreement on definitions and classifications of waste are required to enable the accurate formulation of local, regional and national waste management planning. Waste data, in the form of the types

Waste Treatment and Disposal, Second Edition   Paul T. Williams
© 2005 John Wiley & Sons, Ltd   ISBNs: 0-470-84912-6 (HB); 0-470-84913-4 (PB)

and quantities of wastes generated, are required at National level of the European Union to fulfil obligations for a National Waste Strategy under the 1975 EC Waste Framework Directive.

The European Waste Framework Directive (1975) has at its centre the hierarchy of waste management and the basis of its strategy is to encourage movement up the hierarchy and thereby increase the levels of waste reduction, re-use and recycling. A key element of the strategy is the compilation of accurate data on waste arisings. This enables recycling targets to be set and responses measured, and the diversion of different waste types from disposal, etc to be monitored. Accurate information on the source and composition of wastes also enables strategies for waste minimisation, re-use and recycling to be developed.

The classification of waste is difficult since, in many cases, waste is very heterogeneous and there can be great variation in composition between different loads of waste, but it is necessary for consistency in the description of the waste wherever the waste has arisen and whoever has described it. In addition to accurate definitions of waste, reporting methods also are required to be uniform. For example, waste arisings data across Europe are notoriously inaccurate due to different methods of data collection and quantification, different dates of collection, reporting methods and administrative systems.

It is clear then, from several standpoints, that accurate definitions and classifications of waste are required. However, by its very nature, waste is a heterogeneous material and difficult to describe, define and classify. In many instances the waste will be a mixture of different types or may be on the border between two categories. Also, waste can vary in composition on a daily, monthly and seasonal basis or from location to location. Consequently, when legal or financial outcomes are dependent on the definition of waste then the area is fraught with potential problems. In addition, certain wastes are exempt from environmental taxes such as Landfill Tax, so disagreements concerning the classification of the waste then become financially very important to companies.

A further problem associated with the definition of waste is where it may be recycled and thus become a recycled product or 'good'. The EU Waste Framework Directive defines waste as a 'any substance or object which the holder discards or intends to discard'. For recycled goods, when they change from being a 'waste', with the consequent legislative and handling requirements expected of a waste, to when they become a recycled product or 'good', is in some cases not clear cut. This has significant implications since whether something is a waste or a good implies legislative control and, potentially, has financial implications. The value of a waste or good is determined by gate fees, transport costs, quantities of material involved, technological innovation and the availability of appropriate waste management systems (Laurence 1999). These have a major influence on whether substances are seen as wastes or goods. Laurence (1999) has reviewed the definition of waste in detail from a legal perspective.

2.1.1   The EU Definition of 'Waste'

The definition of 'waste' was originally derived from the EC Waste Framework Directive (Waste Framework Directive 75/442/EEC 1975) as 'any substance or object which the holder discards or intends to discard'. A 'holder' means the producer of the waste or the person who is in possession of it, and 'producer' means any person whose activities produce waste, or any person who carries out pre-processing, mixing or other operations resulting in a change in the nature or composition of this waste. Waste was defined as any substance or object in the categories set out in the original Waste Framework Directive (Waste Framework Directive 1975):

1. Production or consumption residues not otherwise specified below.
2. Off-specification products.
3. Products whose date for appropriate use has expired.
4. Materials spilled, lost or having undergone other mishap, including any materials, equipment, etc., contaminated as a result of the mishap.
5. Materials contaminated or soiled as a result of planned actions (e.g., residues from cleaning operations, packing, materials, containers, etc.).
6. Unusable parts (e.g., reject batteries, exhausted catalysts, etc.).
7. Substances which no longer perform satisfactorily (e.g., contaminated acids, contaminated solvents, exhausted tempering salts, etc.).
8. Residues of industrial processes (e.g., slags, still bottoms, etc.).
9. Residues from pollution abatement processes (e.g., scrubber sludges, baghouse dusts, spent filters, etc.).
10. Machining or finishing residues (e.g., lathe turnings, mill scales, etc.).
11. Residues from raw materials extraction and processing (e.g., mining residues, oil field slops, etc.).
12. Adulterated materials (e.g., oils contaminated with PCBs, etc.).
13. Any materials, substances or products whose use has been banned by law.
14. Products for which the holder has no further use (e.g., agricultural, household, office, commercial and shop discards, etc.).
15. Contaminated materials, substances or products resulting from remedial action with respect to land.
16. Any materials, substances or products which are not contained in the above categories.

However, these broad classifications of waste resulted in the Member States of the EU adopting different notions of waste and establishing different waste lists (Gervais 2002; Laurence 1999). The EU strategy for waste management across the EU was dependent on the accurate statistical data of wastes in different categories and measures of policy success in recycling, waste minimisation, etc. The lack of clarity in waste definitions would therefore impact on the implementation of EU strategies in waste management. This difficulty in the harmonisation of the various terms defining waste across the EU was recognised by the EU Commission as a problem. A 1991 amendment to the 1975 Waste Framework Directive, 91/156/EEC, listed certain broad categories of wastes. The amendment also required the European Commission to draw up a list of wastes belonging to each of the categories listed. This the Commission did in 1994 as a Council Decision by publishing the European Waste Catalogue (94/3/EC 1994). In 1996 the European Environment Ministers adopted by Resolution a revised Waste Strategy (COM (96) 399 Final 1996) which amongst other measures called for a review of the waste definitions and catalogue of wastes. Commission Decision 2000/532/EC in 2000 (Commission

Decision 2000) replaced earlier lists of wastes and hazardous waste in one unified document. There are more than 650 waste categories on the European Waste Catalogue list, but it is not regarded as complete and exhaustive and other waste categories could be added at a later date. The different categories of waste in the European Waste Catalogue are described in Chapter 1.

2.2 Waste Arisings

It has been estimated that 3000 million tonnes of waste are generated throughout Europe each year (European Environment Agency 2003). Manufacturing industry, construction and demolition waste, mining and quarrying and agricultural waste are the main categories of waste generated in Europe. Mining and quarrying waste is the largest single category of waste generated, but is one of the few wastes that is decreasing with time due to a decline in the mining and quarrying activity across Europe. Figure 2.1 shows the percentage distribution of the different categories of waste generated for certain countries in Western Europe (European Commission 2003). In most European countries, the tonnages of waste generated in most waste categories, including municipal solid waste, continue to rise year on year. Figure 2.2 shows the *total* waste arisings in certain countries throughout Europe comparing total waste generation from 1995 to 2000 (European Environment Agency 2003). Difficulties arise in direct comparison of waste generation data due to the different waste classification systems, estimating methods and paucity of data available from country to country. For example, in some countries total waste includes materials which, in other countries, are not defined as wastes at all. This is also

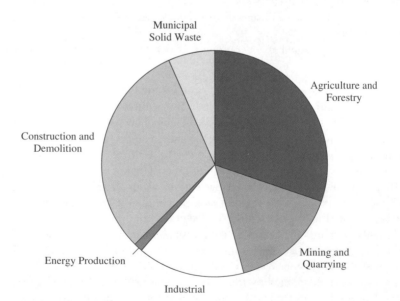

Figure 2.1 *Total waste generated by sector in the EU (15 members 2001). Source: European Commission 2003.*

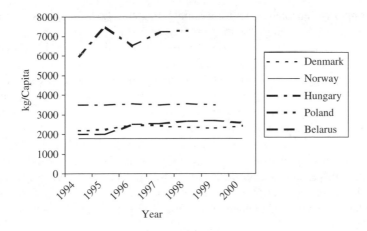

Figure 2.2 *Total waste generation in various countries in Europe. Source: European Environment Agency 2003.*

true of estimations of waste generation from individual countries where estimates for each waste producer sector can vary markedly, depending on the estimator.

2.2.1   Estimating Waste Arisings

Accurate data, concerning estimates of present and future production and composition of different types of waste, is essential for long-term efficient and economical waste management planning. The estimates are used by both planners and waste treatment and disposal engineers to determine the type, size, design and location of waste treatment and disposal facilities. The information is also used to determine the associated transport infra-structure and personnel requirements. Accurate waste arisings data are also required to meet national and international legal and policy obligations. The EC strategies for increased re-use and recycling of waste requires that the composition of the waste is characterised so that potentially suitable materials can be identified. Targets for waste minimisation, re-use and recycling also require accurate data to determine if and when targets are met. A key element of the National Waste Strategy required under the EC Waste Framework Directive for the Member States of the EU is the gathering of waste arising statistics on a regional basis, which will be used at local and regional level for planning future waste management facilities and at National level to implement the Waste Strategy. The estimate of waste arisings is required to allow for sufficient capacity and facilities for waste treatment and disposal for waste as a whole in the region sampled. The requirement under the Strategy is for an annual investigation of waste arisings, waste movements and waste management operations.

More detailed compositional analysis is required in some cases for the design of certain treatment facilities. For example, a knowledge of the chemical composition of waste, including the presence of metals, would be important for composting facilities and the composition of waste in terms of energy content, volatile ash and moisture content, would be important for the design of an incinerator. The physical particle size of the waste may also be an important characteristic, for example, in sizing feeder hoppers for incinerators and in the design of screening systems to segregate waste.

Statistical data on the quantification of waste are usually by weight, although sometimes it may be more appropriate to report the data as volume. For example, plastic bottles for recycling are often reported as volume rather than weight. Classification and quantification of industrial waste may be on an industry-source basis, such as the chemicals, food and petroleum industries, or on generic terms, such as waste solvents and oils from a number of industrial sources. Analysis of the composition of waste may also be based on the source of the waste such that industrial waste might be analysed in terms of chemical composition and clinical and household waste might be either by material types, such as glass, paper metal etc., or by-product types, such as glass containers, tins, magazines, etc. In addition to weight and composition, the energy value, moisture content, volatile content and elemental composition may also be carried out by a series of standard tests.

Whilst detailed classification, quantification and compositional analyses are clearly desirable, it is another matter to obtain accurate data from the producers of the waste. In most countries throughout the world there is no statutory requirement for waste producers to record waste statistics. By its inherent nature waste is heterogeneous and the components of the waste can have a large variance, therefore a large number of samples is required for statistical accuracy. However, for a large waste source population, this may not be possible and therefore representative samples, with their consequent errors, are used. Wastes may also be contaminated with other wastes not necessarily due to the process of the organisation.

There are, however, many problems associated with waste arisings data. Waste, particularly household waste, can vary on a daily, monthly and seasonal basis, on the size of the population, the type of housing stock, etc. A further factor involved in the accuracy of waste arisings data, is that variations in the amount of waste collected can vary depending on the method of waste collection and the size of waste container. This is particularly true for household waste (Pescod 1991–93). Therefore, it should always be borne in mind that waste is a very variable and heterogeneous material.

Two approaches are generally adopted for estimating waste quantity and composition. Either questionnaires are sent to the producers of the waste, or a direct waste analysis of the waste stream is carried out at the point of waste production or at the waste treatment facility (Waste Management Planning 1995; Yu and Maclaren 1995). The questionnaire is distributed to a range of companies, with questions concerning the quantities of waste generated and the composition in relation to a predetermined list of product-based or material-based categories. Other questions which might be included are concerning the seasonality of the waste generated and any waste re-use and recycling schemes in operation. However, in many cases companies do not keep records of the amount of waste that their company generates, let alone the composition of that waste. Consequently, data acquisition systems that rely on questionnaires may not be reliable because the data is not available and reporting of both weight and composition may be estimated by visual observation of the waste stream which is a notoriously inaccurate method. In addition, if there is no obligation to reply to the questionnaire, then the response rate to the questionnaire may be low due to time constraints, apathy and confidentiality issues. Non-response to the questionnaire may induce a bias, such that responding companies may have better waste management and reporting systems than non-respondents.

In the UK, a system used for estimating commercial and industrial waste is based mainly on the questionnaire system, with questionnaires sent by post or via personal visits to the company, representing a sample of selected industrial and commercial sectors in the survey area (Waste Management Planning 1995). To mitigate against possible errors in estimating, a dual approach to collection of data is recommended, by estimating waste arisings from the source and also from the waste management facilities which receive the waste. The questionnaire, which may be up to ten pages in length, covers such areas as the type and quantity of waste arising, the waste collection/disposal method used by the organisation, the types of container used for the waste, the general categories and sub-categories of the waste, a description of the process generating the waste, a detailed description of the waste in terms of the percentage components within it, the physical form of the waste, the proportion of packaging, the transport method, the location of the disposal site and the weight of the waste, etc. The waste data from the producers of the waste should then be matched against the waste received at the waste treatment and disposal facilities, to ensure accuracy of the survey data. The waste treatment and disposal facilities would include incinerators, landfill sites, composting facilities, recycling centres and transfer stations. The data collected from the facility would include the classes and types of waste, the quantities, the recycled output and also would include data on pollution control measures, energy recovery, the remaining useful life of the facility, etc. (Waste Management Planning 1995).

The alternative to the questionnaire system in obtaining waste data is the direct waste analysis method which involves the direct examination of the waste stream. The waste, usually in 90–180 kg samples, is taken directly from the stream and sorted by hand into a set of characteristic material or product categories and weighed (Yu and Maclaren 1995). The direct waste analysis system also has disadvantages. The cost of the method is high because it is so labour intensive and may be more than ten times the cost of a questionnaire. The direct analysis method also represents a sample typically taken during one day which may not be representative of the waste stream on an annual basis.

Figure 2.3 compares the accuracy of the questionnaire and direct waste analysis methods for determining the quantities of industrial, commercial and institutional waste generated in Toronto, Canada (Yu and Maclaren 1995). A good correlation between the estimated weight of waste using the two methods can be seen. However, Table 2.1 shows the rather poorer comparison for the *composition* of the waste using the two methods which is particularly different for paper and paperboard, plastics, textiles and food.

2.2.2   Waste Generation Multipliers

Waste generation multipliers are used for estimating waste from all the generation sources of waste in the region. For example, household waste generation multipliers may be based on the size of the population, while industrial and commercial waste generation multipliers may be based on the number of employees. In the case of agricultural waste, the multiplier might be acreage or number of livestock. For example, Table 2.2 shows typical waste generation factors based on the type of generator (Warmer Bulletin 49, 1996). Use of the correct multiplier in predicting future waste arisings has major

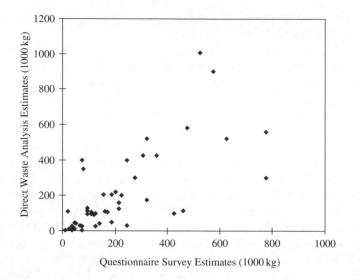

Figure 2.3 *Comparison of municipal solid waste quantity estimates using the 'direct waste analysis' and 'questionnaire' methods. Source: Yu and Maclaren 1995. Reproduced by permission from the International Solid Waste Association.*

Table 2.1 *Comparison of waste composition estimates by direct waste analysis and by questionnaire (weight %)*

Waste type	Direct waste analysis	Questionnaire
Paper	24.7	33.2
Paperboard	22.3	9.0
Ferrous metal	5.9	3.3
Non-ferrous metal	0.9	0.7
Plastics	13.3	6.9
Glass	2.8	8.4
Rubber	0.4	0.5
Leather	0.0	0.0
Textiles	4.5	0.7
Wood	7.5	10.3
Vegetation	1.4	0.4
Fines	0.3	2.2
Special wastes	0.6	0.7
Construction materials	4.6	2.2
Food	10.7	20.9

Source: Yu and Maclaren 1995. Reproduced by permission from the International Solid Waste Association.

implications for future planning, particularly if there is some doubt as to which multiplier to use.

In an attempt to obtain more accurate waste generation multipliers, some surveys have attempted to take into account a wide range of factors, for example, the size of each local population in a region, the type and age of residence occupied, the seasons of the year, the mixture and type of businesses in the area, and also economic data such as industrial

Table 2.2 Typical US waste generation rates by type of generator

Waste generation sector	Average	Units
Single family residential	1.22	kg/person/day
Apartments	1.14	kg/person/day
Offices	1.09	kg/employee/day
Eating and drinking establishment	6.77	kg/employee/day
Wholesale and retail trade[1]	0.009	kg/$ sales
Food stores	0.015	kg/$ sales
Educational facilities	0.23	kg/student/day

[1] Except food stores.

Source: Warmer Bulletin 49, 1996.

output, number of employees, company turnover, etc. (Rhyner and Green 1988; Warmer Bulletin 49, 1996).

Household waste generation multipliers have shown a wide variation depending on the source of the survey, for example, estimates of the arisings of household waste in the US have varied between 1.08 and 1.22 kg/person/day (Rhyner and Green 1988). It has also been shown that more accurate estimates can be produced using waste generation multipliers based on the population size of the local community, for household waste (Table 2.3, Wisconsin USA). Smaller communities produce a lower waste generation/person/day than larger populated communities.

The UK is undertaking a major analysis of household waste in terms of weight, composition and chemical analysis under the National Household Waste Analysis Project (1994, 1995). The Project has tried to produce more accurate waste multipliers and has included socio-economic classifications of households and waste arisings. Whilst there will clearly be differences in the waste amounts collected from multi-storey flats compared with family houses and single-storey houses, etc., this approach does not distinguish between differences in family size, socio-economic and other factors which are included in the UK Project. The system used in the analysis is a classification of all the households in the UK into demographic, housing and socio-economic units. Each unit consists of about 450–550 persons or about 160–180 households. The classification of each group is based on such factors as unemployment, number of cars, age and number of the residents, ownership, number of rooms, toilet facilities, employment type, etc. In addition, a detailed count of the number of houses in each category is also made (National Household Waste

Table 2.3 Comparison of household waste generation multipliers based on the population size of the community

Population	Waste generation multiplier (kg/person/day)
<2500	0.91
2500–10 000	1.22
10 000–30 000	1.45
>30 000	1.63

Source: Rhyner and Green 1988.

Table 2.4   *Waste factor variation between different local authorities for different household groups*

ACORN group	Waste factors (kg/household/week)	
	Range	Average
Agricultural area	17.9–32.4	22.1
High income, modern family housing	9.6–21.5	14.3
Older housing, intermediate income	9.0–14.8	11.8
Older terraced housing	8.8–20.4	12.6
Municipal housing, higher status	10.7–23.3	14.6
Municipal housing, intermediate status	7.2–16.7	12.2
Municipal housing, lower status	7.2–17.6	13.6
Mixed metropolitan areas	5.0–12.3	9.8
High status, non-family areas	7.7–27.1	13.1
Affluent suburban housing	5.4–20.7	14.2
High status retirement areas	5.4–16.6	11.1

Source: Department of the Environment 1991.

Analysis Project 1994, 1995). The results show that waste output *does* vary significantly between household groups as shown in Table 2.4. The results have shown that in many cases the better off houses produce more waste compared to the less well off groups with an average of 15–17 kg/household/week compared to 10–11 kg/household per week respectively.

Waste factors can be calculated for each household group, which enables the prediction of the likely waste arisings from a particular area or for the local municipality as a whole. However, other factors can distort the waste factor, such as the method of waste collection. Local municipalities who use a traditional bin or wheeled bin system, which restrict the waste collected to that in the dustbin container, tend to collect less waste than those using unrestricted waste amounts, such as the plastic sack system. In addition, municipalities using the larger wheeled bins, with their larger capacity, tend to generate more waste than the traditional smaller bins (National Household Waste Analysis Project 1994). In some cases municipalities moving to a wheeled bin system have experienced increases in the waste generated of up to 30% (Pescod 1991–93).

It is not only the contents of the dustbin that should be analysed to obtain an accurate picture of waste generation, since the weight and composition will be influenced by how much recycling is carried out by the residents of the local population. Consequently, household waste analyses should also include waste taken to civic amenities, recycling sites such as bottle banks and scrap metal sites and would also include composting.

Industrial waste multipliers will vary depending on the type of industry or business sector. For example, the food industry produces much more waste than, for example, the printing and publishing industry (Table 2.5, Rhyner and Green 1988). In addition, the particular waste multiplier used from one city to the next will depend on a number of factors, including, variations in manufacturing process, economies of scale of large-scale production and regional differences in the mix of industries included in an industrial sector (Table 2.5, Rhyner and Green 1988).

Table 2.5   *Comparison of industrial waste generation multipliers*

Industrial sector	Industrial waste generation multiplier (tonnes/employee/year)	
	New York	Wisconsin
Food and similar products	5.26	4.42
Textile mill products	0.24	0.76
Paper and allied products	13.06	13.52
Printing, publishing	0.44	1.03
Chemicals and allied products	0.57	7.45
Petroleum and refining	0.00	26.35
Rubber and plastics	2.36	1.01
Stone, clay, glass and concrete	2.18	10.69
Primary metals industry	21.77	6.09
Fabricated metal products	1.54	3.38
Electrical machinery	1.54	2.43
Transportation equipment	1.18	1.18

Source: Rhyner and Green 1988.

Estimation of industrial and commercial waste multipliers for the UK, Holland and Germany are shown in Table 2.6 (Analysis of Industrial and Commercial Waste 1995). Table 2.6 shows the differences which can be obtained using different methods of estimation. In some sectors the differences in the waste multipliers are enormous, with consequent implications for determining the appropriate waste management operations to

Table 2.6   *Industrial and commercial waste multipliers (tonnes/employee/year)*

Industry sector	Holland	Germany	UK Survey[1]	UK Plan[2]
Mining	—	53.4	—	—
Extraction of mineral oil and gas, mineral oil processing	—	24.5	—	—
Public utilities	—	—	9.1	48.5
Metal manufacturing	12.4	55.3	17.0	83.4
Quarrying	—	46.9	0.25	83.4
Manufacture of non-metallic mineral products	13.4	6.3	16.5	83.4
Chemical industry	7.0	15.7	9.5	83.4
Manufacture of other metals goods	1.8	8.7	9.3	3.0
Mechanical engineering	0.7	2.8	4.1	3.0
Manufacture of office machinery	—	1.2	0.1	3.0
Electrical and electronic engineering	0.4	1.6	3.5	3.0
Manufacture of motor vehicles and parts and other transport	0.8	3.7	12.5	3.0
Instrument engineering	0.3	1.0	7.8	3.0
Food, drink and tobacco manufacturing industries	18.4	22.3	13.9	4.8
Textile, leather and leather goods industries	1.3	2.4	3.7	4.8
Timber and wooden furniture industries	6.0	22.3	15.2	4.8

Table 2.6   *Continued*

Industry sector	Holland	Germany	UK Survey[1]	UK Plan[2]
Manufacture of paper and paper products; printing and publishing	5.4	22.7	5.1	4.8
Processing of rubber and plastic	1.7	3.7	5.4	4.8
Other manufacturing industries	0.5	1.4	2.9	4.8
Construction	—	99.2	4.6	25.9

[1] Survey data of companies.
[2] Data from waste regulation authorities waste disposal plans.
Source: Analysis of industrial and commercial waste going to landfill in the UK 1995.

treat and dispose of the waste. This may in part be due to the different ways in which waste data are produced, but highlights the care which must be exercised when using waste generation multipliers and comparisons of waste arisings data between countries and regions.

Estimates of construction and demolition waste have been based on the size of the city as 0.63 kg/person/day for cities greater than 10 000 in population size and 0.14 kg/person/day for cities smaller than 10 000 population. The tonnages of construction and demolition waste will be highly dependent on the type of housing stock and the level of on-site recycling. Commercial waste multipliers have also been estimated in terms of population and range between 0.42 and 2.04 kg/person/day with a median of 1.06 kg/person/day (Rhyner and Green 1988).

2.3 Municipal Solid Waste (MSW)

2.3.1 Municipal Solid Waste Generation

Municipal solid waste is defined as waste collected by, or on behalf of, municipalities. These generally originate from households, commerce and trade, small businesses, office buildings and institutions such as schools, hospitals, government buildings, etc. (European Commission 2003). In some cases, waste from parks and gardens and street-cleaning services are also included. Figure 2.4 shows that municipal solid waste generation per head of population in OECD countries has shown a steady increase from 1990 to 2000 (OECD 2004). Figure 2.4 also shows that North America has a higher per capita generation of municipal solid waste compared to the EU countries (2000), but both regions show a steady increase in waste generation. This increase is linked to a number of factors, including economic growth of OECD countries, since a rise in income of individuals leads to higher rates of consumption of electrical goods and increased packaging waste, etc. The growing trend of urbanisation of the population where there is a movement away from rural areas to urban areas, also tends to increase the per capita generation of waste, since urban populations tend to have higher incomes, higher consumption of goods and, consequently, higher generation of waste compared to rural populations (OECD 2004). However, increased urbanisation of the population leads to a greater

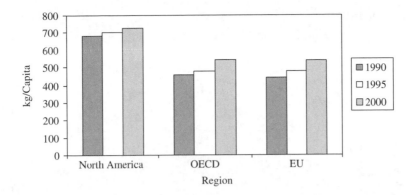

Figure 2.4 *Municipal solid waste generation per capita in OECD Regions. Source: OECD 2004.*

potential for recycling since the population density is increased. Changes in socio-cultural habits may also influence the type and quantities of municipal solid waste generated, for example, by a growth in single-occupancy households and the consequent lifestyle exhibited, or increasing environmental awareness which may also influence per capita waste generation and may develop differently from country to country (OECD 2004)

The increased generation of MSW is reflected in the growth in population shown in Table 2.7 and influences the total increase in MSW generation, which was an 8% increase over the ten-year period, from 530 million tonnes in 1990 to 605 million tonnes in 2000. In addition, households with younger children have been shown to produce more waste than households with older people (OECD 2004). Increase in the population will also lead to an increase in the per capita generation of waste, since the increased population would mean an increase in the number of households which, in turn, increases waste generation.

Table 2.7 suggests there is a strong link between waste generation rates and the economic standing of a country. Figure 2.5 (Stanners and Bourdeau 1995) shows a link between Gross Domestic Product (GDP) and waste production for several countries of the OECD. Increasing economic development, represented by GDP, reflects in an increasing rate of waste production.

Table 2.7 *Growth in generation of municipal solid waste from 1990 to 2000 in relation to population growth and gross domestic product*

	Growth in MSW generation (%)	Growth in GDP (%)	Population growth (%)
North America	13	37	10
Asia–Pacific Region	−11	25	5
OECD Europe	23	23	5
EU (15 countries)	26	23	3
OECD Total	14	28	8

Source: OECD 2004.

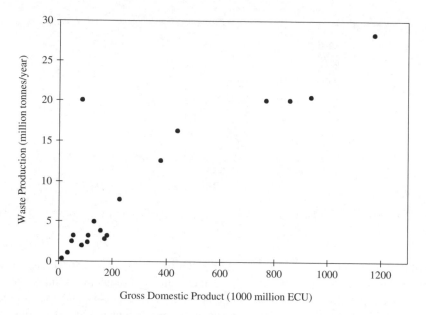

Figure 2.5 *Municipal solid waste production as a function of gross domestic product (GDP). Source: Stanners and Bourdeau 1995.*

Many national and international studies on waste arisings and composition classify municipal solid waste as a specific category of waste. However, comparative data are often difficult to interpret since different types of waste, including in some cases industrial wastes, may be collected by the local authority. Figure 2.6 shows a comparison of municipal solid waste generation per capita per year for various countries (European Commission 2003). These data should be compared with those from the developing world where waste generation weights are very low. For example, the annual per capita production of municipal solid waste in Delhi, India, has been estimated at 136 kg/person/year, in Kathmandu in Nepal it has been estimated at 109 kg/person/year, and in Wuhan, China at 200 kg/person/year (Rushbrook and Finney 1988).

The proportion of municipal solid waste that is generated from households, varies quite widely between countries. For example, in Denmark, Spain, the Netherlands and United Kingdom, the proportion of household waste is approximately 85% of municipal solid waste. However, for Iceland, the figure is 37%, for Finland 40% and for Estonia the proportion of household waste is represented by 32%. The generation of municipal solid waste per head or per capita of population has increased for most European countries and in some cases the increase is quite marked. For example, Spain, Denmark and Finland have shown a recent annual average increase in the per capita generation of municipal solid waste of 5% (European Commission 2003). Whilst most countries of Western Europe have shown an increase in the generation of municipal solid waste, some countries have shown a decrease. For example, the Czech Republic, Hungary and Poland showed a decrease of between 5 and 9% in MSW generation per capita between 1999 and 2001 (European Commission 2003).

Figure 2.6 *Municipal solid waste generated per capita in selected European countries. Source: European Commission 2003.*

2.3.2 Municipal Solid Waste Composition

The main compositional categories of municipal solid waste are paper and cardboard, organic waste such as food and garden (yard) waste, plastics, metals, glass, textiles and other minor fractions of waste. Figure 2.7 shows the average composition of municipal solid waste for several Western European countries (Austria, Belgium, Denmark, Finland, France, Germany, Greece, Ireland, Italy, Luxembourg, the Netherlands, Portugal, Spain, Sweden, United Kingdom, Iceland, Liechtenstein, Norway, Switzerland, Andorra, Monaco, San Marino – European Commission 2003). Also shown, for comparison,

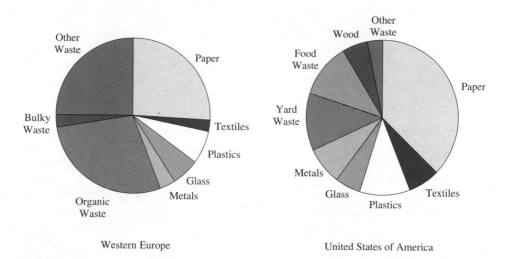

Western Europe   United States of America

Figure 2.7   *Average municipal solid waste composition in Western Europe (See Box 1.5 for countries) and the United States of America. Sources: European Commission 2003; US Environmental Protection Agency, Municipal Solid Waste 2004.*

is the average composition for municipal solid waste for the United States of America (US EPA 2004).

Comparison of municipal solid waste composition can be made between countries. Compositions of municipal solid waste in some European countries are compared in Figure 2.8 (European Commission 2003). In some cases there are large variations in waste compositions, showing that composition of municipal solid waste is very dependent

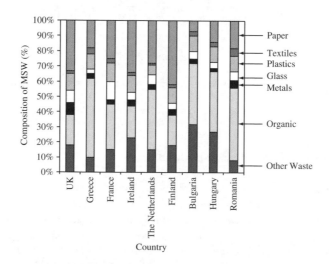

Figure 2.8   *Comparison of municipal solid waste composition for some European countries. Source: European Commission 2003.*

on the local conditions. The comparison of waste compositions from different countries can be difficult, since the methods of reporting and classification and the degree of recycling, all influence the reported composition. The increasing trends in waste management are to waste minimisation, re-use and recycling of waste. This will inevitably influence the quantities and composition of waste. Each town or city produces a different composition of waste, since the inputs will depend on socio-economic factors, types of industry and level of industrialisation, geographic location, climate, level of consumption, collection system, population density, the extent of recycling, legislative controls and public attitudes.

Examples of the influences on the composition of municipal solid waste may be considered. For example, socio-economic conditions would influence the composition of municipal solid waste. Figure 2.9 shows waste compositional data for a group of households in the UK representative of 'better off' and 'less well off' (National Household Waste Analysis Project 1994, 1995). The 'better off' households are classified as 'modern family housing with higher incomes' compared with 'less well off' classified as 'older housing of intermediate status'. The data show clear differences between the two socio-economic groupings of households. For example, waste from 'better-off' households

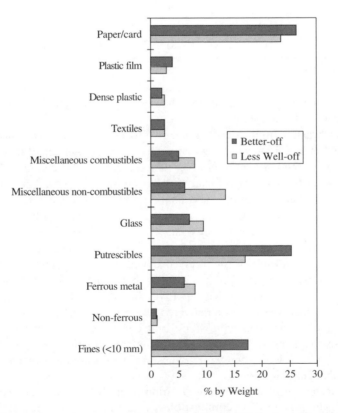

Figure 2.9 *Comparison of waste compositional data for UK 'better off' and 'less well off' household groups. Source: The analysis and prediction of household waste arisings 1991. Crown copyright material is reproduced with the permission of the Controller of HMSO and the Queen's Printer for Scotland.*

Table 2.8   Seasonal variation in US Municipal solid waste composition

Waste component	Autumn	Winter	Spring	Summer	Average
Organics	86.0	86.5	87.7	89.8	87.5
Paper	44.7	45.7	47.5	40.3	44.5
Plastic	6.1	6.5	7.0	6.0	6.4
Yard waste	15.0	15.1	7.4	15.0	13.1
Wood	1.4	0.8	1.0	0.8	1.0
Food	15.2	14.3	18.0	21.5	17.3
Textiles	1.5	1.3	1.3	2.0	1.5
Other organics	2.3	2.9	5.4	4.2	3.7
Inorganics	13.4	12.8	12.2	9.8	12.1
Metals	3.0	3.9	4.0	3.1	3.5
Glass	7.0	7.1	7.1	5.6	6.7
Soil	0.9	0.4	0.0	0.0	0.3
Other inorganics	2.5	1.4	1.1	1.2	1.5
Special wastes	0.5	0.6	0.1	0.3	0.4
Appliances	0.2	0.3	0.0	0.2	0.2
Chemicals	0.1	0.2	0.1	0.1	0.1
Re-useable	0.3	0.1	0.0	0.0	0.1
Total	100.0	100.0	100.0	100.0	100.0

Source: Warmer Bulletin, 49, 1996.

contained more putrescible and garden wastes and also more paper, plastic film and glass than the 'less well-off' households.

In addition, waste composition has been shown to vary with seasons of the year. For example, Table 2.8 shows the waste composition of municipal solid waste (local authority waste) in the US over a one-year period (Warmer Bulletin 49, 1996). Some of the component wastes show significant variation through the year. Short-term seasonal variations of waste are common, for example, organic yard waste (garden waste) are known to increase during the growing season or, where there are influxes of visitors into a tourist area, then the total quantity of waste will also increase.

Month-by-month variations in waste arisings can also be detected. For example, the monthly arisings of residential waste received at landfill sites in Brown County, Wisconsin in the US over a five-year period from 1985–89, showed large swings in the quantities of residential waste being delivered (Rhyner 1992). The data showed increases of 33.4% in May over the average yearly waste delivered, to decreases in February of 23.8% of the average. Such variations can have significant consequences on the manning levels, equipment use, financing, etc., of the facility. Similar fluctuations were seen for commercial, construction and demolition waste, whose industries are particularly influenced by seasonal weather conditions.

The significance of changes in composition over short time spans can be seen in Figure 2.10, which shows the combustible waste fraction of residential solid waste generated in 1989, in the Kita-Ku District of Sapporo, Japan (Matsuto and Tanaka 1993). The collection system in Sapporo is based on separate collections of combustible and non-combustible waste, separated at source, and the disposal system is mainly incineration.

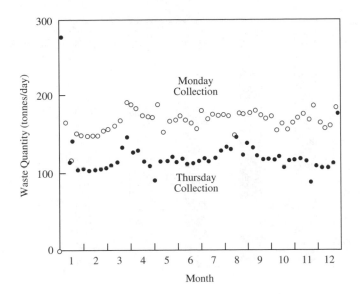

Figure 2.10 *Daily collection of combustible municipal solid waste in Kita-Ku District, Sapporo, Japan. Source: Matsuto and Tanaka 1993. Reproduced by permission of the International Solid Waste Association.*

Collection in Kita-Ku is twice a week, Monday and Thursday, for the combustible fraction, and once a week for the non-combustible fraction. The figure shows the difference in the quantity of the combustible fraction of the waste between two days (Monday and Thursday) in the week. The Monday data obviously contains waste generated at the week-end which is shown to have a higher combustible fraction. Where heat or electrical output from an incinerator has to be guaranteed, then such variations can significantly influence the economic operation of such a facility.

There have been long-term effects which have influenced the composition of municipal solid waste over decades. Figure 2.11 shows the historical change in household or domestic waste arisings in the UK from 1879 to 1990 (The Open University 1993). The dominant component of waste, in the late part of the nineteenth and early part of the twentieth centuries in the UK, was dust and cinder, derived largely from the residues of the household coal fire. The dramatic fall in the contribution of dust and cinders arose from the move away from coal fires, during the late 1950s and 1960s, to the more convenient central heating systems, fired by oil and gas. One of the major factors influencing the move way from coal was the introduction of the Clean Air Act of 1956 which sought to reduce the smoke and sulphur dioxide pollution from residential and also industrial sources. This lead directly to the replacement of the living room coal fire by gas and electric fires. Vegetable and putrescible waste had grown from about 8% of domestic waste to about 28% in 1965, after which the percentage contribution has remained fairly constant at between 22 and 25%. The biggest increases in percentage contribution to the domestic dustbin composition have been paper and board, metals, glass and plastics. These wastes are associated with the increases in packaging, newspapers and magazines, plastic bottles, tin cans, etc., reflecting our change to a consumer society (The Open University 1993).

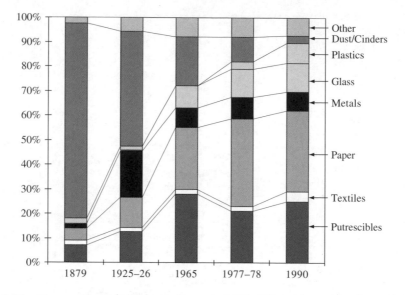

Figure 2.11   *Composition of UK domestic waste 1879–1990. Source: The Open University 1993.*

Such long-term analysis of waste data is not common. However, similar long-term analysis of municipal solid waste composition from New York City in the US (Table 2.9) shows that, in the early part of the 20th century, ash and cinders from coal combustion for heating and cooking was the dominant waste collected, as was also found for the UK (Walsh 2002). The progressive decrease in ash content, shown in Table 2.9, reflects the increasing use of oil and natural gas fuels with low ash content, as replacement fuels for coal. The increase in the paper fraction of municipal solid waste was due in part to the

Table 2.9   *Household waste composition from 1905 to 1989 for New York City*

	1905 (%)	1939 (%)	1971 (%)	1978 (%)	1989 (%)
Food refuse	13.4	17.0	15.6	17.8	14.1
Ash	79.9	43.0	2.8	1.5	2.3
Paper	5.0	21.9	35.5	32.9	34.7
Plastic	—	—	2.7	8.8	9.9
Metal	0.2	6.8	11.1	13.3	5.3
Glass	0.2	5.5	23.1[a]	9.4[b]	5.5
Textiles	1.0	—	3.9	5.7	5.2
Hazardous	—	—	—	—	0.4
Rubber and leather	0.1	—	3.5	—	0.2
Wood	0.1	2.6	1.2	4.5	2.4
Yard refuse	—	—	0.7	4.7	4.7
Miscellaneous	—	3.2	—	1.5	15.2

[a] Includes ceramics and stones.
[b] Reported as combined glass and ceramics.
Source: Walsh D.C. 2002.

increased use of packaging in the US after the First World War. Plastics are not reported for New York City in 1905 and 1939 since they were not widely used, but the steady rise in the latter half of the 20th century is evident, reflecting the replacement of conventional products and packaging with lighter weight plastics. It has also been suggested that the relative increases in metals and glass were due to the increasing use of metal food containers and non-returnable glass bottles, up to 1971. The move away from non-returnable bottles and metal cans to plastic substitutes, is suggested as the reason for the decline in the metal and glass contents of municipal solid waste in New York City from 1978 to 1989 (Walsh 2002).

The composition of municipal solid waste also has other factors to be considered, particularly where the waste is to be combusted in an incinerator. Incinerator operators are concerned with such aspects as the energy (calorific) value of the waste, the elemental (ultimate) analysis and the proximate analysis of the waste. Proximate analysis is the ash, moisture and combustible fraction of the waste, in some cases the combustible fraction is subdivided into the volatiles and fixed carbon content. The calorific value may be reported as the as-received figure (gross calorific value), corrected for the moisture content (net or dry calorific value) or corrected for the moisture and ash content (dry, ash-free calorific value). All these factors are important in the design, operation and pollutant emissions of the incinerator. For example, the incinerator grate, ceramic lining, furnace chamber and boilers would be designed to cope with a certain calorific value of waste. If the calorific value changed potentially, rapid corrosion of the boiler or ceramic lining could occur. In addition, the heat output from the plant would also be affected. Similarly, the incinerator flue gas clean-up system would be designed to cope with certain pollutants at estimated concentrations. If the nature of the waste changed, then the resultant pollutants and their concentration could mean that authorised limits might be exceeded. For example, an increase in chlorinated waste in the incoming stream could lead to increases in hydrogen chloride or dioxins and furans. The typical calorific value of municipal solid waste in the UK is approximately 8500 kJ/kg (Royal Commission on Environmental Pollution 1993). This figure compares with about 30 000 kJ/kg for a typical coal and 42 000 kJ/kg for a typical fuel oil. Estimation of calorific value and ultimate and proximate analyses of municipal solid waste, are inherently inaccurate due to the heterogeneous nature of the waste. In addition, the majority of analyses of wastes are standard tests derived from solid fuel analyses, such as coal, where the fuel is more homogeneous and sample sizes are typically one gram. Obtaining a representative one-gram sample of a dustbin lorry full of waste is very difficult, even if strict sampling procedures and processing of the sample to a representative ground sample, are followed. Larger-scale (1 kg) instruments to determine the calorific value of waste have been reported, specifically designed for municipal solid waste analysis (Daborn 1988).

Table 2.10 shows some typical properties of municipal solid waste (Williams 2002) giving proximate, elemental analysis, including metals and calorific value analyses of municipal solid waste (Waste Management Paper 28, 1992; Kaiser 1978; Buekens and Patrick 1985). Some components have a high calorific value, for example, plastic material at approximately 40 000 kJ/kg, and, although present in low concentration in the waste, will have a significant influence on the overall calorific value. The volatile, ash and moisture content will vary considerable depending on the source of the waste. For example, commercial waste with a high plastic content will produce a high evolution

Table 2.10   *Typical properties of municipal solid waste*

Composition	Wt%	Elemental analysis	Wt%
Paper/board	33.0	Carbon	21.5
Plastics	7.0	Hydrogen	3.0
Glass	10.0	Oxygen	16.9
Metals	8.0	Nitrogen	0.5
Food/garden	20.0	Sulphur	0.2
Textiles	4.0	Chlorine	0.4
Other	18.0	Lead (ppm)	100–2000
		Cadmium (ppm)	1–150
Proximate analysis	**Wt%**	Arsenic (ppm)	2–5
		Zinc (ppm)	400–1400
Combustibles	42.1	Copper (ppm)	200–700
Moisture	31.0	Chromium (ppm)	40–200
Ash	26.9	Mercury (ppm)	1–50

Sources: Williams 2002; European Commission 2004.

of combustible volatiles with negligible ash and moisture. In addition, the day-to-day rainfall influences how wet the waste is and consequently affects the measured moisture content. The ash content of municipal solid waste is high compared, for example, to a bituminous coal which typically would be less than 10 wt%. The elemental analysis of municipal solid waste has most influence on the treatment method which, for most countries, is either landfill or incineration. The sulphur and chlorine content of the waste will influence the emissions of sulphur dioxide and hydrogen chloride if the waste is incinerated. Similarly, the content of hazardous heavy metals such as cadmium, lead, chromium and copper, will influence emissions to air and land from incineration and would require control. Since landfill leachate may at some stages of the biodegradation process be acidic, the heavy metals may be dissolved in the leachate and be transported into ground water. Consequently, leachate monitoring and control for heavy metals is normally required for municipal solid waste landfills.

More detailed analyses of the components of municipal solid waste are available in some countries. For example, Table 2.11 shows a breakdown of the components of UK household waste, represented by the northern UK town of Warrington (National House-hold Waste Analysis Project 1994). The data is useful, particularly for recycling and re-use strategies, such as the economic potential of recycling plastic bottles, glass and metal cans, which can be more easily and accurately assessed. In addition, the potential of the waste in terms of combustibility in an incinerator may also be assessed. Table 2.11 also shows that the more readily recyclable newspapers and magazines make up the bulk of the paper and card composition of household waste, with a figure of 27.4 wt%. The plastic components of waste are almost 10 wt% in total from the Warrington analysis, representing a major and increasing component of the waste. Interestingly, disposable nappies make up a significant proportion of the composition of household waste and also have a significant contribution to the overall calorific value of the waste. Putrescible material (represented by food and garden waste) and fines also have a certain combustible fraction. The glass content of the waste approaches 10 wt% and, together with the metals component, represents a significant proportion of the waste which can be readily recycled.

Table 2.11 Typical composition of UK household waste (Warrington 1992, Housing type – modern family housing, higher incomes)

Category	Sub-category	Composition		
		Weight percent		kg/household
Paper/card	Newspapers		8.6	
	Magazines		4.3	
	Other paper	27.4	6.8	4.4
	Liquid cartons		0.6	
	Card packaging		3.3	
	Other card		3.8	
Plastic film	Refuse sacks		0.3	
	Other plastic film	3.8	3.5	0.6
Dense plastic	Clear beverage bottles		0.6	
	Coloured beverage bottles		0.1	
	Other bottles	6.0	1.0	1.1
	Food packaging		1.6	
	Other dense plastic		2.7	
Textiles	Textiles	1.9	1.9	0.3
Miscellaneous	Disposable nappies		7.1	
Combustibles	Other misc. combustibles	10.3	3.2	1.7
Miscellaneous Non-combustibles	Miscellaneous Non-combustibles	4.6	4.6	0.8
Glass	Brown glass bottles		0.4	
	Green glass bottles		1.5	
	Clear glass bottles	9.7	2.2	1.7
	Clear glass jars		1.8	
	Broken glass		3.8	
Putrescible	Garden waste		3.6	
	Other putrescibles	21.7	18.1	3.6
Ferrous metal	Beverage cans		0.8	
	Food cans		4.2	
	Batteries	6.1	—	1.0
	Other cans		0.3	
	Other ferrous		0.8	
Non-ferrous Metals	Beverage cans		0.4	
	Foil	1.7	0.5	0.3
	Other non-ferrous metals		0.8	
Fines	10 mm fines	6.8	6.8	1.1
Total	100.0	100.0	100.0	16.6
Collection details:	Number of properties	231	Bulk moisture	31.9
	Total weight collected	3840 kg	Ash	16.15%
	Mean weight/household	16.6	Gross CV	10.93 MJ/kg

Source: National Household Waste Analysis Project 1994. Crown copyright material is reproduced with the permission of the Controller of HMSO and the Queen's Printer for Scotland.

Coupled with the analysis of the components of municipal solid waste are analyses of the properties of the various components of municipal solid waste (Tables 2.12, 2.13 and 2.14) show that the calorific value, ultimate and proximate analyses of even different sorts of paper, food and garden waste, can be very different (Kaiser 1978). Other

Table 2.12   Proximate analysis of municipal solid waste components

Component	Moisture	Volatiles	Fixed carbon	Ash
Paper/paper products				
Paper – mixed	10.24	75.94	8.44	5.38
Newsprint	5.97	81.12	11.48	1.43
Corrugated boxes	5.20	77.47	12.27	5.06
Plastic coated paper	4.71	84.20	8.45	2.64
Waxed milk cartons	3.45	90.92	4.46	1.17
Junk mail	4.56	73.32	9.03	13.09
Food/garden waste				
Vegetable food waste	78.29	17.10	3.55	1.06
Meat scraps (cooked)	38.74	56.34	1.81	3.11
Fried fats	0.00	97.64	2.36	0.00
Lawn grass	75.24	18.64	4.50	1.62
Leaves	9.97	66.92	19.29	3.82
Green logs	50.00	42.25	7.25	0.50
Evergreen shrubs	69.00	25.18	5.01	0.81
Flowering plants	53.94	35.64	8.08	2.34
Wood and bark	20.00	67.89	11.31	0.80
Household waste				
Leather shoe	7.46	57.12	14.26	21.16
Rubber	1.20	83.98	4.94	9.88
Upholstery	6.90	75.96	14.52	2.62
Polystyrene	0.20	98.67	0.68	0.45
PVC	0.20	86.89	10.85	2.06
Linoleum	2.10	64.50	6.60	26.80
Rags	10.00	84.34	3.46	2.20
Vacuum cleaner dirt	5.47	55.68	8.51	30.34

Source: Kaiser 1978.

Table 2.13   Ultimate analysis (dry) of municipal solid waste components (%)

Component	Carbon	Hydrogen	Oxygen	Nitrogen	Sulphur
Paper/paper products					
Paper – mixed	43.41	5.82	44.32	0.25	0.20
Newsprint	49.14	6.10	43.03	0.05	0.16
Corrugated boxes	43.73	5.70	44.93	0.09	0.21
Plastic coated paper	45.30	6.17	45.50	0.18	0.08
Waxed milk cartons	59.18	9.25	30.13	0.12	0.10
Junk mail	37.87	5.41	42.74	0.17	0.09
Food/garden waste					
Vegetable food waste	49.06	6.62	37.55	1.68	0.20
Meat scraps (cooked)	59.59	9.47	24.65	1.02	0.19
Fried fats	73.14	11.54	14.82	0.43	0.07
Lawn grass	46.18	5.96	36.43	4.46	0.42
Leaves	52.15	6.11	30.34	6.99	0.16
Green logs	50.12	6.40	42.26	0.14	0.08

Evergreen shrubs	48.51	6.54	40.44	1.71	0.19
Flowering plants	46.65	6.61	40.18	1.21	0.26
Wood and bark	50.46	5.97	42.37	0.15	0.05
Household waste					
Leather shoe	42.01	5.32	22.83	5.98	1.00
Rubber	77.65	10.35	—	—	2.00
Upholstery	47.10	6.10	43.60	0.30	0.10
Polystyrene	87.10	8.45	3.96	0.21	0.02
PVC	45.14	5.61	1.56	0.08	0.14
Linoleum	48.06	5.34	18.70	0.10	0.40
Rags	55.00	6.60	31.20	4.12	0.13
Vacuum cleaner dirt	35.69	4.73	20.08	6.26	1.15

Source: Kaiser 1978.

Table 2.14 *Calorific Values (CV's) of municipal solid waste components (kJ/kg)*

Component	As received	Dry	Moisture/ash free
Paper/paper products			
Paper – mixed	15 750	17 530	18 650
Newsprint	18 550	19 720	20 000
Corrugated boxes	16 380	17 280	18 260
Plastic coated paper	17 070	17 910	18 470
Waxed milk cartons	26 350	27 290	27 660
Junk mail	14 160	14 830	17 210
Food/garden waste			
Vegetable food waste	4170	19 230	20 230
Meat scraps (cooked)	17 730	28 940	30 490
Fried fats	38 300	38 300	38 300
Lawn grass	4760	19 250	20 610
Leaves	18 490	20 540	21 460
Green logs	4870	9740	9840
Evergreen shrubs	6270	20 230	20 750
Flowering plants	8560	18 580	19 590
Wood and bark	19 570	19 940	20 140
Household waste			
Leather shoe	16 770	18 120	23 500
Rubber	25 930	26 230	29 180
Upholstery	16 120	17 320	17 800
Polystyrene	38 020	38 090	38 230
PVC	22 590	22 640	23 160
Linoleum	18 870	19 240	26 510
Rags	15 970	17 720	18 160
Vacuum cleaner dirt	14 790	15 640	23 060

Source: Kaiser 1978.

components that may occur in domestic waste can also significantly influence the properties of the waste, such as rubber and plastic waste.

The chemical analysis of municipal solid waste for a comprehensive range of metals and non-metals has been undertaken in the UK (National Household Waste Analysis Project 1995). Table 2.15 shows a detailed chemical analysis of four waste samples associated with the household groups represented by (a) modern family housing, higher incomes, (b) older terraced housing, (c) council estates – category II and (d) affluent suburban housing. In addition to ultimate and proximate analysis, Table 2.15 also shows detailed analysis of halides and heavy metals. The table shows the variability of chemical

Table 2.15   *Analysis of bulk waste samples as received from household group classifications in Leeds, UK*

Element	Sample			
	a	b	c	d
Moisture content (%)	37.8	41.4	38.7	37.0
Ash (%)	27.35	14.7	27.28	29.17
Gross calorific value (MJ/kg)	6.94	8.95	7.73	7.76
Fixed carbon (%)	4.59	5.96	4.49	4.1
Volatile matter (%)	30.24	38.32	29.5	29.4
Carbon (%)	20.04	22.76	19.74	20.38
Hydrogen (%)	2.93	3.23	2.73	2.82
Nitrogen (%)	0.40	0.67	0.89	0.72
Sulphur (%)	0.06	0.09	0.05	0.06
Chlorine (%)	0.25	0.31	0.21	0.97
Bromide (%)	0.01	0.01	0.01	0.01
Fluoride (%)	0.01	0.01	0.01	0.01
Silicon (%)	4.13	1.53	4.04	3.74
Mercury (ppm)	0.03	0.12	0.02	0.03
Sodium (ppm)	11 783	2220	7977	12 022
Magnesium (ppm)	2721	1737	2056	1908
Aluminium (ppm)	12 243	14 920	8828	8532
Potassium (ppm)	2277	2788	4185	3643
Calcium (ppm)	13 116	7821	17 251	11 723
Chromium (ppm)	48	24	37	61
Manganese (ppm)	383	374	372	572
Iron (ppm)	26 609	3217	73 066	11 127
Nickel (ppm)	24	17	28	105
Copper (ppm)	100	71	44	41
Zinc (ppm)	149	240	201	313
Arsenic (ppm)	2.4	1.8	6.4	10.0
Molybdenum (ppm)	2.8	2.4	1.95	2.75
Silver (ppm)	0.73	0.3	0.62	4.58
Cadmium	0.4	0.5	0.59	1.81
Antimony (ppm)	1.8	0.68	3.4	3.7
Lead (ppm)	84	33	41	247

a – High income, modern family housing; b – Older terraced housing; c – Municipal housing, intermediate status; d – Affluent suburban housing.

Source: National household waste analysis project 1994. Crown copyright material is reproduced with the permission of the Controller of HMSO and the Queen's Printer for Scotland.

analyses of waste from different sources. For example, the ash content shows a greater than 100 wt% variation depending on the type of household sampled. Other elements showing a wide variation in the four samples given are aluminium, calcium, chromium, iron, nickel, arsenic, cadmium and mercury.

2.3.3   Treatment and Disposal Options for Municipal Solid Waste

Figure 2.12 shows the average management of MSW across the OECD countries for 1995 compared with 2000 (OECD 2004). It is clear that the majority of municipal solid waste is landfilled, followed by waste incineration either with or without energy recovery, with some recycling and composting. The trends from 1995 to 2000 show that there is a movement away from landfill with a significant fall from 64 to 58% of MSW going to landfill. There was a corresponding growth in incineration, with energy recovery of 2%, increased recycling of 2% and increased composting of 2%. The data in Figure 2.12 reflect the average of treatment methods throughout the OECD. However, individual countries have a wide variation in how their MSW is managed. For example, in Japan, 77.4% of MSW is incinerated, 5.9% landfilled and 16.7% recycled, whereas in the USA, about 57% of MSW is landfilled, 28% is recovered/recycled/composted and 15% is incinerated.

Landfill is chosen as the most suitable option in most cases because of its low cost, its ready availability and its applicability for a wide range of wastes. Landfill also has advantages where, for example, holes produced from quarries and mineral workings are infilled with waste to produce restored landscapes. The biodegradable fraction of household waste will biodegrade within the landfill to produce a liquid leachate and landfill gas, composed largely of methane and carbon dioxide. In addition, landfill gas produced from the normal biodegradation of the organic waste in the site can be utilised for energy recovery. Inert wastes such as construction and demolition waste will normally not decompose and produce pollutants and will generate reduced levels of landfill gas. Because of the production of methane and carbon dioxide from municipal solid waste degradation which are both 'greenhouse gases', the EC, through the EC Waste Landfill Directive, is to limit the proportion of biodegradable waste going to landfill.

The promotion of a sustainable waste management strategy throughout the European Community has led to the initiation of targets to minimise the production of waste by the use of newer technologies and processes, to minimise the proportion of waste, particularly

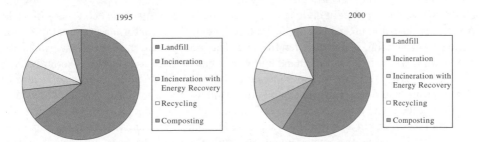

Figure 2.12   *Management of municipal solid waste in OECD member countries. Source: OECD 2004.*

biodegradable waste, going to landfill and also to encourage re-use and recycling of waste. A large proportion of the waste landfill sites designed and engineered today have energy recovery as an essential component of the system in order to derive economic benefit from the disposal of the waste.

Incineration utilises the energy content of the waste to produce steam from boilers, which in turn is used for district heating systems or to produce electricity for sale or in combined heat and power systems. In addition, incineration reduces the waste to about 10% of its original volume and to 30% of its original weight. The bottom ash residue or ash from the incinerator is either recycled as secondary building aggregate or landfilled. However, the costs of incineration are high. The costs derive from the high plant costs since the incinerator consists not only of the combustion unit to burn the waste but also a very high-efficiency, high-cost, gas clean-up system. High efficiencies of clean-up are required to meet stringent EC emission limits. A typical modern incinerator might have, for example, electrostatic precipitators to remove particulate material, a wet or dry lime gas scrubbing system to remove acid gases such as hydrogen chloride and sulphur dioxide, and a bag filter system with activated carbon additive to remove fine particulate, heavy metals and dioxins.

Recycling of municipal solid waste might be via segregation of the recyclable materials at source or segregation at a materials recycling facility. Municipal solid waste has a number of components which can be readily segregated. A major component of municipal solid waste is packaging waste, consisting of paper and board, aluminium cans, glass, plastics and steel food cans from household waste, and metal and plastic drums, wooden pallets and board and plastic crates and containers from the commercial and industrial sectors. The EC Directive on Packaging and Packing Waste, set a target for recovering 50–65% with a recycling target of 25–45%. The EC Directive requires the implementation of national programmes for the prevention of packaging waste and the principle of producer responsibility means that the packaging industry must contribute to the costs of recycling and recovery.

Other treatment and disposal options for municipal solid waste include composting and anaerobic digestion. Composting is the aerobic biological degradation of biodegradable organic waste such as garden and food waste by micro-organisms. Whilst anaerobic digestion is the biological degradation of the organic components of the waste by different groups of micro-organisms, which thrive under anaerobic conditions. Composting is a relatively fast biodegradation process, taking typically about 4–6 weeks to reach a stabilised product. Small-scale household composting has been carried out for many years and large-scale composting schemes, using organic waste collected from parks, and household garden waste collected from civic amenity sites, derive benefits from economies of scale. Anaerobic digestion takes place in an enclosed, controlled reactor under conditions similar to those operating in a mature landfill site. The biodegradation of the waste produces a product gas which is rich in methane and which can be used to provide a fuel or act as a chemical feedstock. The gas also contains carbon dioxide. In addition, the solid residue arising from anaerobic digestion can be cured and used as a fertiliser. The biodegradable fraction of the waste requires separation from the other components of the waste. Biodegradation takes place in a slurry of the separated organic waste and micro-organisms.

Figure 2.13 shows the management of municipal solid waste in selected European countries (European Commission 2003). It is clear that landfill is the major option for waste disposal in most European countries. In particular, the countries of Central and

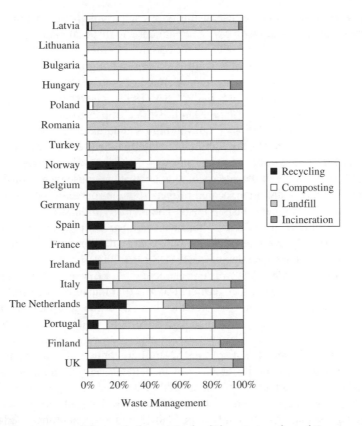

Figure 2.13 *Treatment and disposal of municipal solid waste in selected European countries. Source: European Commission 2003.*

Eastern Europe such as Lithuania, Bulgaria, Hungary and Poland, where the waste treatment option is almost exclusively waste landfill. The countries of Western Europe (Austria, Belgium, Denmark, Finland, France, Germany, Greece, Ireland, Italy, Luxembourg, the Netherlands, Portugal, Spain, Sweden, United Kingdom, Iceland, Liechtenstein, Norway, Switzerland, Andorra, Monaco and San Marino) show a more mixed use of the various waste management options. Indeed, there has been a recent trend in the decline of waste landfill in Western Europe (European Commission 2003) and increased recycling. For example, Germany has approximately 45% of recycling and composting. Similarly, Norway, Belgium and the Netherlands have significant recycling of waste. Incineration of waste is also a major option throughout Western Europe for waste treatment, but is almost non-existent in Eastern Europe.

Individual country characteristics in many cases dictate the options chosen for municipal waste treatment and disposal. For example, the Netherlands is a densely populated country with a limited landfill capacity and so incineration, waste reduction and recycling are the preferred options. Waste management policies are based on waste reduction and recycling to minimise the amount of material that ultimately is destined for landfill, and the main route for waste disposal is incineration either with or without energy recovery.

Municipal waste disposal is the responsibility of each municipality in Japan and each is responsible for the preparation of a waste treatment and disposal plan.

By comparison, the USA generates about 230 million tonnes of municipal solid waste each year, of which the majority is landfilled, representing about 57%. Approximately 15% is incinerated and 28% is recovered, recycled or composted (US EPA 2004). There are approximately 2300 landfill sites in the USA (1999). The recent trend in the USA has been towards a fewer number of larger landfills, due in part to the introduction of more stringent waste landfill regulations requiring high standards of site lining, monitoring of gas and leachate and post closure liabilities, which has led to increased costs. This has consequently led to the closure of some sites and the move to larger landfill sites allows economies of scale to offset the site regulation costs. One consequence of the closure of landfill sites is that they are often located further from the source of the waste production, and therefore involve higher transport costs. The majority of landfill sites are owned and operated by local government, but a significant number are privately owned. The USA has 102 municipal solid waste incinerators with energy recovery (1999) with the capacity to burn up to 96 000 tonnes per day of waste (US EPA 2004).

In North America, Canada's waste treatment and disposal is primarily managed through municipal and regional government and is dominated by the option of landfill, due in part to the ready availability of suitable cheap land. In addition, there are relatively cheap sources of energy via petroleum, hydroelectricity and nuclear electricity which lowers the incentive for energy from waste schemes such as incineration.

The situation in Japan is that, of the 52 million tonnes of municipal solid waste generated each year, the majority, at 77.4% is incinerated, 5.9% is landfilled and 16.7% is recycled (Statistical Handbook of Japan 2003). Japan is a highly mountainous and volcanically active country with only approximately 10% of the land suitable for residential purposes and consequently there is little land available for waste landfill.

Box 2.1 describes the management of municipal solid waste in China.

Box 2.1
The Management of Municipal Solid Waste in China

Of the 1.2 billion citizens of China, approximately 208 million people are located in 666 cities. These cities generate about 140 million tonnes of municipal solid waste each year. Before 1980 almost all of the municipal solid waste was dumped without any concern for the environment. The current main municipal solid waste treatment option is waste landfill, which accounts for more than 70% of the total, followed by composting at 20%, with waste incineration playing a minor role. This has left a legacy of water and air pollution emanating from this poor management of waste. The last decade has seen technological improvement in the management of wastes, including in some cases, improved sanitary landfill design, increased recycling and recovery technologies. However, in most cities in China the main approach to municipal solid waste management is centralised dumping and low technology waste landfill. Recycling

Continued on page 93

Continued from page 92

of waste is conducted by waste collectors either at source or at the disposal site, however, statistical data on recycling is poor. Composting is of the mechanical or simple high-temperature type and there are less than 10 municipal solid waste incinerators in the whole of China, representing about 2% of MSW. The waste management industry in China is characterised by inadequate investment, out-of-date technical expertise and poor education of personnel.

The management of municipal solid waste is organised at national level by the Ministry of Construction. For each of the 666 cities of China, the management of municipal solid waste is the responsibility of the Environmental Sanitary Departments who organise the management, collection, transportation, and disposal of the waste. Management of environmental pollution is the responsibility of the Environmental Protection Departments.

The regulation of municipal solid waste in China:

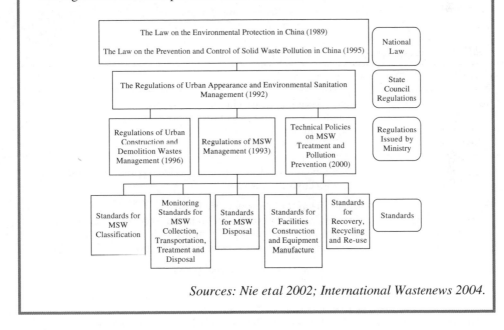

Sources: Nie et al 2002; International Wastenews 2004.

2.4 Hazardous Waste

Hazardous waste is a term used throughout the world to describe waste which is dangerous or difficult to keep, treat or dispose of, and may contain substances which are corrosive, toxic, reactive, carcinogenic, infectious, irritant, or otherwise harmful to human health and also which may be toxic to the environment. Within the countries of Europe, the highest proportion of hazardous waste is derived from manufacturing industry (European Commission 2003)

In 1991, the European Commission under an amendment (91/689/EEC) to the 1975 Waste Framework Directive (Council Directive 75/442/EEC; Waste Framework

Directive 1975) identified the properties of hazardous waste which make it hazardous such as that it is explosive, oxidising, flammable, irritant, harmful, toxic, carcinogenic, corrosive, infectious, etc. The amendment listed various categories of hazardous waste such as pharmaceuticals, wood preservatives, inks, dyes, resins, tarry materials, mineral oils, etc. This was followed, in 1994, by the introduction by the European Commission of a list of over 200 different types of hazardous waste which were listed in Council Decision 94/904/EC (1994).

Commission Decision 2000/532/EC in 2000 (Commission Decision 2000) replaced earlier lists of wastes, including hazardous wastes, in one unified document, the European Waste Catalogue. There are more than 650 waste categories on the list, each with its individual number code, divided into different categories of waste according to their source. The hazardous wastes within each category are highlighted in the following text by an asterisk. Table 2.16 shows examples of hazardous wastes in various waste categories of the European Waste Catalogue (Commission Decision 2000).

Table 2.16   *Examples of hazardous wastes in the European Waste Catalogue*

Code	Waste category
06	**Wastes from inorganic chemical processes**
06 01 01*	Sulphuric acid and sulphurous acid
06 01 02*	Hydrochloric acid
06 02 02*	Calcium hydroxide
06 01 02*	Ammonia
06 13 01*	Inorganic pesticides, biocides and wood preserving agents
06 13 02*	Spent activated carbon
09	**Wastes from the photographic industry**
09 01 01*	Water based developer and activator solution
09 01 03*	Solvent based developer solutions
09 01 05*	Bleach solutions and bleach fixer solutions
10	**Inorganic wastes from thermal processes**
10 01 04*	Oily flyash
10 03 01*	Tars and other carbon containing wastes from anode manufacture
10 04 04*	Flue gas dust (wastes from lead thermal metallurgy)
10 06 03*	Flue gas dust (wastes from copper thermal metallurgy)
12	**Wastes from shaping and surface treatment of metals and plastics**
12 01 06*	Waste machine oils containing halogens
12 01 10*	Synthetic machining oils
12 03 02*	Steam degreasing waste
13	**Oil wastes**
13 01 01*	Hydraulic oils containing PCBs or PCTs
13 01 08*	Brake fluid
13 02 02*	Non-chlorinated engine, gear and lubricating oils
13 02 03*	Chlorinated engine, gear and lubricating oils
13 03 05*	Mineral insulating and heat transmission oils
13 04 01*	Bilge oils from inland navigation

16	**Wastes not otherwise specified in the list**
16 02 09*	Transformers and capacitors containing PCBs or PCTs
16 02 12*	Discarded equipment containing free asbestos
16 04 01*	Waste ammunition
16 04 02*	Fireworks waste
16 04 03*	Other waste explosives
16 06 01*	Lead batteries
16 06 02*	Nickel–Cadmium batteries
18	**Wastes from human or animal health care**
18 01 08*	Cytotoxic and cytostatic medicines
18 01 10*	Amalgam waste from dental care
19	**Wastes from waste treatment facilities**
19 01 13*	Flyash containing dangerous substances (waste incineration)
19 01 15*	Boiler dust containing dangerous substances (waste incineration)
19 01 17*	Pyrolysis waste containing dangerous substances (waste pyrolysis)
20	**Municipal wastes**
20 01 19*	Pesticides
20 01 21*	Fluorescent tubes and other mercury-containing waste
20 01 31*	Cytotoxic and cytostatic medicines

Source: Commission Decision 2000/532/EC in 2000.

A range of inorganic and organic compounds which are hazardous to health or which may pose a physical hazard, are shown in Table 2.17 (Woodside 1993). The hazardous materials are derived from a variety of industries. Many industries use hazardous materials in their processes and consequently may be generated as part of the waste stream. Contaminated soils may also be designated as hazardous waste by containing hazardous materials such as heavy metals, pesticides or by being contaminated with tars, oils and other organic materials from old industrial sites such as old gasworks.

It has been estimated that about 62 million tonnes of hazardous waste are generated each year throughout Europe (European Commission 2003). This equates to approximately 47 million tonnes per year generated by Western European countries (Austria, Belgium, Denmark, Finland, France, Germany, Greece, Ireland, Italy, Luxembourg, the Netherlands, Portugal, Spain, Sweden, United Kingdom, Iceland, Liechtenstein, Norway, Switzerland, Andorra, Monaco, San Marino) and approximately 15 million tonnes per year from Central and Eastern European countries. Figure 2.14 shows the generation of hazardous waste in relation to different countries of Europe (European Environment Agency 2002(a); European Environment Agency 2003). As with many waste statistics, data comparisons between countries are difficult due to the different classifications of hazardous waste used by different countries. In many countries, the tonnages of hazardous waste generated are small compared with the total tonnages of waste arising, but because of their associated hazard, represent difficult and expensive wastes to be treated.

Table 2.17   *Examples of hazardous chemicals and physical hazards*

[A] Health hazard	Chemicals that create health hazards
Carcinogen	Aldrin, formaldehyde, ethylene dichloride, methylene dichloride, dioxin
Toxic	Xylene, phenol, propylene oxide
Highly toxic	Hydrogen cyanide, methyl parathion, acetonitrile, allyl alcohol, sulphur dioxide, pentachlorophenol
Reproductive toxin	Methyl cellosolve, lead
Corrosive	Sulphuric acid, sodium hydroxide, hydrofluoric acid
Irritant	Ammonium solutions, stannic chloride, calcium hypochlorite, magnesium dust
Sensitizer	Epichlorohydrin, fibreglass dusts
Hepatotoxin	Vinyl chloride, malathion, dioxane, acetonitrile, carbon tetrachloride, phenol, ethylenediamine
Neurotoxin	Hydrogen cyanide, endrin, mercury, cresol, methylene chloride, carbon disulphide, xylene
Nephrotoxin	Ethylenediamine, chlorobenzene, dioxane, acetonitrile, hexachlonaphthalene, allyl alcohol, phenol, uranium
Blood damage	Nitrotoluene, benzene, cyanide, carbon monoxide
Lung damage	Asbestos, silica, tars, dusts
Eye/skin damage	Sodium hydroxide, ethylbenzene, perchloroethane, allyl alcohol, nitroethane, ethanolamine, sulphuric acid, liquid oxygen, phenol, propylene oxide, ethyl butyl ketone
[B] Physical hazard	**Chemicals that create the hazard**
Combustible liquids	Fuel oil, crude oil, other heavy oils
Flammable materials	Gasoline, isopropyl alcohol, acetone, butane spray cans
Explosives	Dynamite, nitroglycerine, ammunition
Phyrophoric	Yellow phosphorus, white phosphorus, superheated toluene, silane gas, lithium hydride
Water reactive	Potassium, phosphorus pentasulphide, sodium hydride
Organic peroxides	Methyl ethyl ketone peroxide, dibenzoyl peroxide, dibutyl peroxide
Oxidisers	Sodium nitrate, magnesium nitrate, bromine, sodium permanganate, calcium hypochlorite, chromic acid

Source: Woodside 1993. Copyright © 1993. Reprinted with permission of John Wiley & Sons Inc.

Table 2.18 shows the main industrial categories which are the sources of hazardous waste in several European countries (European Environment Agency 2002(b)). The manufacturing industry is the main source of hazardous waste for most countries and regions of Europe. This is particularly the case for Finland, Germany and Norway, where more than 75% of the hazardous waste arisings come from the manufacturing sector. However, for other countries, including Austria and the Netherlands, hazardous waste from the manufacturing industries represents less than 30%. Within the category of manufacturing industry are a wide variety of different industrial types which contribute to the overall hazardous waste category for the manufacturing industry. For example, in Finland the dominant source of hazardous waste comes from the manufacture of refined petroleum products, at 30.6%, and the manufacture and processing of basic metals, at 35.3%. In Norway, and to some extent Germany, the main manufacturing industry responsible for the production of hazardous waste is the chemical industry at 78.4% and 25.2%, respectively. The main differences in the production of hazardous waste from the various

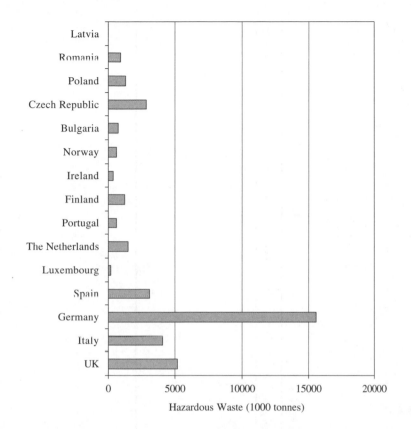

Figure 2.14   *Hazardous waste generation in selected countries of Europe. Source: European Commission 2003.*

industrial sectors lies in the differences in the range and extent of industrial activity. Wastes from commercial and service activities, which includes shipping and repair of motor vehicles, are large contributors to waste generation in the Netherlands and Portugal. Wastes from refuse disposal activities are significant sources of hazardous waste, where there is a large waste incineration activity generating significant tonnages of hazardous flyash, for example, in Austria.

One category of hazardous waste is that of healthcare waste or clinical waste. The 2000 European Waste Catalogue (Commission Decision 2000/532/EC 2000) lists the various categories of wastes according to the different economic sectors (designated as 'chapters') of Europe. Chapter 18, 'wastes from human or animal health care and/or related research', lists various hazardous and non-hazardous healthcare wastes, each with its individual six-digit code number. For example, hazardous clinical waste from human and animal health care is mostly included under, 'waste whose collection and disposal is subject to special requirements in view of the prevention of infection' (codes 18 01 03* and 18 02 02*).

Healthcare waste includes waste from hospitals, doctors' and dental surgeries, health centres, nursing homes and veterinary surgeries. For example, it might include human or

Table 2.18 Hazardous waste generation in terms of industrial category

Waste source category	Austria (%)	Finland (%)	Germany (%)	The Netherlands (%)	Norway (%)	Portugal (%)
Manufacturing	21.2	79.9	79.2	26.5	82.5	56.9
Agriculture, fishing, quarrying	5.7	0.2	1.2	3.1	5.4	1.9
Electricity, gas, water supply	20.7	0.8	13.8	—	0.3	11.9
Construction	1.4	—	5.4	1.8	0.4	3.5
Commercial and service activities	18	—	—	41.3	5.4	21.8
Public services and health care	1.4	—	0.4	—	0.6	3.9
Refuse disposal	19.0	15.1	0.1	—	0.5	—
Activity not stated	12.6	4	—	27.3	4.9	0.1

Source: European Environment Agency 2002(b).

animal tissue, blood or other body fluids, excretions, drugs or other pharmaceutical products, swabs, dressings, syringes, needles or other sharp instruments. It is waste which, unless rendered safe, may prove hazardous to persons coming into contact with it.

Some wastes arising from the rest of the health care sector may be separated from clinical wastes and be classified as household or commercial waste, depending on where in the health care sector the waste has arisen. For example, newspapers, magazines, flowers and office waste, would clearly be non-hazardous and could be classified as commercial waste. Similarly, waste from nurses' accommodation and nursing homes would be classified as household. This reduces the amount of waste to be handled under the 'hazardous health care waste' definition, which requires more expensive treatment and disposal. However, careful segregation of the different categories of waste are required to prevent the mixing of hazardous and non-hazardous healthcare wastes.

The type and quantity of waste produced will vary according to the institutional type. For example, geriatric and long-stay hospitals produce about 1.0 kg of waste/bed/day, while general short-stay hospitals produce about 3.1 kg/bed/day and large teaching hospitals, with hostel accommodation and administrative centres, can have waste generation rates of over 6.0 kg/bed/day. The composition of the waste will also change with the institution. For example a higher proportion of human tissue will be present in general hospital waste, whereas there may be none from a geriatric-type hospital. The plastic content of hospital waste is high compared with municipal solid waste. Typical hospital waste has a calorific value of about 15 MJ/kg (Wallington and Kensett 1988) compared with municipal solid waste at about 8.5 MJ/kg and a typical coal at 28–32 MJ/kg.

A further category of hazardous waste, which has significance for local authority waste management, is household hazardous waste. This includes such materials as garden pesticides and herbicides, paints, medicines, oils, batteries, solvents and other materials which put human health or the environment at risk because of their chemical or biological nature. The waste in this category comprises about 1–5 kg/household/year, representing less than 1% of the total domestic waste but it does represent a disproportionate risk. Whilst for most European countries, household hazardous waste is less than 1% of municipal solid waste, the consequent tonnages are significant. For example, it is estimated that France produces 260 000 tonnes per year of household hazardous waste, Germany 390 000 tonnes, Italy 254 000 tonnes, Spain 143 000 tonnes and the UK 252 000 tonnes per year (European Commission – Directorate General Environment 2002(a)). In addition, the household hazardous waste, whilst occurring in small quantities, occurs over a large number of locations, adding to the difficulties of management. Also, it has been noted that some household hazardous wastes such as pesticides and paints may be stockpiled for many years and then the obsolete products discarded in large quantities as a single consignment. Such concentrated releases of hazardous waste into the waste streams can create high risks to health during waste collection and treatment, and also to the environment if not handled correctly (European Commission – Directorate General Environment 2002(a)). Much household hazardous waste is intermixed with other household waste and mostly collected together, and may represent a disproportionate risk to human health and the environment unless correctly handled (Warmer Bulletin 50, 1996). In addition, the options for treatment and disposal of such wastes may be limited. For example, high levels of heavy metals may rule out the option of composting the waste, since the final product would have heavy metal concentrations outside regulated levels. Similarly, the presence of persistent pesticides could result in groundwater contamination if the mixed waste were landfilled (White et al 1995). Such wastes may be separated by the householder and collected from the household either on demand or via yearly or twice-yearly collections, or at civic amenity sites. Once sorted, such a concentration of hazardous wastes would be classified as hazardous, with the consequent regulation of its storage, transport, handling and treatment (White et al 1995).

The European Commission has proposed that household hazardous waste be separately collected from households and other municipal sources (ENDS Report 265, 1997). The Commission also requires manufacturers to mark such products with a special logo informing consumers not to dispose of them with ordinary household waste. Several categories of such wastes are listed, for example, paint, mineral and synthetic oils and their filters, medicines, aerosols, bleaches, batteries, solvents, adhesives, etc. Many European countries, such as Austria, Denmark, Finland, Germany, Luxembourg and the Netherlands, already have systems for the separate collection and/or delivery of at least some household hazardous waste (ENDS Report 265, 1997; European Commission – Directorate General Environment 2002(a)).

2.4.1   Treatment and Disposal Options for Hazardous Waste

There are a range of treatment and disposal options for hazardous waste which depend on the type of hazard involved. These include landfill, storage, recovery, thermal, biological and physico-chemical treatment methods (Freeman 1998).

Recovery methods involve separation of recoverable materials from the hazardous waste. Such methods include, for example, distillation, ion exchange, solvent extraction, membrane separation, catalytic extraction, etc. (Freeman 1998). The recovered material then has the potential to be recycled. Recovery has the advantages of reducing waste disposal costs and, potentially, reduction in the use of raw materials. It is carried out either on or off-site.

Disposal of hazardous waste in waste landfills is agreed as the option of last resort in EC countries and should only be used when all possibilities of treatment have been exhausted (European Commission 2003). The extent of hazardous waste landfill is dependent on an individual country's national policy on waste treatment and disposal. The landfill designs used by European countries are highly engineered containment landfills which use natural and synthetic polymer barriers to contain the waste and prevent leachate moving beyond the site boundary. A 'double' or 'multiple' liner containment system would be typical. Containment systems such as this require that there is minimal interaction of the barriers with the waste which would deteriorate the barrier system. The liner systems used a complex and often contain a number of leachate drainage systems, separated by impermeable synthetic barriers.

Under the Council Directive (1999/31/EC) landfills are categorised into those accepting non-hazardous waste (such as municipal solid waste), hazardous waste and inert waste. Each type of designated landfill can only accept the particular waste for which it is designated. Consequently, only hazardous wastes are permitted in hazardous waste landfill sites. Some types of waste are not permitted to go to landfill at all and include liquid waste, flammable waste, explosive or oxidising wastes, infectious clinical waste or hospital waste.

Co-disposal of hazardous waste into landfill has been used as a common route to dispose of hazardous waste. The principle behind co-disposal is that the landfill site acts as a bioreactor in which the micro-organisms in the municipal solid waste breakdown the components of the hazardous waste. The hazardous waste is deposited into the municipal solid waste in the landfill site where the waste is in an advanced stage of biodegradation. There is some evidence that the leachate derived from such co-disposal sites is, in fact, quite similar to a municipal solid waste site leachate (Waste Management Paper 26F, 1996). The co-disposal of hazardous waste is regarded throughout Europe as a poor disposal option and the EC Waste Landfill Directive seeks to eliminate the practice through designation of landfills into those accepting hazardous, non-hazardous and inert wastes. The drawback of co-disposal has centred on possible interaction of the wastes to produce a toxic product which may harm the population or environment.

Incineration of hazardous waste with energy recovery would be the preferred option for the sustainable disposal of hazardous waste. However, the flue gas emissions from the incinerators requires extensive clean-up using a variety of systems such as electrostatic precipitators, scrubbers and bag filters to remove the potentially highly toxic pollutants. The flue gas treatment systems are expensive and, consequently, disposal costs via incineration are high and can represent between 10 and 50 times the equivalent cost of landfill, depending on the degree of hazard associated with the waste. However, incineration for certain types of waste such as liquid organic waste is regulated as the only option in some countries. Incineration of the waste is regarded as a thermal treatment process. Hazardous waste incinerators have used rotary kilns as the preferred design, whilst other designs have included fluidised bed and vortex or other spray combustor-type incinerators (Dempsey and Oppelt 1993; LaGrega et al 1994). Rotary kiln technology involves a continuously rotating ceramically lined cylinder in which the waste is combusted. The kiln is tilted at an angle and the waste moves down the kiln until complete burn-out of the waste is achieved. Where hazardous waste is combusted, a secondary chamber is also involved, which combusts the derived gaseous products.

Incineration of certain types of hazardous waste, blended with other fuels in cement kilns, is also practised in some countries, utilising the high combustion temperatures (>1400 °C) and long residence times in the rotary kiln of the process, to destroy the waste. In addition, when chlorinated or fluorinated wastes are combusted, the large mass of alkaline clinker from the process, absorbs and neutralises the acidic stack gases. The utilisation of wastes by the cement industry has mutual benefits in that the cement kiln process in very energy intensive and the organic wastes that are often used have high calorific values (Holmes 1995; Benestad 1989).

The European Commission Waste Incineration Directive (Council Directive 2000/76/EC) introduced in 2000, represents a single text which covers the incineration of hazardous and non-hazardous wastes. The Directive concerns the incineration of all types of waste, including municipal solid waste, hazardous waste, sewage sludge, tyres, clinical waste, waste oils and solvents, etc. The 2000 EC Waste Incineration Directive repeals the earlier EC Directive on incineration of hazardous waste (Council Directive 94/67/EC).

Biological treatments use microbiological organisms to breakdown the components of the waste into less hazardous products. The ability of the organisms to break down the organic waste is dependent on the types of organic compounds present in the waste material. For example, easily biodegradable compounds include alkenes, alcohols and aldehydes;

intermediate biodegradable compounds are alkanes, aromatic compounds and nitrogen-containing compounds; and difficult wastes to degrade are those containing halogenated compounds. Some inorganic compounds can also be treated in biological systems, for example, cyanides and metals such as lead and arsenic.

Biological treatment systems are categorised as either aerobic or anaerobic. Aerobic systems are oxygenated and the microbiological organisms convert the organic components of the hazardous waste into carbon dioxide and water. Anaerobic systems are devoid of oxygen and the organisms convert the waste into methane and carbon dioxide. The aerobic activated sludge process is widely used for the treatment of industrial wastewaters containing organic wastes and also for sewage sludge. The process of waste and microbiological interaction takes place in suspension and the reactants are continuously mixed in a bioreactor. The solid suspension may separate out as a sludge and would be recycled back to the bioreactor to maintain a suitable concentration. The sludge would have a typical residence time of 20–30 days in the bioreactor, after which it is removed for treatment by de-watering, thickening and final disposal, which is typically via landfill or incineration (Woodside 1993; La Grega et al 1994; Kim and Qi 1995).

Other aerobic treatment systems are, for example, the supported sludge processes, where the biological organisms are supported on some sort of solid substrate in a bioreactor and the organic waste material is slowly passed over the solid bed of biological reactant. Because the residence time in the bioreactor is less than that of the activated sludge process, the organic waste may have to be recycled through the reactor several times to achieve the required low concentrations of organic material in the effluent.

Anaerobic processes, utilise a different type of microbiological organism, since the reacting environment is devoid of oxygen. The bioreactor used is similar to the activated sludge process used in sewage sludge treatment works, where the waste and organisms interact in suspension. The organisms biodegrade the organic components of the waste via a two-step process, firstly converting the waste to organic acids, alcohols, carbon dioxide and water followed by breakdown of the acids and alcohols to carbon dioxide and methane.

Physical, chemical and physico-chemical treatment processes have all been used for hazardous waste (Woodside 1993; LaGrega et al 1994; Kim and Qi 1995). Physical treatment includes carbon adsorption, fractional distillation, solvent extraction and sedimentation. Carbon adsorption uses activated carbon with a very high active surface area, typically $1000–2000 \, m^2/g$, to physically adsorb the organic material from solution or wastewater. The waste components interact with the activated carbon either in suspension in a reactor or as fixed filter beds of activated carbon, through which the waste stream flows. Distillation relies on the different boiling points of organic material in the waste. Solvent wastes can be fractionated and separated into different purified organic compounds, or organic material can be separated from wastewater streams. Similar in principle is evaporation, where volatile compounds can be separated from non-volatile compounds. Sedimentation is the separation of hazardous suspended fine solid material from solution by means of gravity.

Chemical treatment includes: neutralisation of acidic or alkaline wastes to produce an acidically neutral solution; concentration of the waste from solution by precipitation of the hazardous components; removal of inorganic material, such as heavy metals, from solution by ion exchange resins, which selectively remove such components; and oxidation/

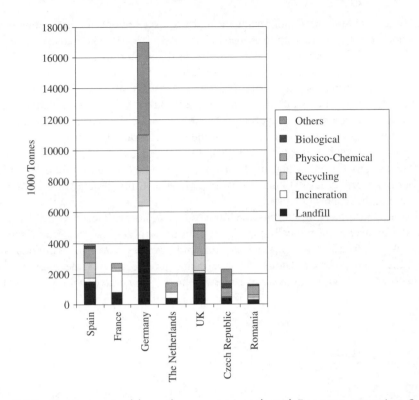

Figure 2.15 *Management of hazardous waste in selected European countries. Source: European Commission 2003.*

reduction reactions to produce less hazardous components or less volatile waste by adding or subtracting electrons. Chemical treatments, which include incineration, have also included: thermal processes such as wet-air oxidation which involves treatment of organic wastes at high temperatures and pressures in the presence of steam; pyrolysis, the thermal degradation of organic wastes at high temperature in the absence of oxygen; and vitrification or solidification where the wastes, such as contaminated soils and sludges, are combined with silica-containing material at high temperature to form a glass.

Solidification and stabilisation are physico/chemical treatment methods. The treatment involves mixing liquid waste with solids to produce a waste material which is more easily handled, can be landfilled and is less susceptible to leaching. The treatment may be physical adsorption or chemical interaction, depending on the type of waste and the type of solid material. For example, lime or cement-based treatment of toxic metal ions has been used (Woodside 1993; LaGrega et al 1994).

In some countries, long-term storage of hazardous waste has been used as an option for the management of hazardous waste, with the waste being stored in 55-gallon drums. Clearly this is not regarded as a sustainable option for waste treatment. A range of reviews are available for the different options available for the treatment and disposal of hazardous waste (Freeman 1998; Visvanathan 1996; Woodside 1993).

Figure 2.15 shows the major treatment and disposal options for hazardous waste in certain countries across Europe (European Commission 2003). The management options include a range of different treatments and disposal and there is no over-reliance on one method over another, for most countries in Europe. There is significant recovery and recycling of hazardous waste in Europe, where recovery also covers energy recovery from hazardous waste incineration. Spain, Germany and the UK have significant operations involved in the recovery and recycling of hazardous waste. Between 1997 and 2001, the amount of hazardous waste landfilled in Europe, has generally shown an increase. This is not a consistent trend, for example, in Turkey, Norway, the UK, Italy, Denmark and Romania, the proportion of landfilled hazardous waste has decreased. In Spain, Finland, UK, Germany, Bulgaria, Estonia, Romania and Estonia this amounted to a significant proportion of the disposal method.

Because of the potentially offensive nature of hazardous healthcare waste, the route for disposal is incineration, typically in a rotary kiln or pyrolytic incineration system. Non-hazardous healthcare wastes arising as household wastes from, for example, nursing homes, are similar to other domestic wastes and are treated via the municipal waste routes. Non-hazardous healthcare waste may be co-incinerated with municipal solid waste.

2.5   Sewage Sludge

Sewage sludge is formed as a by-product of the different treatment stages of raw sewage from domestic households, but may also include industrial and commercial effluent (European Commission 2002(b)). Consequently, the sludge composition can vary considerably depending on the main source of the sewage. Sewage sludge contains constituents such as organic matter, nitrogen, phosphorus, potassium and, to some extent, calcium, magnesium and sulphur which are of agricultural value. However, the sludge also contains pollutants such as heavy metals, organic pollutants and pathogens which are environmentally harmful. Where industrial sewage systems contribute to the domestic waste loadings, significantly higher concentrations of heavy metals such as lead, zinc and copper or high levels of soluble organic matter may result (Dean and Suess 1985; Hall 1992; Digest of Environmental Statistics 1996; Hudson 1993; Try and Price 1995). The heavy metals, in particular, may be carried through the treatment processes and end up in the final sewage sludge. Where such sludges are produced, this limits the disposal options, since application to agricultural land via landspreading may not be allowable within existing EC regulations. The sewage is mostly water but, after treatment, the particulate and colloidal matter is concentrated to form sewage sludge. Treatment of the raw sewage involves effective separation of the suspended organic matter from the liquid stream by settlement in primary, secondary and tertiary stages. The aim of such treatment is to reduce the water content, the potential for fermentation and to reduce the presence of pathogens (European Commission 2002(b)).

The biological treatment of the sewage sludge utilises aerobic digestion in either the activated sludge process or the supported sludge process. The activated sludge process involves the interaction of the sewage with organic micro-organisms in suspension in a bioreactor. The suspension is continuously stirred and aerated by bubbling air through the bioreactor and the settled sewage sludge is recycled until it has been fully biodegraded. The supported sludge process relies on the biological micro-organisms being supported on a solid substrate contained in a bioreactor, through which the sewage wastewater is trickled. The supported sludge system has a shorter residence time than the activated sludge process and the sludge may therefore be recycled through the bioreactor to ensure sufficiently low concentrations of pollutants. The settled solid material is removed for final treatment, which is anaerobic digestion where the sludge is heated to about 35 °C for approximately two weeks until the organic material is broken down, to reduce pathogen levels and also odours. Mechanical de-watering takes place to reduce the water content of the sludge product. The final material is termed 'treated sewage sludge' and this is either recycled by spreading on the land for agricultural use, landfilled, or incinerated (European Commission 2002(b); Bruce et al 1989; Wheatley 1990; Hall 1992). Rather than biological treatment of raw sewage sludge, the sludge may be chemically stabilised by the addition of alkaline material such as lime (Environment Agency 1998).

The tonnages of sewage sludge on a dry-weight basis, are shown in Figure 2.16 for selected countries throughout Europe (European Commission 2003; European Environment Agency 2002(a)). The tonnages when expressed on a dry-weight basis are low compared with the total generation of all the wastes in a country. However, the generated sludges contain up to 96% water content, thus considerably increasing the actual tonnage to be treated. Interestingly, there are significant variations in the generation of sewage sludge expressed on a per capita basis, with some countries such as Denmark, Finland, Germany, Luxembourg, Portugal and Ireland having per capita generation rates of sewage sludge of more than 30 kg/person, whilst other countries, such as Belgium, Greece and France have less than 20 kg/person.

The composition of sewage sludge is shown in Table 2.19 (Frost 1991; Yokoyama et al 1987). The sewage sludge has a high ash content, typically between 35 and 40% of the solid content. This has great significance for incineration processes and the efficient operation of the incinerator and for clean-up of the flue gases which will contain high

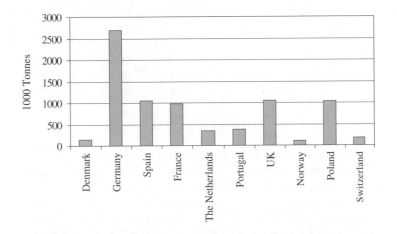

Figure 2.16   *Generation of sewage sludge as weight of dry solid in selected European countries. Sources: European Commission 2003; European Environment Agency 2002(a).*

Table 2.19   *Typical composition and properties of sewage sludge*

Property	Typical value
Calorific value	21.3 MJ/kg (dry, ash free)
Ash content	37%
Composition of combustible fraction (dry, ash free)	
Carbon	53.0%
Hydrogen	7.7%
Oxygen	33.5%
Nitrogen	5.0%
Sulphur	0.8%
Organic composition (dry, ash free)	
Crude protein	30.0
Crude fat	13.0
Crude fibre	33.0
Non-fibrous carbohydrate	24.0

Sources: Frost 1991; Yokoyama et al 1987.

concentrations of particulate and associated heavy metals from the sludge. However, the calorific value is moderately high and sewage sludge can be successfully incinerated, even if there is a requirement for supplementary fuel. The sewage sludge would have to be dried to some extent to remove the water to a level to enable the formation of a combustible fuel or other combustible, with the addition of a supplementary fuel, such as natural gas. The nitrogen content of the sludge is high, representing the nutrient value of the sludge when it is used for land spreading, however, high concentrations of nitrogen can lead to high concentrations of polluting nitrogen oxides during the incineration processes.

Of more concern in the, environmentally acceptable, disposal of sewage sludge is the potential concentration of toxic inorganic and organic compounds. Table 2.20 shows the ranges of potentially toxic chemicals which have been shown to occur in raw sewage sludge (Hall 1992; Dean and Suess 1985). Discharges from industrial sources to sewer are subject to their own legislative control and, consequently, heavy metal pollutants should be present in low concentration. Domestic sewage, whilst containing many metals in significant concentrations, does not approach the amounts from industrially derived sewage. Organic compounds occur naturally in sewage sludge. However, those listed in Table 2.20 are synthesised compounds, which may be resistant to the biodegradation sewage treatment process and may persist in the treated sludge and wastewater.

2.5.1   Treatment and Disposal Options for Sewage Sludge

The main treatment and disposal options for sewage sludge throughout Europe is via recycling, incineration and landfill. Recycling involves the spreading of the sludge or sludge-derived material onto land, to utilise the agricultural fertilising benefits of the sludge constituents. Figure 2.17 shows the treatment and disposal of sewage sludge in selected European countries in 1998 (European Environment Agency 2002(a)). The data for the UK in Figure 2.17 are for 2000 (DEFRA 2004(a)). Prior to the introduction of the Urban Wastewater Treatment Directive, approximately 29% of UK sewage sludge and

Table 2.20   *Concentration ranges of potentially toxic contaminants found in sewage sludge (mg/kg of dry solid)*

Inorganic contaminant[1]	Concentration	Typical domestic concentration
Cadmium	2–1500	5
Copper	200–8000	380
Nickel	20–5000	30
Zinc	600–20 000	515
Lead	50–3600	120
Mercury	0.2–18	1.5
Chromium	40–14 000	50
Molybdenum	1–40	4
Arsenic	3–30	3
Selenium	1–10	2
Boron	15–1000	50
Fluorine	60–40 000	200
Organic contaminant[2]	**Concentration**	
Phthalic acid esters	1 100	
Polycyclic aromatic hydro carbons (PAHs)	0.01–50.0	
Polychlorinated biphenyls (PCB)	0.16–9.11	
Dieldrin (pesticide)	0.018–3.90	
Lindane (pesticide)	0.025–0.410	
Aldrin (pesticide)	0.02–0.24	
Dichlordipheyltrichloroethane (DDT) (pesticide)	0.02–0.80	
Dioxins and furans[3] (ng/kg)	20–200	

Sources: Hall 1992[1]; Dean R.B. and Suess M.J. 1985[2]; Environment Agency 1998[3].

about 8% of Spanish sewage sludge was disposed of at sea. This was banned under the Directive, with subsequent diversion to incineration, landfill and agricultural use. The largest producer of sewage sludge in Europe is Germany, with the largest proportion being recycled as agricultural use. In fact, for most countries, agricultural use in the form of land spreading is the preferred option for the management of sewage sludge

Whilst land spreading may not, at first, appear to be the best option for sewage, the sludge in fact contains valuable nutrients such as nitrogen, phosphorus and organic matter in high concentration and is a suitable supplement to other fertilisers. Typical application rates of sewage sludge to land are about 50–100 tonnes/hectare (Hall 1992; Try and Price 1995). The sludge is applied to land after being treated via biological, chemical or heat treatment. The use of untreated sewage sludge on land is prohibited, unless the sludge is injected or worked into the soil. In some cases the sewage sludge may be composted, which stabilises the sludge due to the addition of vegetal material added as a co-composting material during the composting process (European Commission 2002(b)). The use of sewage sludge in forestry areas has also been undertaken (Hall 1992; Try and Price 1995). Where the source of the sewage has been derived from a largely industrial area, the resultant sewage sludge may contain high concentrations of heavy metals or other chemicals, which are resistant to the biological treatment processes. Consequently, the use of such sewage sludges as land spreading may not be appropriate and may exceed legislative limits.

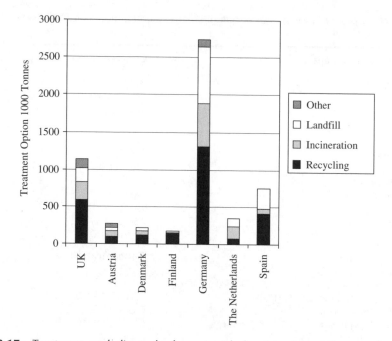

Figure 2.17 *Treatment and disposal of sewage sludge in selected European countries.
Sources: European Environment Agency 2002; DEFRA 2004(a) (UK data).*

The regulation of sewage sludge in agricultural use is covered by EC Directive 86/278/
EEC (1986) and seeks to promote the use of sewage sludge use on land but with more
control on the impact on the environment. The Directive states that the use of sewage
sludge in agriculture must not impair the quality of the soil and agricultural products. In
addition, other EC Directives covering the quality of soil (EC Directive 75/440/EEC) and
groundwater (80/68/EEC) ensure their protection from indiscriminate use of sewage
sludge for agricultural use (Environment Agency 1998). A key requirement of the Directive, for use of sewage sludge in agriculture, is that a specified time must elapse before the
use of the land to grow crops, which may be eaten raw. In addition, a certain period of
time must elapse between land spreading and the use of the land for pasture and subsequent grazing or harvesting of forage. The use of sewage sludge on fruit crops and vege-
tables during the growing season is prohibited. In addition, Member States of the EC are
required to keep records on the quantities, composition and properties of the sludge and
details of where it was spread on the land. There is a requirement to test the sludge and
soil to limit the concentration of certain heavy metals, namely, cadmium, mercury, chromium, zinc, nickel, copper and lead. The control of the build-up of heavy metals in soils
is achieved by, setting the maximum quantities of sludge that can be used each year and
ensuring that the metal levels in the sludge are not exceeded. The levels of nitrogen are
also subject to limitation.

The use of anaerobic digestion of sewage sludge is common in some countries in
Europe and produces a high calorific value gas, composed largely of methane and carbon
dioxide, for use as a fuel. Typically about 50% of the organic matter in sludge is readily

biodegradable to produce a gas, the other 50% is composed of, for example, lingo-cellulosic material which only breaks down very slowly (Environment Agency 1998). In addition, some trade effluents may have high proportions of material that is toxic to the bacteria, such as heavy metals and toxic organic chemicals, and may stop the anaerobic digestion process by killing the bacteria. The process operates anaerobically using various micro-organisms, such as hydrolytic acid-forming bacteria, acetogenic bacteria and methanogenic bacteria, to degrade the organic material.

Composting of sewage sludge is carried out under aerobic conditions in order to develop a product suitable for agricultural or horticultural use. It has been applied to non-digested sewage sludge for example, in Italy and France and to digested sewage sludge in the Netherlands (European Commission 2002(b); European Environment Agency 1997). In order to produce a suitable compost, certain nutrients, such as nitrogen and phosphorus, are required. The high water content of sewage sludge means that the sludge has to be partially dried or co-composted with another suitable organic source.

Landfill is used as a disposal option for sewage sludge, usually where there is no suitable agricultural land nearby for application, or where the sludge poses a pollution hazard to the soil because of high concentrations of pathogens or heavy metals. The legislation covering landfill sites and the associated landfill-permitting systems will limit the type and quantities of waste, including sewage sludge, allowed to be accepted by the landfill. The disposal of sewage sludge in landfill sites by co-disposal with municipal solid waste is encouraged in some cases since the high organic content of sewage sludge promotes the biological breakdown of other co-disposed wastes. The sewage sludge, as an organic waste, will degrade in the landfill by a range of biological and physico-chemical processes to produce leachate and landfill gas composed largely of methane and carbon dioxide. The constraints on the landfilling of biodegradable wastes set out in the EC Waste Landfill Directive (1999) seeks to limit the amount of biodegradable waste going to landfill quite markedly. This will encourage redirection to other treatment options for sewage sludge.

The Waste Incineration Directive (Council Directive 2000/76/EC), introduced in 2000, concerned the incineration of all types of waste, including sewage sludge. The emissions from the incineration of sewage sludge will be controlled and regulated by the provisions of this EC Directive. The sludge has a high ash content, which can range between 20 and 50%, and which remains as a residue after incineration, requiring disposal. The ash is generally landfilled because it will contain a high concentration of toxic heavy metals, indeed, in some European countries the ash is regarded as toxic waste and may be subject to special regulation (Hall 1992). In addition, the sewage sludge has a very high water content and regulations for the incineration of sewage sludge require that the water content of the sludge be reduced to 70% or less, using mechanical de-watering and thermal drying treatments, resulting in increased processing costs (Frost 1991). Alternatively, support fuel can be added to aid incineration. The dry solid content of the sewage sludge, where support fuel is not required, is termed the autothermic solids content. It is estimated that 15% of European sewage sludge is incinerated (European Environment Agency 1997).

The technologies used for sewage sludge incineration are based on multiple-hearth designs, fluidised bed incinerators, rotary kiln furnaces, electric furnaces and cyclone furnaces (Environment Agency 1998). The increasingly common option for sewage sludge incineration design is the fluidised bed, using sand as the fluidised bed material, heated to approximately 850 °C (European Environment Agency 1997). The plants involve a de-watering stage utilising centrifuge, filter belt presses, filter plates or membrane press systems. The solids content of the sludge is increased to between 24 and 95%, depending on whether supplementary fuel is used. The heat generated is usually used to preheat the incoming combustion air or to pre-dry the sewage sludge. Where excess heat is generated, this may be used for steam generation for use either as electricity generation or for heating purposes. Since the incinerators are subject to regulation of emissions, clean-up of the flue gases, in particular, requires expensive treatment. Such clean-up may include electrostatic precipitators and scrubbers to remove particulates and heavy metals. Dioxin emissions have also been detected from sewage sludge incineration, but at very low levels (A Review of Dioxin Emissions in the UK 1995; Hudson 1993).

Co-incineration of sewage sludge with municipal solid waste is also an option, where the sewage sludge is dried to 65% dry solids to enable correct feeding of the sludge to the incinerator grate (European Environment Agency 1997). For sludges with a higher water content of approximately 20–30% dry solids, the sludge is pumped under high pressure to the combustion zone of the burning municipal solid waste. Other novel treatment processes for sewage sludge have included gasification and water oxidation.

2.6   Other Wastes

2.6.1   Agricultural Waste

Agricultural waste consists of organic material, such as manure from livestock, slurry, silage effluent and crop residues. Figure 2.1 shows that agricultural and forestry waste is one of the major waste categories generated throughout Europe. Individual countries within Europe show a variation in arisings of agricultural waste due to the different extents of agriculture within the economy and differing farming methods. Examples of agricultural waste arisings in Europe are: Spain, estimated at 114 million tonnes/year; France, 377 million tonnes/year; the UK, 87 million tonnes/year; and Finland, 25 million tonnes/year (European Commission 2003).

The majority of the agricultural waste is landspread and some is used as animal feed or for composting. The organic waste is high in nutrients and provides a substitute for commercial fertilisers. Consequently, landspreading is regarded as the best practicable environmental option. Landspreading of organic agricultural waste is generally beneficial and has resulted in increased yields as well as helping to maintain high organic carbon and nitrogen levels. Figure 2.18 shows the typical land area needed for spreading the wastes arising from certain animals (Waste Management Planning 1995).

Poultry litter, consisting of a mixture of bird droppings and wood shavings, is a particular form of agricultural waste that has received interest due to its high generation rates and its high calorific value and therefore its potential use as a fuel rather than for landspreading. In addition, application of broiler litter to land has been shown to adversely affect soil chemistry in that increased levels of phosphorus, nitrogen, potassium, calcium and magnesium have been detected with the potential for significant

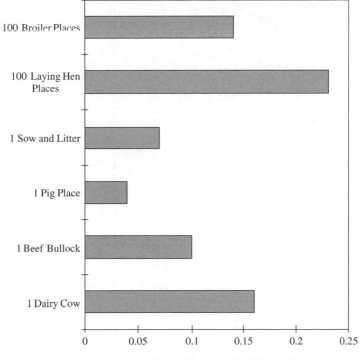

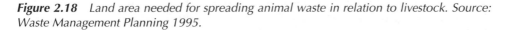
Land Area Needed for Spreading Wastes (Hectares)

Figure 2.18 *Land area needed for spreading animal waste in relation to livestock. Source: Waste Management Planning 1995.*

environmental impact. The combustion systems are used to generate either heat for use as space heating within the farm, including for the poultry themselves, or using poultry litter to generate electricity (Dagnall 1993).

2.6.2 Industrial (Manufacturing) Waste

It is estimated that 740 million tonnes of waste from the manufacturing industries is generated each year throughout Europe (European Environment Agency 2003). Table 2.21 shows the breakdown of the main industrial sectors which generate the industrial waste for selected European countries. The wide range of manufacturing industries, generating a wide range of wastes with different compositions, results in a broad range of potential management options, including recycling, recovery and disposal. However, there is very little information available as to the waste treatment and disposal methods used by the manufacturing industry. Figure 2.19 shows the management of industrial non-hazardous waste in selected European countries. In the countries shown, there is significant recycling of the waste, and the disposal method is dominated by waste landfill.

Table 2.21 Industrial waste generation in selected European countries in relation to industrial sector

Industrial waste category	Italy	The Netherlands	Portugal	UK	Czech Republic	Poland	Romania	Turkey
Food, beverages, tobacco	6874	10271	865	7203	1525	9058	1254	3904
Textile and leather industries	1124	85	1211	1010	134	182	1084	233
Wood and wood products	837	275	1141	1064	294	1213	2716	46
Paper and paper products	1061	829	615	2265	568	1485	99	127
Printing and publishing	440	332	102	1935	50	18	17	11
Refineries	187	584	11	5500	96	223	53	15
Chemical industries	2080	1466	181	4425	896	5034	684	1796
Rubber and plastics industries	507	187	55	1339	129	110	86	6
Non-metallic mineral products	5356	826	1972	2217	667	1709	1282	837
Basic metal industries	4405	2458	1312	9108	3805	37358	2480	5185
Fabricated metals industry	3570	860	335	6482	1341	918	2463	659
Other manufacturing industry	1981	832	558	1282	253	438	379	19

Source: European Commission 2003.

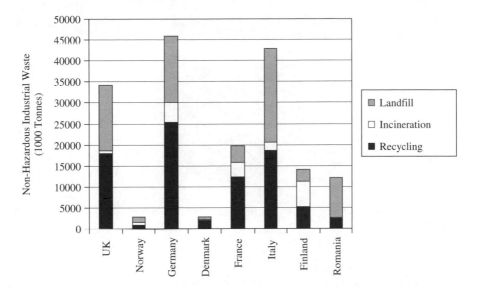

Figure 2.19   *Non-hazardous industrial waste management in selected European countries (UK data includes hazardous industrial waste). Source: Environment Commission 2003.*

2.6.3   Construction and Demolition Waste

Construction and demolition waste is a term used for a variety of wastes and includes waste arising from the construction or demolition of buildings or civil infrastructure, soil, rocks and vegetation arising from land levelling or civil works, and waste materials associated with road maintenance (European Commission 1999). Extremely large tonnages of demolition and construction industry waste arise throughout Europe each year, of the order of 1000 million tonnes/year, representing 32% of the total European waste arisings (European Environment Agency 2003). The waste consists of soil, brick, plaster, metal work, concrete, glass, tiles, wood, plastic, etc., and is generally bulky and inert. Figure 2.20 shows the typical proportions of the components of construction and demolition waste (Warmer Bulletin 47, 1995). Typically soil, stones and clay form the largest percentage of the composition. The routes for management of construction and demolition waste in Europe are mainly landfill and recycling, with some incineration. Construction and demolition waste contains a high proportion of concrete, bricks and tiles which are well suited to crushing and recycling as a substitute aggregate for engineering fill and road sub-bases (European Commission 1999). Landfill also includes a significant proportion of the waste which is used to provide aggregate for access roads to the landfill, and landfill site construction engineering in order to build the embankment walls of landfill 'cells' and for cover and final site capping.

The proportion of construction and demolition waste which is recycled is in some cases high. For example, in Denmark, 90% of such waste is recycled, in Germany 83%, the Netherlands 87%, for Ireland and Italy more than 40% and for the UK about 50%

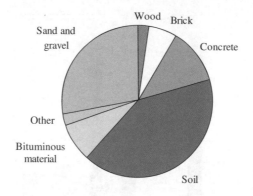

Figure 2.20   *Typical composition of construction and demolition waste. Source: Warmer Bulletin 47, 1995.*

(European Environment Agency 2002(a); DEFRA 2004(b)). Recycling involves use of the construction and demolition waste as low-grade bulk fill, and construction site engineering within the construction industry. For example, materials may be used for hard core for roadways, landscaping and car parks. Such use involves the crushing, removal of metals and size grading of the waste. Higher level use of the waste is not often possible due to the poor quality and heterogeneous nature of the material, compared with primary aggregates used in the industry.

Construction and demolition waste may contain hazardous materials such as asbestos, paints and coatings, adhesives, mineral fibre insulation, damp-proofing chemicals, resins, plasterboard, etc. (European Commission 1999). Such materials may be hazardous due to their historical use such as asbestos insulation, lead-based paints and tars, whilst others may be hazardous due to modern building techniques, where resins, adhesives, sealants and coatings themselves are hazardous or are made up on-site using hazardous materials. Surplus or empty containers containing residues of hazardous material may also end up in the construction and demolition waste.

Inter-country comparisons of construction and demolition waste show that Germany has a huge generation of demolition waste of the order of 250 million tonnes per year, Spain generates 22 million tonnes/year, Italy 24 million tonnes/year, the UK 72 million tonnes/year, Finland 35 million tonnes/year, and France 24 million tonnes per year (European Commission 2003). This compares with the US which generates about 140 million tonnes of construction and demolition waste each year. However, as with most waste statistics, care should be taken in direct comparison of waste arisings data since, for example, for construction and demolition waste, there are differences in the definitions used. One reason for the high generation rates of construction and demolition waste in Germany, is that they include excavated soil and stones in their definition, whereas other countries do not. In addition, other countries may not include waste bricks and concrete in statistical data if they are used directly as construction material for roads and paths, or as filling material at the construction site (European Environment Agency 2002(a)).

2.6.4 Mines and Quarry Waste

Mines and quarrying wastes are one of the largest categories of waste generated throughout Europe. The wastes consist mainly of colliery spoil from coal mines, and other mining and mineral industry materials such as china clay wastes and slate wastes and are generally inert mineral materials. The large majority of mines and quarry wastes are disposed of as large-scale above ground open tips, close to the mine or quarry, which are then later landscaped and restored.

In many European countries, waste from mining and quarrying is not subject to environmental or waste management legislation (European Environment Agency 2003). Consequently, the reliability and availability of waste arisings data is poor. Comparison with other countries of Europe shows a wide variation in waste arisings from the mining and quarrying industries. For example, Germany produces 68 million tonnes/year, Spain 70 million tonnes/year, France 75 million tonnes/year, whilst Norway produces only 9 million tonnes/year and the Netherlands, Austria and Portugal each produce less than 1 million tonnes/year (Eurostat 1996).

2.6.5 Power Station Ash

Approximately 50 million tonnes/year of coal ash generated from coal-fired electricity production power stations was estimated to be produced in the EU in the 1990s (European Environment Agency 2003). This has changed in recent years as the coal-fired power industry has declined in favour of natural gas-fired power stations, and the increasing use of renewable energy. Power station ash is generated from the ash in coal which amounts to about 15% of the original coal. The coal is first pulverised to a small grain size before use in the power station. The ash from the furnace chamber is coarse-grained and termed 'coarse slaggy clinker', the fine-grained material captured in the flue gas clean-up system is termed 'pulverised fuel ash'.

More than 70% of the coal-fired power station ash is recycled for use in the construction industry as a concrete mix ingredient, in blended cement, for use as structural fill in road building, as a lightweight aggregate and also for use as a constituent in building blocks. The remainder is landfilled, usually close to the power station. The fine-grained, pulverised fuel ash is pumped into settling lagoons in solution or conditioned with water then landfilled.

2.6.6 Waste Tyres

Tyres are composed of vulcanised rubber in addition to the rubberised fabric with reinforcing textile cords, steel or fabric belts and steel-wire reinforcing beads (Williams and Besler 1995; Williams et al 1995). Typical tyre compositions are shown in Table 2.22 (Ogilvie 1995). A number of different natural and synthetic rubbers and rubber formulations are used to produce tyres. Other components in the tyre include: carbon black, which is used to strengthen the rubber and aid abrasion resistance; extender oil, which is a mixture of aromatic hydrocarbons and serves to soften the rubber and improve workability; sulphur which is used to cross-link the polymer chains within the rubber and also to harden and

Table 2.22   *Approximate proportions of components in tyres*

Car tyres

Component	Steel-braced radial	Textile-braced radial	Cross-ply
Rubber compound	86	90	76
Steel	10	3	3
Textile	4	7	21

Truck tyres

Component	All-steel radial	Cross-ply
Rubber compound	85	88
Steel	15	3
Textile	<0.5	9

Composition of tyre rubber compound

Component	Weight (%)
Rubber hydrocarbon	51
Carbon black	26
Oil	13
Sulphur	1
Zinc oxide	2
Others	7

Sources: Ogilvie 1995; Williams et al 1995.

prevent excessive deformation at elevated temperatures; an accelerator, typically an organo-sulphur compound, added as a catalyst for the vulcanisation process; zinc oxide and stearic acid used to control the vulcanisation process and to enhance the physical properties of the rubber.

Approximately 250 million car and truck tyres are scrapped each year in Europe representing about 3 million tonnes by weight of tyre. The worldwide generation of scrap tyres is estimated at 1000 million tyres/year. The generation of waste tyres are linked to the generation of end-of-life vehicles. The European Union has attempted to control the management of end-of-life vehicles and tyres through the 2000 End-of-Life Vehicle Directive (Council Directive 2000/53/EC). The Directive stipulates the separate collection of tyres from vehicle dismantlers and encourages the recycling of tyres. In addition, the 1999 Waste Landfill Directive (Council Directive 1999/31/EC) seeks to ban the landfilling of tyres by 2006. These two Directives will inevitably impact on the management of tyres throughout the European Union. The main routes for management of waste tyres (2002) are landfill, energy recovery, material recycling, retreading and export (Figure 2.21, Archer et al 2004).

Landfilling of tyres is declining as a disposal option, as they are specifically banned from waste landfills after 2006 to comply with the 1999 Waste Landfill Directive. In addition, tyres do not degrade easily in landfills, they are bulky, and can cause instability within the landfill, indeed, many landfill sites refuse to take tyres. In addition, they can be a breeding ground for insects and a home for vermin.

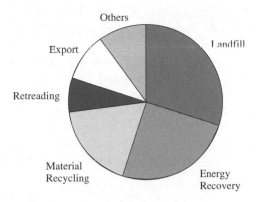

Figure 2.21   *Waste tyre management in Western Europe. Source: Archer et al 2004.*

Energy recovery utilises the high calorific value of tyres, about 32 MJ/kg, and currently accounts for about 25% of the total tyre waste arisings in Western Europe. Waste tyre incineration plants operate successfully in Germany, Italy and also the USA and Japan (Archer et al 2004). Cement kilns have been used for the disposal of tyres by combustion to offset the high energy costs associated with cement production. Most of the major cement manufacturers throughout the world use tyres as a fuel source. The tyres are fed directly to the rotary kiln of the process where very high temperatures of the order of 1500 °C ensure complete burnout of the tyre. The metal and ash components of the tyre are incorporated into the cement clinker. Waste tyres are also used in other energy intensive industries, such as the pulp and paper industry.

Materials recycling of waste tyres includes the production of rubber crumb which is derived from tyres, via shredding and grinding of the rubber part of the tyre to produce a fine-grained product for use in such applications as children's playgrounds, sports surfaces, carpet backing and other applications, such as absorbents for oils and hazardous and chemical wastes. There is increasing use of granulated tyres in cement and asphalt applications, for example, in road and pavement construction. Rubber reclaim consists of devulcanisation of the rubber by temperature and pressure, using various reclaiming chemicals and solvent treatments to produce a rubber which can be used in low-grade applications. Such applications include cycle tyres, conveyor belts and footwear.

Retreading of tyres, where a new rubber tread is bonded to the surface of worn tyres, accounts for about 10% of annual tyre consumption. The process involves grinding of the tyre surface to produce a clean surface onto which the new rubber is bonded. Car tyres can be retread once only, but truck tyres may be retreaded up to three times. Other routes for the management of scrap tyres includes dumping and stockpiling. This is unsightly and has the potential for accidental fires or arson resulting in high pollution emissions to the atmosphere and water courses.

In the USA in 2001 it was estimated that the production of scrap tyres was 280 million tyres (US Rubber Manufacturers Association 2002). In addition, it is estimated that there is a stockpile of 500 million tyres awaiting disposal (Scrap Tyre Management Council

1999). The main methods of scrap tyre disposal in the USA are energy recovery in cement kilns, pulp and paper works, tyre incinerators and boilers which accounts for 40.9% of the arisings of scrap tyres or some 114 million tyres. About 10% of tyres are landfilled, while 14.2% are used in civil engineering applications including drainage aggregates, bulk fill aggregates, insulation and landfill engineering. Approximately 11.7% of tyres are recycled in ground rubber applications and 5.3% exported (US Rubber Manufacturers Association 2002). The route for the remaining 17.9% includes miscellaneous methods and a significant proportion (12.5%) where the disposal route is unknown. In the USA, retread tyres are not considered in many scrap tyre data statistics, since they are deemed as still to be in use and are only finally scrapped when they cannot be retread further.

2.6.7   End-of-Life Vehicles

Approximately 15.5 million cars are scrapped in Western Europe (Austria, Belgium, Denmark, Finland, France, Germany, Greece, Ireland, Italy, Luxembourg, the Netherlands, Portugal, Spain, Sweden, United Kingdom, Iceland, Liechtenstein, Norway, Switzerland, Andorra, Monaco, San Marino) and Central and Eastern Europe each year (European Environment Agency 2004). Scrap vehicles or end-of-life vehicles are regulated by the European Union through the 2000 EU End-of-Life Vehicle (ELV) Directive (Council Directive 2000/53/EC). The approach of the EU to the problem of vehicle waste is through minimising waste in the production process, coupled with increased recycling and re-use of the waste. The Directive has therefore set targets for the recovery of scrap vehicles (cars and vans) and requires Member States of the EU to re-use and recover 85wt% of the average vehicle weight by 2006, increasing to 95wt% by 2015. The Commission makes the distinction, in the Directive, between targets for re-use and recycling which must make up 80wt% of the 2006 target and 85wt% of the 2015 target. The remaining percentage is a recovery target which is mainly achieved through incineration with energy recovery.

   The number of end-of-life vehicles are predicted to increase over the next decade to reach almost 19 million cars in Western Europe and Central and Eastern Europe by 2015 (Figure 2.22, European Environment Agency 2004). The management of end-of-life vehicles is currently a high proportion of recycling. Vehicles have a high steel and aluminium content representing about 75% of the weight of the vehicle which are readily recycled. The remainder is composed of plastics, rubber and other components which are currently disposed of to waste landfill or incineration. With the move to lighter components to improve fuel economy, the proportion of non-metallic components is likely to increase. However, the wide range of non-metallic components used in vehicles inhibits their effective recycling. Vehicle manufacturers are responsible for the management of the waste arising from the automotive industry under the 'Producer Responsibility' principle. Consequently, manufacturers are developing design and construction techniques and systems which allow the easy dismantling of the vehicle, to aid recycling.

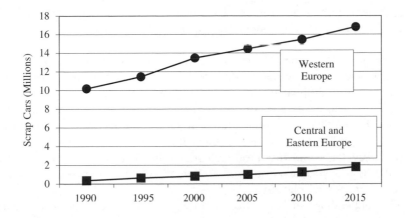

Figure 2.22   *Predicted estimate of scrap car arisings in Western Europe (Austria, Belgium, Denmark, Finland, France, Germany, Greece, Ireland, Italy, Luxembourg, the Netherlands, Portugal, Spain, Sweden, United Kingdom, Iceland, Liechtenstein, Norway) and Central and Eastern Europe (Bulgaria, Cyprus, Czech Republic, Estonia, Hungary, Latvia, Lithuania, Malta, Poland, Romania, Slovakia, Slovenia, Turkey). Source: European Environment Agency 2004.*

2.7   Waste Containers, Collection Systems and Transport

It has been estimated that the collection of solid waste involves a cost of more than 50% of the total solid waste budget and is very labour intensive. The type of container used to store the waste generated from households and commercial and industrial premises, depends on the frequency and efficiency of collection, the amount of waste, the type of housing, the density of the collected wastes, the collection vehicle type, the vehicle usage and the manpower and economics relating to the container and the collection system (Scharff and Vogel 1994). The correct size of waste container is important since it has been shown that the use of non-standard containers is the greatest cause of litter (Pescod 1991–93).

Household waste containers include traditional metal or plastic dustbins, wheeled bins and plastic sacks. The capacity of the household waste storage container depends on how many collections are made per week. With a general increase in the amount of waste generated, there has been a tendency to use bigger or more containers. This also has the potential to reduce manning costs by less frequent waste collections. Another factor dictating size of container and frequency of collection is climate; in cooler areas such as Northern Europe, where odour from the degradation of the waste occurs more slowly, the frequency of collection may be once or twice per week. Consequently, the container must be able to store a full week's volume of waste. In warmer climates, such as Southern Europe, collections of waste have to be more frequent, often daily, to minimise nuisance from odours and fly breeding (Pescod 1991–93). Consequently, the waste container can be much smaller than one used for weekly collection. The use of non-standard containers is the greatest cause of litter. The size of the waste container and the frequency of collection has been shown to influence the quantity of household waste placed in the container. For example, experience has shown that a twice-weekly collection produces more waste than a weekly one and that, where wheeled bins are used compared to traditional smaller metal or plastic dustbins, then increases in waste have been up to 30% (Pescod 1991–93). The use of larger wheeled bins has also resulted in large bulkier items being placed in the larger containers. In some European cities, recycling initiatives have led to more than one container which segregate the waste into recyclable and non-recyclable fractions. The segregation of waste also reduces the non-recyclable portion of household waste and therefore reduces final waste treatment and disposal costs.

Commercial buildings such as offices and shops, but also larger household areas such as blocks of flats, and institutions such as schools, require much larger container capacities than single households. A typical waste container for such buildings would be about $0.85 \, m^3$ capacity. Industrial and trade wastes and large commercial buildings require even larger containers and use waste 'skips' of up to $30 \, m^3$ capacity (Pescod 1991–93).

Some recycling schemes involve the 'bring' system where the recyclable materials such as paper, glass, metal tins and plastic bottles, are placed in large containers sited throughout the community. These are collected periodically, using specialised vehicles, and transferred to the materials recycling facility.

Waste collection vehicles have compaction rams to reduce the volume of waste. The vehicle has a team of up to six collectors, depending on the type and capacity of vehicles used and the method of collection. The typical compaction collection vehicle is rear loading and involves a team of collectors with a rear bin lift. The waste is then compacted using a packing plate, under hydraulic pressure, to a compaction ratio of about 6:1. Side loading vehicles are designed for a one or two-man operation, where the bins are loaded via a bin lift under the control of the driver. Rear loading demountable collection vehicles are common in Europe where the collected waste is compacted on the vehicle as usual during the local collection round, but the container is demountable. The container is then transferred to a transfer station where the collection vehicle picks up a new container body (Bates 2004). The containers are then transferred by large vehicle, rail or barge to the final disposal site (Institute of Wastes Management 1997). A major advantage is the reduction in waste transport costs by reductions in the number of vehicles travelling to the disposal site and manning levels of personnel travelling with the vehicle.

There has been a trend throughout Europe over the last 30 years, for municipal solid waste collection vehicles to become larger in size, rising from 12–14 tonnes to 30 tonnes (Bates 2004). That trend has been coupled with an increase in complexity and higher compaction ratios. However, that increase in size has raised issues of manoeuvrability in congested streets, road safety issues, noise, and the environmental impact of such large trucks. One development is the use of 'satellite' collection vehicles that collect waste from households in small side streets and operate in conjunction with a larger vehicle on the main street. The satellite vehicle periodically empties its waste into the larger main vehicle (Bates 2004). There has also been a recent trend towards the re-emergence of the smaller 12–14 tonne truck, but with more manoeuvrability, lower noise pollution and with the flexibility to be used in a range of waste management situations (Bates 2002).

Comparison of collection systems used for municipal solid waste in European cities has shown that they are highly efficient, with a combination of containers, vehicles, personnel and logistics individually suited to the local conditions such as population density, residential structure or traffic conditions (Scharff and Vogel 1994). The countries compared in the survey were Berlin, Budapest, Copenhagen, Munich, Paris, Stockholm, Vienna and Zurich. The size of collection container for household waste was calculated, in most instances, on a figure of 120 l volume of waste arising per household. Collection frequencies range from Paris, with a daily emptying of containers or a daily collection of waste bags from each collection point, to two to three times per week in other cities, with a once-per-week collection being the minimum acceptable. Traffic congestion, number of collection vehicles, population density collection frequency and the treatment and disposal used for the waste, greatly influence the daily distance travelled by the collection vehicles. For example, Paris has 500 collection vehicles with a population density four times higher than the other cities surveyed and achieves a daily travelling distance of between 20 and 52 km. Budapest, however, with a similar waste quantity to Paris but a different collection organisation and traffic situation, uses 175 collection vehicles and achieves a daily travelling distance of between 53 and 175 km (Scharff and Vogel 1994). The size of waste containers used in the eight cities surveyed ranged from 30 l volume to over 1100 l. The waste collection efficiencies for waste collected per day were found to be greater by up to four times for the larger containers, compared to smaller ones.

The USA use a range of collection containers for municipal solid waste collection (Bonomo and Higginson 1988). Residential containers range from plastic or paper bags, 90 l volume metal cans, to 350–500 l volume plastic wheeled carts which are transferred to the collection point. Larger containers are used for multi-family or apartment blocks and may have compaction facilities or require automated dedicated vehicles for collection. Commercial waste is collected in 6000 l volume plastic or metal containers and up to 30000 l for commercial waste from, for example, shopping centres. Recycling involves source-separated waste which is collected using compartmentalised trucks. Transport of the waste involves trucks which may go directly to the disposal site or to transfer stations where the waste is compacted and sent in larger trucks for treatment or disposal. The movement of waste by rail and barges has also been used in the USA.

The collection of household waste in Japan is mostly from collection sites serving between 10 and 40 properties, although in some cases individual household collection may occur (Waste Management in Japan 1995). In Tokyo, for example, there are some 240 000 waste collection points. Waste is segregated by the householder into combustible and non-combustible types. The combustible fraction of the waste is the largest and is collected more frequently than the non-combustible waste. In some areas, waste is collected up to four times per week, as the very narrow streets within the cities and the segregation of waste, require frequent visits by very small vehicles. In Tokyo there are some 3000 waste collection vehicles of either 1.2 tonnes or 2.4 tonnes capacity.

The collection systems of the developed world should be contrasted with the systems in the developing world. In urban areas of India, municipal solid waste is collected from individual households by various means, including push carts, animal drawn carts and tricycle trolleys (Sundaravadivel and Vigneswaran 2002). The collected waste is transferred to designated large bins where it is collected by truck for transfer to the disposal site. In the smaller towns, waste collection and treatment is less organised. For example, many households do not have regular collection but instead, concrete pits placed on street corners are used as bins for waste disposal. Where these concrete bins are not provided, then the waste is dumped in piles at the street corner. The waste from street corners and pits is transferred to collection points where it is transported by trucks or mini-lorries to dumpsites. In many cases, the waste is not collected regularly or disposed of environmentally correctly and often unauthorised dumpsites are used (Sundaravadivel and Vigneswaran 2002).

An innovative collection system for municipal solid waste, which consists of a fully enclosed vacuum system and a system of underground pipes, has been developed for offices, theme parks, residential areas and nursing homes (Envac 2004). The waste is thrown into a normal inlet, either indoors or outdoors. Sorting at source, for the recycling of components, is handled by using one inlet for each fraction, with up to four separate inlets and containers for each category of waste. The system consists of a number of collection points, linked together by piping that transports the waste to a central collection station. When a refuse bag is deposited into an inlet within an office block, hospital or residential area, it is temporarily stored in a chute on top of a discharge valve. At regular intervals the waste collected at the inlet point is automatically emptied, the control system switches on the fans and a vacuum is created in the network of pipes. An air inlet valve is opened to allow transport air to enter the system. One by one, the discharge valves below each of the chutes are opened and the refuse bags fall down by gravity into the horizontal network of pipes and are sucked to the collection station. The refuse enters the collection station via a cyclone that separates the refuse from the air. The refuse falls down into a compactor, which compacts the refuse in the sealed container. The transport air then passes through dust and deodorant filters and a silencer (Envac 2004). For source-separated waste, required for recycling, there is an additional inlet and container for each category of refuse. The control system directs a diverter valve to convey each category of sorted waste into the correct container. When the containers are full, trucks collect them for emptying and further transportation to recycling facilities, incinerators, composting plants or waste landfills. Such systems have been used for a range of applications, with examples in Sweden, Spain and Portugal (Envac 2004).

References

Analysis of Industrial and Commercial Waste Going to Landfill in the UK. 1995. Department of the Environment, HMSO, London.

Archer E., Klein A. and Whiting K. 2004. The Scrap Tyre Dilemma: Can Technology Offer Commercial Solutions? Waste Management World, January, James & James Science Publishers Ltd, London.

A Review of Dioxin Emissions in the UK. 1995. Environment Agency (Her Majesty's Inspectorate of Pollution). Department of the Environment, HMSO, London.

Bates M. 2002. Moving Waste Transport up a Gear. Waste Management World, July–August.

Bates M. 2004. Collection and Transport Special. Waste Management World, January–February.

Benestad C. 1989. Waste Management and Research, 7, 351–361.

Bonomo L. and Higginson A.E. 1988. International Overview on Solid Waste Management. Academic Press, London.

Bruce A.M., Colin F. and Newman P.J. 1989. Treatment of sewage sludge. Elsevier Applied Science, London.

Buekens A. and Patrick P.K. 1985. Incineration. In, Solid Waste Mangement. Selected Topics, Suess M.J. (Ed.). World Health Organization, Regional Office for Environmental Health Hazards, Copenhagen, Denmark.

COM (96) 399 Final. 1996. Communication from the commission on the review of the community strategy on waste management. Official Journal of the European Communities, Brussels, Belgium.

Commission Decision. 2000, 2000/532/EC Commission Decision of 3 May 2000, Official Journal of the European Communities, L226/1, 6.9.2000, Brussels, Belgium.

Council Directive 75/440/EEC. 1975. Quality required of surface water intended for the abstraction of drinking water in the Member States. Official Journal of the European Communities, L194, Brussels, Belgium.

Council Directive 75/442/EEC. 1975. Waste Framework Directive. Official Journal of the European Communities, L194, Brussels, Belgium.

Council Directive 80/68/EEC. 1980. Protection of groundwater against pollution caused by certain dangerous substances. Official Journal of the European Communities, L020, Brussels, Belgium.

Council Directive 86/278/EEC. 1986. Protection of the environment, and in particular of the soil, when sewage sludge is used in agriculture. Official Journal of the European Communities, Brussels, Belgium.

Council Directive 94/67/EC. 1994. Incincration of Hazardous Waste. Official Journal of the European Communities, Brussels, Belgium.

Council Directive 1999/31/EC. 1999. Landfill of Waste. Official Journal of the European Communities, L182, Brussels, Belgium. European Commission, 1999.

Council Directive 2000/53/EC. 2000. End-of-life Vehicles. Offical Journal, L269, Brussels.

Council Directive 2000/76/EC. 2000. Inceneration of Waste. Official Journal of the European Communities, L332, Brussels, Belgium.

Daborn G.1988. In, Brown A., Evemy P. and ferrero G.L. (Eds.). Energy Recovery Through Waste Combustion. Elsevier Applied Science, London.

Dagnall S.P. 1993. Poultry Litter as a Fuel in the UK: A Review. Energy Technology Support Unit, Harwell, Oxfordshire.

Dean R.B. and Suess M.J. 1985. Waste Management and Research, 3, 251–278.

DEFRA. 2004(a). Sewage treatment in the UK: Implementation of the EC urban waste water treatment Directive, Re-use and disposal of sewage sludge. Department of Environment, Food and Rural Affairs, UK.

DEFRA. 2004(b). Construction and Demolition Waste, 1999 and 2001, Key facts about waste and recycling. Department of Environment, Food and Rural Affairs, UK.

Dempsey C.R. and Oppelt E.T. 1993. Incineration of Hazardous waste: A critical Review Update. Air and Waste, 43, 25–73.

Department of the Environment. 1991. The analysis and prediction of household waste arisings: Controlled Waste Management Report, CWM/037/91. Department of the Environment, HMSO, London.

Digest of Environmental Statistics. 1996. Department of the Environment, HMSO.

ENDS Report 265. 1997. Environmental Data Services Ltd, Bowling Green Lane, Report 265, February, 41–42, London.

Envac Centralsug. 2004. Waste Travel Underground. Information Data, Stockholm, Sweden.

Environment Agency. 1998. A Review of Sewage Sludge Treatment and Disposal Techniques, Project P2/064/1. Environment Agency, Bristol, UK.

European Commission. 1999. Construction and Demolition Waste Management Practices and their Economic Impacts. European Commission DGXI, Brussels.

European Commission – Directorate General Environment. 2002(a). Study on Hazardous Household Waste with a main Emphasis on Hazardous Household Chemicals. European Commission, Brussels.

European Commission. 2002(b). Disposal and Recycling Routes for Sewage Sludge. Synthesis Report. European Commission, DG Environment-B2, Luxemburg.

European Commission. 2003. Waste Generated and Treated in Europe, European Commission. Office for Official Publications of the European Communities, Luxembourg.

European Commission. 2004. Integrated pollution prevention and control. Draft reference document on the best available techniques for waste incineration, European Commission, Brussels, Belgium.

European Environment Agency. 1997. Sludge Treatment and Disposal: Management Approaches and Experiences. European Environment Agency, Copenhagen, Denmark.

European Environment Agency. 2002(a). Review of selected waste streams: sewage sludge, construction and demolition waste, waste oils, waste from coal fired power plants and biodegradable municipal solid waste. European Environment Agency, Copenhagen.

European Environment Agency. 2002(b). Hazardous Waste Generation in EEA Member Countries: Comparability of Classification Systems and Quantities. European Environment Agency, Copenhagen.

European Environment Agency. 2003. Europe's Environment: The Third Assessment. European Environment Agency, Copenhagen.

European Environment Agency. 2004. Generation of Waste from End-of-life Vehicles. Indicator Fact Sheet, European Environment Agency, Copenhagen.

Eurostat Year Book. 1996. Office for Official Publications of the European Communities, Luxembourg.

Freeman H.M. 1998. Standard Handbook of Hazardous Waste Treatment and Disposal. McGraw-Hill, New York.

Frost R. 1991. Incineration of sewage sludge, University of Leeds Short Course 'Incineration and Energy from Waste'. The University of Leeds, Leeds, UK.

Gervais C. 2002. An overview of European Waste and Resource Management Policy. Forum for the Future, London.

Hall J.E. 1992. Treatment and use of sewage sludge. In, The Treatment and Handling of Wastes, Ed. Bradshaw A.D., Southgwood R. and Warner F. Chapman & Hall, London.

Holmes J. 1995. The UK Waste Management Industry 1995. Institute of Wastes Management, Northampton, 1995.

Hudson J. 1993. Sewage sludge incineration. In, Incineration an Environmental Solution for Waste Disposal. IBC Technical Services Ltd (London), Conference, February, Manchester.

Institute of Wastes Management. 1997. Refuse Collection Systems. Institute of Wastes Management, Northampton, UK.

International Wastenews. 2004. International Solid Waste Association, Copenhagen. International Wastenews 1, p20.

Kaiser E. 1978. In, Combustion and Incineration Processes. Niessen, W.R. Dekker Press, New York.

Kim B.J. and Qi S. 1995. Water Environment Research, 67, 560–570.

LaGrega M.D., Buckingham P.L. and Evans G. J. 1994. Hazardous Waste Management. McGraw-Hill Inc., New York.

Laurence D. 1999. Waste Regulation Law. Butterworths, London.

Matsuto T. and Tanaka N. 1993. Waste Management and Research, 11, 333–343.

National Household Waste Analysis Project – phase 2 volume 1. 1994. Report on composition and weight data Report No. CWM 082/94. Wastes Technical Division, Department of the Environment, HMSO, London.

National Household Waste Analysis Project – phase 2 volume 3. 1995. Chemical analysis data. Report No. CWM/087/94. Wastes Technical Division. Department of the Environment, HMSO, London.

Nie Y., Li J., Niu D., Bai Q. and Wang H. 2002. Waste Management World, March–April.

OECD. 2004. Addressing the Economics of Waste. Organisation of Economic Cooperation and Development.

Ogilvie S.M. 1995. Opportunities and Barriers to Scrap Tyre Recycling. AEA Technology, National Environmental Technology Centre, Abingdon, HMSO.

Pescod M.B. (Ed.) 1991–93. Urban Solid Waste Mangement. World Health Organization, Copenhagen, Denmark.

Rhyner C.R. 1992. Waste Management and Research, 10, 67–71.

Rhyner C.R. and Green B.D. 1998. Waste Management and Research, 6, 329–338.

Royal Commission on Environmental Pollution. 1993. 17th Report, Incineration of Waste. HMSO, London.

Rushbrook P.E. and Finney E.E. 1988. Planning for Future Waste Management Operations in Developing Countries. Waste Management and Research, 6, 1–21.

Scharff C. and Vogel G. 1994. Waste Management and Research, 12, 387–404.

Scrap Tyre Management Council. 1999, US.

Stanners D. and Bourdeau P. 1995. Europe's Environment. The Dobris Assessment, European Environment Agency, Copenhagen.

Statistical Handbook of Japan. 2003. Statistics Bureau, Ministry of Public Management, Home Affairs, Posts and Telecommunications. Japan Statistical Association, Japan.

Sundaravadivel M. and Vigneswaran S. 2002. Sustainable MSW Management in Developing Countries. Waste Management World, November–December, 60–64.

The analysis and prediction of household waste arisings, Report No. CWM/037/91, 1991. Wastes Technical Division, Department of the Environment, HMSO, London.

The Open University. 1993. Course T237: Environmental Control and Public Health, Units 8–9 Municipal Solid Wastes Management. The Open University, Milton Keynes.

Try P.M. and Price G.J. 1995. Sewage and industrial effluents. In, Hester R.E. and Harrison R.M. (Eds.), Waste Treatment and Disposal, Issues in Environmental Science and Technology, 3, The Royal Society of Chemistry, London.

US EPA. 2004. Municipal Solid Waste. United States Environmental Protection Agency, Washington, US.

US Rubber Manufacturers Association. 2002. US Scrap Tyre Markets: 2001. Rubber Manufacturers Association, Washington, US.

Visvanathan C. 1996. Hazardous Waste Disposal. Resources, Conservation and Recycling, 16, 201–212.

Wallington J.W. and Kensett R.G. 1988. Energy recovery from refuse incineration in the National Health Service. In, Engineering for Profit from Waste. Proceedings of the Institution of Mechanical Engineers, Conference, 1988–1, Coventry, March.

Walsh D.C. 2002. Urban Residential Refuse Composition and Generation Rates for the 20th Century. Environmental Science and Technology, 36, 4936–4942.

Warmer Bulletin 47. 1995. Warmer Bulletin Fact Sheet; Construction and Demolition Waste. Journal of the World Resource Foundation, Tonbridge, Kent.

Warmer Bulletin 49. 1996. Journal of the World Resource Foundation, High Street, Tonbridge, Kent, UK, May.

Warmer Bulletin 50. 1996. Household Hazardous Waste. Journal of the World Resource Foundation, Tonbridge, Kent.

Waste Framework Directive. 1975. Council Directive 75/442/EEC, OJ L 194, 25.7.75.

Waste Management in Japan. 1995. The Institute of Wastes Management, Department of Trade and Industry, Overseas Science and Technology Expert Mission Visit Report. IWM, Northampton.

Waste Management Paper 28. 1992. Recycling. Department of the Environment, HMSO, London.

Waste Management Paper 26F. 1996. Landfill Co-disposal (Draft). Department of the Environment, HMSO, London.

Waste Management Planning. 1995. Principles and Practice. Department of the Environment, HMSO.

Wheatley A. 1990. Anaerobic Digestion: A Waste Treatment Technology. Elsevier Applied Science, London.

White P.R., Franke M. and Hindle P. 1995. Integrated Solid Waste Management. Blackie Academic and Professional, London.

Williams P.T. 2002. Issues in Environmental Science and Technology. Royal Society of Chemistry, UK, 18, 141–170.

Williams P.T. and Besler S. 1995. Fuel, 74, 1277–1283.

Williams P.T., Besler S., Taylor D.T. and Bottrill R.P. 1995. Journal of the Institute of Energy, 68, 11–21.

Woodside G. 1993. Hazardous Materials and Hazardous Waste Mangement: a Technical Guide, Wiley Interscience, New York.

Yokoyama S., Murakami M., Ogi T., Kouchi K. and Nakamura E. 1987. Liquid fuel production from sewage sludge by catalytic conversion using sodium carbonate. Fuel, 66, 1150–1155.

Yu C.-C. and Mclaren V. 1995. Waste Management and Research, 13, 343–361.

3

Waste Recycling

Summary

This chapter is concerned with waste recycling and also discusses waste reduction and re-use. Municipal solid waste and industrial and commercial waste recycling in Europe are reviewed in detail. Examples of recycling of particular types of waste, i.e., plastics, glass, paper, metals and tyres are also discussed. Economic considerations of recycling are discussed, together with consideration of life-cycle analysis of waste recycling.

3.1 Introduction

Article 174 of the Treaty, which established the European Community, states that Community policy shall aim at a high level of environmental protection (European Commission 2000). That policy was based on the principles that preventative action should be taken, that environmental damage should, as a priority, be rectified at source, and that the polluter should pay (European Commission 2000). The European Community Environmental Action Programmes also call for significant changes in current patterns of development, production, consumption and behaviour in order to achieve sustainable development. Indeed the Fifth Environmental Action Programme (1993–2000) covers the management of waste in detail and advocates a reduction in the wasteful consumption of natural resources and the prevention of pollution, applying the principles of waste prevention, recovery and safe disposal of waste.

Waste Treatment and Disposal, Second Edition   Paul T. Williams
© 2005 John Wiley & Sons, Ltd   ISBNs: 0-470-84912-6 (HB); 0-470-84913-4 (PB)

The producer responsibility under the 'polluter pays' principle, as laid down in the EC Treaty, has the underlying theme of making responsible for environmental pollution, those who have the possibility of improving the situation. Producers are then deemed to have the incentive to develop approaches to design and manufacture of their products to ensure the longest possible product life and, in the event that it is scrapped, the best methods for recovery and disposal.

The hierarchy of waste management, places waste reduction at the top, followed by re-use, recycling, recovery and finally disposal. Waste reduction is synonymous with waste prevention or waste minimisation and has been defined as any technique, process or activity which either avoids, eliminates or reduces a waste at its source, usually within the confines of the production unit (Crittenden and Kolaczkowski 1995). This is sometimes known as 'clean technology' or producing 'clean' products. Re-use is defined as any operation by which a product or package is designed to accomplish, within its life cycle, a minimum number of trips or rotations, and is then refilled or used for the same purpose for which it was conceived (Council Directive 1994). Recovery and recycling are referred to as separate operations by the European Commission, and many European Directives on waste have separate targets for waste recovery and waste recycling. In particular, recovery of waste by energy recovery is classed separately and there is a clear distinction made between recycling and recycling which specifically excludes energy recovery (Council Directive 1994).

Waste Reduction   Waste reduction is both environmentally and economically beneficial both to society as a whole, and to businesses and the community. A large proportion of the costs of waste treatment and disposal are borne either by business or community-charge payers. However, recent surveys have shown that many companies do not regard waste management as a key business area and many do not even measure waste costs. In addition, it has been found that smaller companies have less regard for waste minimisation and recycling as an issue, than do larger companies (European Environment Agency 2003). Waste treatment and disposal by industry has often been regarded as an 'end of pipe' process of disposal rather than waste reduction within the manufacturing process, which may also reduce costs and increase process efficiency. Environmental and cost savings arise from less waste being processed with, for example, savings in energy costs, waste storage space, transport costs, administrative costs, and lower emissions to air, water and onto land. Reducing waste in the production process in industry can also reduce the amount of raw material inputs, in addition to final disposal costs. Reduction in waste costs need not always equate with reduction in waste amount, but reduction in the toxicity of waste will also reduce costs due to the lower costs of treatment, compared with more hazardous wastes (Crittenden and Kolaczkowski 1995).

Waste reduction at source involves good practice, input material changes, product changes and technological changes (Crittenden and Kolaczkowski 1995). Good practice can be as simple and effective as good housekeeping, maintaining production procedures, minimising spillages and proper auditing of input and final destination of raw materials. There are many examples of where a change in input material has resulted in a reduction of waste in the production process. For example, the replacement of organic solvents with water-based types, or the replacement of hazardous solvents, such as benzene and chlorinated organic solvents, with less hazardous solvents. Product changes can be used to reduce the production of waste or materials used within the manufacturing process, but also when the consumer eventually disposes of the product. Examples include the reduction in the weight of packaging, for example, the thickness and therefore weight of drinks cans have shown a reduction of more than 50% since the 1970s. Plastic items, such as carrier bags and yoghourt pots, have similarly shown a reduction in thickness and therefore weight. Technological changes have included retrofitting and development of cleaner processes. A change in technology can mean a fundamental change in the process, a change in the process conditions, a change to an automated system or re-engineering of the process.

Waste Re-use   After waste reduction or waste minimisation measures have been introduced to reduce the production of waste from municipal, commercial or industrial sources, another option is waste re-use. Re-use involves using a product or package more than once or re-using it in another application. Examples include re-using plastic supermarket carrier bags, glass milk bottles, re-treading partly worn tyres and re-using car parts via car scrap merchants. Re-using a product or packaging extends the lifetime of the material used and therefore reduces the waste quantity requiring treatment and disposal. In addition, other savings for the environment include the costs of producing the replaced item which would include energy, materials and transport costs. Re-use of materials can also take the form of new use applications in a different environment than that of the original function, for example, the use of tyres for securing covers on silage mounds and for boat/dock fenders (Making Waste Work 1995).

The re-use of beverage bottles was very common until the 1980s using a 'deposit refund' system, where a small charge was made on the bottle, which was refunded when the bottle was returned. These schemes were widely used and cost-effective, since collecting and washing the returned bottles was more economic than manufacturing new ones. However, with the introduction of new materials and production technology and changes in consumer preferences, the deposit refund system has declined. Some schemes have still survived, such as the traditional re-use of milk bottles and, with a more environmental public attitude, business has responded with the introduction of new schemes. Some supermarkets now encourage the re-use of plastic carrier bags, often with the incentive of supporting charities. Other countries have extended the deposit refund scheme for products such as batteries (Denmark and the Netherlands), disposable cameras (Japan) and even car bodies (Sweden and Norway). Such schemes have mixed success; where the consumer inconvenience of returning the item is not outweighed by financial gain then there is a lower rate of return. Re-use schemes where industry or business have full control of the product or packaging are easier to implement. For example, transit packaging such as polystyrene chips, and cardboard used to transport items between different destinations within the same company, can be re-used several times.

Re-use of glass bottles, for example, means that savings are made in raw materials and energy used. However, re-use also has an environmental impact. The re-use of the glass bottle involves the energy used in collection and transportation, with their associated effects. In addition, large amounts of detergents, water and chemicals are required to ensure adequate cleaning, in order to meet hygiene regulations. A further factor is that re-useable containers need to be more robust than a single-use container, which means a thicker or heavier container involving an increased use of resources.

3.2   Waste Recycling

Recycling is the collection, separation, clean-up and processing of waste materials to produce a marketable material or product. Recycling can take place within the manufacturing process, such as in the paper industry, where surplus pulp fibres, mill off-cuts and damaged paper rolls are recycled back into the pulping process. Alternatively, recycling takes place at the post-consumer stage where paper can be collected separately or extracted from the waste and can then re-enter the paper making process. The advantages of using recyclable materials are that there is a reduced use of virgin materials, with consequent environmental benefits in terms of energy savings in the production process and reduced emissions to air and water and onto land. There has been a significant increase in the levels of recycling in most countries worldwide (OECD 2004). Recycling may not always be the best environmental or economic option for a particular type of waste and a full analysis of the processes involved in recycling versus treatment and disposal should therefore be made.

Figure 3.1 shows a comparison of the recycling rates for various waste streams in the United Kingdom, Germany and Finland (DEFRA 2004; European Environment Agency 2003; European Environment Agency 2002). The potential for recycling certain types of waste can be seen in the figure. For example, construction and demolition waste has a recycling rate of more than 80% in Germany, 50% in the UK and almost 30% in Finland. Similarly, industrial waste shows high rates of recycling. Both of these waste streams have the direct financial incentive of the waste producer in recycling the waste. Sewage sludge recycling is mainly recycling via spreading on land. Hazardous waste and municipal solid waste show the lowest rates of recycling. However, the potential for increasing the amount of municipal solid waste recycled is shown by the high rates of over 35%, obtained by Germany.

Within the European Union, the principle of producer responsibility extends to certain waste sectors, for example, packaging, end-of-life vehicles, waste electrical and electronic equipment. Each of these sectors is covered by a European Directive which sets targets

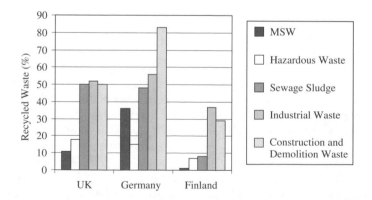

Figure 3.1 *Percentage recycling of various wastes in the UK, Germany and Finland. Sources: DEFRA 2004; European Environment Agency 2003; European Environment Agency 2002.*

for the recycling and recovery (energy recovery) of these waste streams. The Directives seek to increase the share of responsibility in dealing with the waste which arises from the products in these categories. The producers bear some of the costs of waste disposal in that they must introduce measures to re-use and recycle waste in order to meet the imposed targets. The incentive therefore encourages waste producers to reduce the waste arising, in order to minimise the costs of waste management.

Across Europe, a range of economic instruments have been introduced to encourage diversion away from waste landfill and incineration and to develop more waste recycling. For example, economic instruments such as landfill taxes, incineration taxes, direct waste charging schemes and tradeable waste allowances, have been introduced by many European countries. Increasing the costs of waste disposal to the manufacturer, through increased waste disposal costs, or the setting of regulatory recycling targets, encourages the development of low-waste-producing processes and is an incentive to recycle. Underlying this approach is the 'polluter pays' principle, which was introduced in the Fifth Environmental Action Programme (1993), in which the producer of waste has the incentive to develop production processes that minimise waste or increase the levels of recycling in order to minimise their costs.

It should also be borne in mind that the original 1975 EC Waste Frame Directive, defined waste as 'any substance or object which the holder discards or intends to discard'. That is, nobody wants the material. But recycling the waste material does not redefine it as a material that someone now wants. Recycling is susceptible to the market price and price stability for the recycled material. In this respect the recycled material is in direct competition with the virgin product. The market is also a two-way process, the industrial user of the recycled material requires a secure supply of material and confidence that the consumer market for the processed recycled material will be stable and will grow in the future, since investment costs in the necessary recycling processing plant are high, with long pay-back periods.

Whilst there are a range of regulations and measures in place to encourage recycling, there is still a voluntary sense about recycling and, in most sectors of the economy, recycling remains relatively low. However, the increasing influence of the European Union through a range of Directives on waste management, which encourage recycling through the principle of producer responsibility, will inevitably lead to a more regulated environment and an increase in waste recycling.

3.2.1 Municipal Solid Waste Recycling

The composition of municipal solid waste is shown in Figure 3.2 for several countries across Europe (European Commission 2003). The theoretically recyclable components of municipal solid waste include paper and board, plastics, glass, metals and organic or putrescible materials, made up of garden and food wastes, which are suitable for composting. The municipal solid waste for the four countries illustrated in Figure 3.2 shows a wide variation in composition, particularly in the percentage of paper and board products and organic waste. The United Kingdom and Finland have paper and board as the highest proportion of municipal waste at 33% and 42%, respectively, whereas Greece and Bulgaria are dominated by organic waste at 52% and 40%, respectively. However, in some cases it is not possible to recycle some of the wastes due to contamination. Figure 3.3 shows an estimation of the

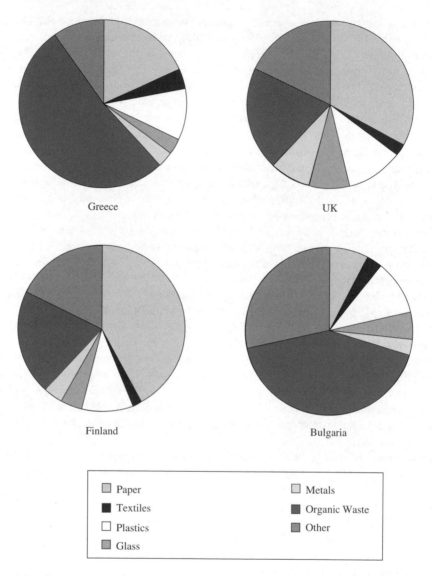

Figure 3.2   *Composition of municipal solid waste in selected European countries. Source: European Commission 2003.*

potentially recyclable components of household waste (Waste Management Paper 28, 1992). Approximately 60% of all household waste in the form of paper, plastics, textiles, glass, metals and organic waste, is potentially recyclable after discounting the contaminated materials.

In some cases it is not technically feasible or economically desirable to recycle all the components of waste. In most European countries, the municipal solid waste category 'paper and board' consists of mainly newsprint made up of newspapers and magazines,

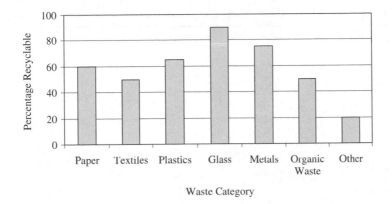

Figure 3.3 *Potentially recyclable components of household waste. Source: Waste Management Paper 28, 1992.*

which are readily recycled back into the pulp and paper industry, provided that they are collected as clean paper before contamination with organic waste (Manser and Keeling 1996). In addition, the category includes packaging materials which may be contaminated with foodstuffs and may also contain plastic or aluminium foil coatings which are less easily recycled. Plastic in municipal solid waste includes dense plastic beverage bottles and food containers, which may be segregated at source or at the recycling facility, but it also contains plastic film and plastic bags which are not easy to handpick or source segregate. Glass waste is in the form of glass bottles and jars of different colours, some of which are more economically desirable than others. For example, in some countries, clear glass is more valuable than coloured glass. Metals includes a whole host of different metals that are found in municipal solid waste to some extent, however, aluminium drinks cans and steel food cans are the most common and are readily segregated and recycled.

Detailed analysis of the municipal solid waste producing sectors can highlight the potential areas where recycling of specific materials would be most beneficial. Table 3.1 shows average waste composition of municipal solid waste from the US in terms of the production source of the waste (Warmer Bulletin 49, 1996). The opportunities for recycling various components of the waste streams associated with each source are very different. For example, paper recycling can be identified as a potential recycling opportunity from office buildings, family residencies and schools, while plastic waste and glass may come from hotels.

Whilst the recyclable materials are found in household waste, they are present in a very heterogeneous matrix and the segregation of the materials is one of the major factors involved in waste recycling. Two types of system exist to reclaim the materials separately, the 'bring' and the 'collect' systems (Hogg and Hummel 2002; Eunomia 2003; White et al 1995; ERRA 1994). The 'bring' systems involve the segregation of recyclable materials, for example, paper, plastic and glass bottles, metals and textiles, from household waste by the public and delivery to a centralised collection site. The sites may be bottle and paper banks situated at the local supermarket, civic amenity sites for the disposal of many types of material or the local scrap merchant. This system has the advantage of being low in capital costs, easily accessible and can also provide

Table 3.1 *Average US waste composition in relation to waste production source*

Waste component	Multi-family residence	Hotel	Schools	Office building	Shopping centres
Mixed paper	10.3	11.3	28.6	43.0	11.8
Newsprint	33.7	11.5	2.5	3.5	6.7
Corrugated board	10.0	27.5	26.1	21.5	37.4
Plastic	8.6	13.7	6.9	3.7	8.0
Yard waste	5.3	0.5	0.0	0.0	2.8
Wood	1.7	0.7	0.1	8.5	3.6
Food waste	13.0	4.9	6.7	2.2	12.4
Other organics	7.4	7.5	0.1	1.2	1.9
Total combustibles	90.0	77.6	71.0	83.6	84.6
Ferrous	2.5	6.9	8.2	0.6	3.6
Aluminium	1.1	2.7	1.5	0.4	0.7
Glass	5.9	11.9	6.2	0.8	2.7
Other inorganics	0.3	0.9	13.1	14.6	8.6
Total non-combustibles	9.8	22.4	29.0	16.4	15.6
Total	100	100	100	100	100

Source: Warmer Bulletin 1996.

an easy method of segregating clean readily marketable materials. However, the take-up of the schemes by the public can be low. In addition, the sites can become unsightly with litter spillage and can be an attraction for vandalism. The collected materials are taken either directly to a materials reprocessor or to larger recycling facilities where further processing takes place to sort, clean and grade the materials before transfer to the reprocessing plant.

The 'collect' systems involve house-to-house kerbside collection of designated recyclable materials, source separated by the householder and placed in separate containers. There are a number of varieties of the 'collect' system. For example, the recyclable materials such as paper, plastics and metal cans are all placed in one container, therefore the mixture has to be sorted, either by processing equipment or by hand at the central materials recycling facility. Alternatively, the materials may be sorted at the kerbside by the collector. More sophisticated systems involve the separation of the recyclable materials into several containers or sections of a container by the householder for separate collection. The latter two systems require a more elaborate collection vehicle to collect the separated waste streams. The advantages of the 'collect' system include convenience for the householder and higher recovery rates of recyclable materials. However, collection costs are higher in that separate collections or purpose-built vehicles, with separate enclosures, are required (Eunomia 2003). In addition, costs associated with the sorting of the materials and transport to the reprocessing facility will be extra. However, these are offset by the income from the sale of the recycled materials.

Throughout Europe, a range of 'bring' and 'collect' systems are used for the recycling of materials from municipal solid waste (ACCR 2000). Austria has a mixture of kerbside collect systems, 'bring' systems, container parks, and central depots. Belgium has door-to-door collect schemes for paper, glass and packaging and 'bring' systems. France has door-to-door collect schemes in both rural and urban areas as well as 'bring' schemes. Spain has a system of road container collect systems and the United Kingdom has door-to-door collect systems for co-mingled recyclable material and on-vehicle separation, coupled with a 'bring' system.

Individual materials achieve high rates of reclamation, with glass collection rates using the 'collect' system being up to 71% and paper 67% (Atkinson et al 1993; Atkinson and New 1992, 1993). Generally 'bring' systems require more motivation and effort on behalf of the householder and usually achieve lower levels of reclamation. The exception to this is glass where the 'bring' system is well established and can achieve recovery rates of over 60% (Eunomia 2003; Atkinson and New 1992).

As on alternative to the bring and collect systems, there are centralised materials recycling facilities, where the household waste is brought to a central plant for reclamation and recycling (ERRA 1994). The waste can be segregated into recyclable materials, partially segregated or completely unsegregated (Warmer Bulletin 1995; Warmer Bulletin 91, 2003; ERRA 1994). The number of components in the waste will be influenced by the degree of pre-segregation, which will influence the sorting and separation technology or manpower required for the materials recycling facility. Unsegregated materials recycling facilities are designed to process household, commercial and industrial wastes. Inevitably the materials are contaminated, for example, with broken glass, foodstuffs, etc., and recovery rates of recyclable materials are low, of the order of 15% from municipal solid waste (Warmer Bulletin 91, 2003). However, where one type of material from a specific source is involved, rates can be much higher, for example, general commercial office waste may contain levels of paper of over 80% by weight (Waste Management Paper 28, 1992). Segregated material streams handled by materials recycling facilities process a set, typically between three and eight components, of particular materials which may be separated or mixed. Such facilities handle 'clean' waste and consequently contamination levels are low and recovery rates are high.

The design of a typical unsegregated municipal solid waste recycling facility is shown in Figure 3.4 (Warmer Bulletin 1995). The example is for a facility which can recover ferrous metals, high-density polyethylene (HDPE) and polyethylene terephthalate (PET) plastics, aluminium and several grades of paper, i.e., corrugated, newspaper and high-grade paper. The design would include mechanical and manual separation processes. At the input stage, large quantities of pre-segregated materials, such as old corrugated cardboard, would not require processing and therefore can be fed directly to the baler. The stages of separation include trommel screening, magnetic separation, and manual sorting. Manual sorting is necessary to separate different types of plastic and different coloured glass. However, the trend is towards an increase in mechanisation of the process. An unsegregated, municipal solid waste materials recycling facility would recover approximately 15% of the waste stream as usable materials. The remaining 85% is largely organic and can be used to produce a fuel (refuse derived fuel, RDF), which is converted to compost or landfilled (Warmer Bulletin 1995).

Figure 3.5 shows a materials recycling facility used to handle a mixture of containers composed of different coloured glass, aluminium and ferrous cans, and plastic containers composed of high-density polyethylene (HDPE) and polyethylene terephthalate (PET) (Warmer Bulletin 59, 1998). Whilst the separation processes are very similar, the pre-segregation of the materials before they arrive at the facility, means that very high recovery rates are possible, compared with recycling facilities for unsegregated materials. Such

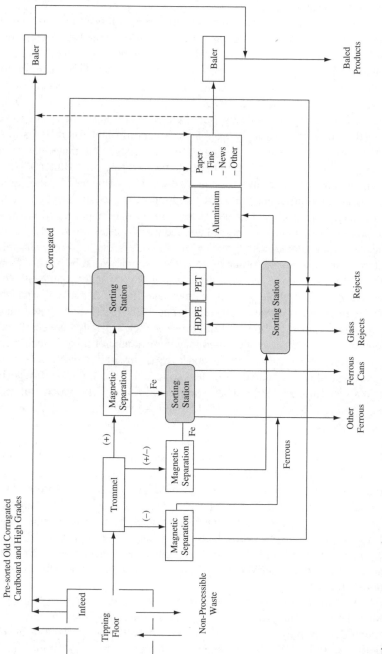

Figure 3.4 *Design of an unsegregated municipal solid waste recycling facility. Source: Warmer Bulletin, Materials Reclamation Facilities 1995. Reproduced by permission of R.C. Strange.*

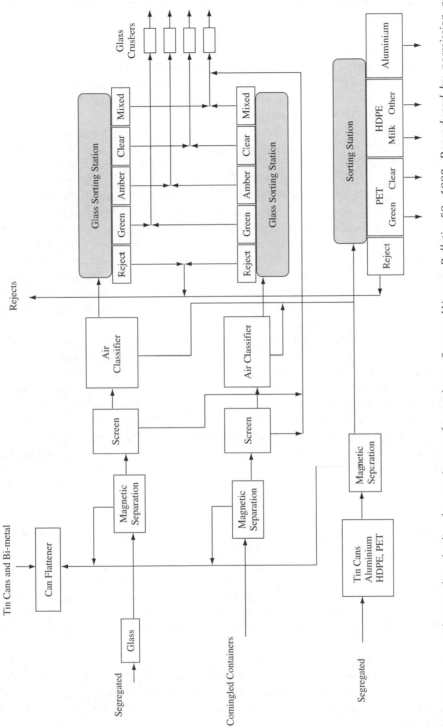

Figure 3.5 *Materials recycling facility for a mixture of containers. Source: Warmer Bulletin 59, 1998. Reproduced by permission of R.C. Strange.*

systems can achieve recovery rates of 90% of the incoming segregated waste material in the form of usable end products.

In some cases, the materials recovery facility may be designed with a compost or a refuse derived fuel (RDF) as the end material, with other recyclable materials as the by-products. For composting from municipal solid waste, the stages of sizing, metal and glass removal, would be essential to provide a compost suitable for marketing (Warmer Bulletin 59, 1998). More successful schemes have relied on source-separated organic wastes, such as garden waste, which are processed at centralised composting facilities.

An important aspect of materials recycling, which is often not considered, is the occupational health risks of the workers employed to sort, often by hand, the municipal solid waste in the materials recycling facility. The potential risks to the workforce include accidents from handsorting due to broken bottles, sharps and tin cans, exposure to electromagnetic fields, used for sorting metals, and chemical hazards from garden and household chemicals, paint vapours and batteries (Gladding 2002). Biological hazards result from the vapours arising from biodegradation and also airborne bacteria and fungi and, potentially, viruses from the collection and sorting of household waste. Gladding (2002) has reviewed the potential health effects of emissions reported in source-segregated and unsegregated, mixed, municipal solid waste recycling facilities. Reported adverse symptoms in the workforce who are working in materials recycling facilities included pulmonary disorders, organic dust-like symptoms, gastrointestinal problems, eye inflammation and irritation of the skin and upper airways.

The number of municipal solid waste recycling schemes has risen markedly over the last decade, particularly across Europe and North America and there have been significant improvements in the overall recycling rates for municipal solid waste. For example, the municipality of Aarhus in Denmark has a population of 282 000 people and achieves an overall recycling rate, for all the different categories of waste, of over 64%. The figure for household waste recycling is only 27%, but, the overall figure of 64% for Aarhus is reached by including garden waste composting at close to 100%, industrial and commercial waste recycling at 56% and construction and demolition waste at 93% (Aumonier and Troni 2001). The collection system is based on a network of street containers for the recycling of glass and paper, with a density of one container for approximately every 430 people. In addition, there are civic amenity sites throughout the municipality where householders can dispose of paper, cardboard, glass, demolition waste, plastics, bottles, garden waste, combustible materials, electronic scrap and hazardous waste. It is estimated that the participation rate in the Aarhus recycling scheme is 98%.

A further example of a successful recycling scheme is that of Wiesbaden, Germany. The city has a population of 267 000 and it is estimated that, of the 91 000 tonnes of municipal solid waste produced each year, 43 500 tonnes are recycled, representing a recycling rate of 48% (Aumonier and Troni 2001). Wiesbaden, like other German cities, operates the DSD 'green dot' recycling scheme, which relies on households sorting their packaging waste into three fractions which are glass, sorted by colour and taken to bottle banks, paper and board, which is sorted to one bin, and aluminium, steel and plastics cartons and composites which are sorted and placed in a yellow bin or bag. The recycling scheme is based on collection of recyclable material from households, a network of containers at recycling points and recycling centres. Organic waste is collected and composted at a central facility.

An example of waste recycling from the United Kingdom is project Integra in the county of Hampshire. Hampshire is divided into three regions with a total population of 1.6 million generating approximately 850 000 tonnes per year of waste (Project Integra 2004). The recycling rate for municipal solid waste is over 24%. The majority of waste is collected by kerbside collection schemes which cover over 90% of the households in Hampshire. In addition, the waste is recycled through household waste recycling centres and composting centres. The main wastes recycled were ferrous metals, board, glass, paper, non-ferrous metals, batteries, textiles, oil, electrical goods and plastics. The collected materials are taken to materials recycling facilities where they are sorted and processed.

There are over 9000 kerbside recycling programmes in the USA serving approximately 50% of the population (US EPA 2004). In addition, there are over 12 000 'bring' recycling sites. It is estimated that, in the USA, there are approximately 480 materials recycling facilities. Overall recycling rates have been estimated at 29.7%, which includes composting (2001). For individual components of the waste stream, the recycling rates are higher. For example, the recycling rate for newspapers is 60.2%, for steel cans 58.1%, yard (garden) trimmings 56.5%, aluminium beer and soft drink cans 49.0%, scrap tyres 38.6% and plastic soft drink bottles 35.6%. In some states, with residential kerbside recycling programmes, it is a requirement by law to recycle selected materials and in such areas the material recovery rates tend to be significantly higher than in areas operating voluntary schemes (Everett and Peirce 1993). There has been a recent trend downwards in the number of landfill sites with current estimates of 55.7% (2001) of municipal solid waste going to landfill. In addition the number of landfill sites has fallen from over 6326 in 1990 to 1858 in 2001, with a trend towards smaller numbers of large landfill sites. Closed landfills are being replaced with waste transfer stations, with the opportunity of separating and recovering materials. Incineration takes place at 97 municipal waste incinerators, which accounts for almost 15% of the total of 230 million tonnes of municipal solid waste generated in the USA each year (US EPA 2003).

An example of a city-wide recycling scheme in the USA, is that of New York. The population of New York is over 7.3 million people who are densely housed, mainly in apartment blocks. The New York City Department of Sanitation is responsible for collection and treatment of the city's waste and recycling is organised by the Department's Bureau of Waste Prevention, Re-use and Recycling. Processing of the collected recyclable materials is carried out by private sector companies under contract to the Department of Sanitation. The municipal solid waste recycling rate for New York is 22%, but there are variations between districts of between 5 and 35% recycling rate (Aumonier and Troni 2001). The city recycles mixed paper, newspaper, magazines, catalogues, phone books, corrugated cardboard, milk and juice cartons, metals, plastic bottles and jugs, glass bottles and jars and aluminium foil. The kerbside collection programme is the largest in the USA, providing a collection service to all 3 million households and also to the commercial sector (Aumonier and Troni 2001).

One form of materials recycling facility is the refuse derived fuel (RDF) plant where the main end product is the production of a fuel in the form of the combustible fraction of municipal solid waste. The plant would normally process unsegregated municipal solid waste. Consequently, a refuse-derived fuel plant seeks to concentrate the combustible fraction of the waste by removal of non-combustible materials, such as glass and metals. Further processing to remove materials of low calorific value, such as putrescibles, and

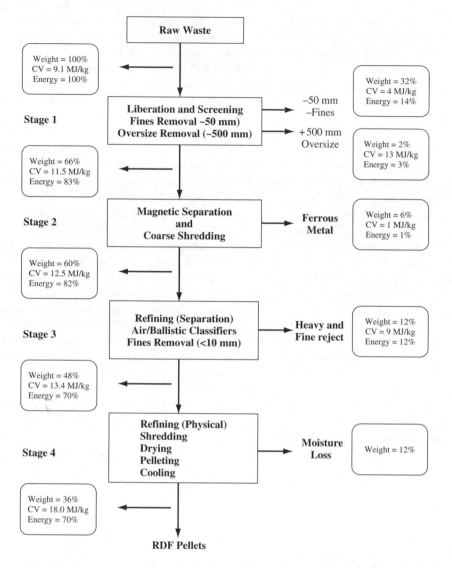

Figure 3.6 *Production process to produce refuse-derived fuel from municipal solid waste. Source: Barton 1989.*

very fine material, increases the calorific value of the residual product which consists of paper, plastics, textiles and other combustible material. Figure 3.6 shows the processing stages in the production of a typical refuse-derived fuel, which is in a compressed pelletised form, of size approximately 1.8 cm diameter by 10 cm long (Barton 1989). The major steps involved are preliminary liberation, where bags of waste are mechanically opened, then size screening, magnetic separation and coarse shredding, a refining separation stage and finally a series of processes to control the physical characteristics of the fuel for ease of combustion. The figure shows the weight and energy content as 100% and the various stages show how the weight and energy content decrease with processing. The final refuse-derived fuel pellets decrease in weight to 36% of their original value, but contain 70% of their original energy content. The 36% conversion of municipal solid waste to refuse-derived fuel is an average value since, depending on the type of process used to produce the refuse-derived fuel, the conversion rate can vary between 23 and 50% by weight (Gendebien et al 2003). The raw waste has a typical calorific value of 9.1 MJ/kg whilst the processed refuse-derived fuel pellets have a calorific value of 18.0 MJ/kg. Some plants do not produce pellets, but instead produce a coarse refuse-derived fuel which requires less processing and has a similar calorific value to pelletised fuel. Table 3.2 shows a comparison of the characteristics of different refuse-derived fuel pellets with a typical bituminous coal (Rampling and Hickey 1988; Jackson 1988; Barton 1989). The refuse-derived fuel has a 30% lower calorific value, lower bulk density, higher ash content and higher volatile content, than coal.

Refuse derived fuel plants have been developed throughout Europe and North America. For example, the production of refuse-derived fuel is well established in Austria, Finland, Germany, Italy, the Netherlands and Sweden (Gendebien et al 2003). Table 3.3 shows the number of plants and production of refuse-derived fuels for certain countries in Europe (Gendebien et al 2003). Refuse-derived fuel may be produced not only from municipal solid waste but also from other wastes which might be included in the process. For example, in Finland, refuse-derived fuel is produced from source-separated household waste, waste from commerce and industry and construction and demolition waste. In Austria, Germany and Italy, refuse-derived fuel is produced from municipal solid waste, waste

Table 3.2 *Analysis of typical refuse derived fuel pellets compared with a typical bituminous coal*

Analysis	Refuse Derived Fuel	Coal
Calorific value	18.7	27.2
Proximate analysis (wt%)		
Volatile matter	67.5	25.9
Fixed carbon	10.2	55.5
Ash	15.0	10.2
Moisture	7.3	8.4
Ultimate Analysis (wt%, dry ash free)		
Carbon	55.0	83.5
Hydrogen	7.3	5.4
Oxygen	35.9	8.4
Nitrogen	0.6	1.7
Chlorine	0.9	0.04
Fluorine	0.01	trace
Sulphur	0.3	1.0
Lead	0.02	0.003
Cadmium	0.0008	0.00001
Mercury	0.0002	0.00007
Bulk density (kg/m³)	600.0	900.0

Sources: Rampling and Hickey 1988; Jackson 1988; Barton 1989.

Table 3.3   *Refuse derived fuel production in European countries*

Country	Number of plants	Production (1000 tonnes/Year)
Austria	10	70
Finland	12	90[1]
Germany	14	330
Italy	16	300
The Netherlands	13	700
United Kingdom	3	90

[1] Maximum.

Source: Gendebien et al 2003.

wood, commercial waste, industrial waste, sewage sludge, etc. In the United Kingdom, refuse-derived fuel is produced from raw municipal solid waste or source-separated municipal solid waste.

Refuse-derived fuel is used in dedicated refuse-derived fuel incinerators, in paper mill furnaces, in cement kilns, power plants and also in district heating plants. The main type of combustion system for refuse-derived fuel is the fluidised bed, where the waste-derived fuel is fed to a bubbling bed of hot sand (typically at 800 °C) where the waste combusts. Refuse-derived fuel may also be incinerated or co-incinerated with other fuels in small-scale plants for power or district heating. For example, Finland co-incinerates refuse-derived fuel with biomass waste, such as tree bark, sawdust, etc. The combustion of refuse-derived fuel in cement kilns is also a significant route for energy recovery utilisation.

3.2.2   Industrial and Commercial Waste Recycling

Industrial waste recycling includes direct recycling, where waste material is recycled back into the manufacturing process in-house within the factory. For example, broken and mis-shapen glass would be routinely re-melted in the production process. Similarly, plastic off-cuts and scrap are also recycled during the manufacturing process. Other industrial sources of waste are routinely recycled within the industry. Agricultural waste is mostly landfilled or used as animal feed and, consequently, the material does not enter the general waste management process. Similarly construction and demolition waste is often recycled on-site as aggregate or ballast in the construction of new buildings. The majority of materials which would include soil, sub-soils, concrete, bricks, timber, plaster, etc., are used for bulk fill, such as aggregates for road construction projects. The iron and steel industry also has a significant proportion of in-house recycling for waste metals.

Other industrial wastes are commonly recycled, but indirectly, as post-consumer waste. Commercial and industrial wastes are by their nature very variable in composition. Commercial waste would include waste from shops, offices, restaurants and institutions such as schools. Office waste contains a high proportion of waste paper, whilst restaurants will have high proportions of putrescible waste, but also glass, metal cans and plastic packaging. Industrial waste will be heterogeneous in its composition, and depends on the individual business and type of on-site production. Many larger companies have separate waste collection and disposal arrangements, which may include recycling. Figure 3.1

shows examples of the percentage of industrial category wastes recycled in the United Kingdom, Germany and Finland. For certain wastes, such as construction and demolition waste, the rates can be high, being over 80% in Germany. Similarly, industrial waste recycling is over 50% in some countries, including the United Kingdom and Germany.

3.3   Examples of Waste Recycling

3.3.1   Plastics

Plastic polymers make up a high proportion of waste and the volume and range used is increasing dramatically. The two main types of plastic are thermoplastics, which soften when heated and harden again when cooled, and thermosetts, which harden by curing and cannot be re-moulded. Thermoplastics are by far the most common types of plastic comprising almost 80% of the plastics used in Europe and they are also the most easily recyclable. Table 3.4 shows typical applications of the main plastic types (Warmer Bulletin 1992). The consumption of plastics in Western Europe (Austria, Belgium, Denmark, Finland, France, Germany, Greece, Ireland, Italy, Luxembourg, the Netherlands, Portugal, Spain, Sweden, United Kingdom, Iceland, Liechtenstein, Norway, Switzerland, Andorra, Monaco, San Marino) is of the order of 38 million tonnes per year (2004) and the majority is used in the production of plastic packaging, household and domestic products, electrical and electronic goods. There is also significant consumption of plastics for the building and construction industry and automotive industry. Figure 3.7 shows the distribution of plastics end-use in Western Europe by sector (APME 2004).

Table 3.4   *Primary applications of plastics*

Plastic type	Typical application
[1] Thermoplastics	
High density polyethylene (HDPE)	Bottles for household chemicals, bottle caps toys, housewares
Low density polyethylene (LDPE)	Bags, sacks, bin liners, squeezy bottles, cling film, containers
Polyvinyl chloride (PVC)	Blister packs, food trays, bottles, toys, cable insulation, wallpaper, flooring, cling film
Polystyrene (PS)	Egg cartons, yoghurt pots, drinking cups, tape cassettes
Polyethylene terephthalate (PET)	Carbonated drinks bottles, food packaging
Polypropylene (PP)	Margarine tubs, crisp packets, packaging film
[2] Thermosets	
Epoxy resins	Automotive parts, electrical equipment, adhesives
Phenolics	Appliances, adhesives, automotive parts, electrical components
Polyurethane	Coatings, cushions, mattresses, car seats
Polyamide	Packaging film
Polymethylmethacrylate	Transparent all weather electrical insulators
Styrene copolymers	General appliance mouldings

Source: Warmer Bulletin 32, 1992.

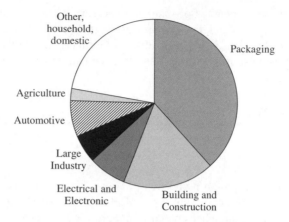

Figure 3.7   *Plastics end-use in Western Europe by sector. Source: APME 2004.*

Table 3.5   *Consumption and waste management of plastics by sector for Western Europe (1000 tonnes)*

Sector	Consumption (1000 t)	Waste (1000 t)	Recycling (1000 t)	Recovery (1000 t)	Disposal (1000 t)
Agriculture	953	286	161	0	125
Automotive	2669	851	61	35	755
Building and Construction	6710	530	58	0	472
Industry	5969	4130	1418	441	2271
Electrical and Electronic	2783	854	34	4	816
Household	19039	13324	1087	4103	8139
Total	38123	19980	2819	4583	12578

Source: APME 2004.

It is estimated that only about 50% of the plastics produced in Western Europe each year are available for collection and recycling, the remainder accumulating in the environment for long-term use such as in building works including plastic window frames, pipes, electrical wiring, etc. Of this collectable waste plastic, about 15%, representing about 3 million tonnes, is recycled in Western Europe. A further 23% of the plastics available for collection is incinerated with other wastes, mainly as municipal solid waste and the 'recovery' of the plastic in such cases is via energy recovery while the remainder is disposed of, mainly to landfill. Table 3.5 shows the main industrial/commercial and municipal waste sectors which produce the waste plastic and the routes for recycling, energy recovery and disposal (APME 2004). The majority of waste plastic arises from packaging. However, the EC Packaging and Packaging Waste Directive, sets targets for the recovery and recycling of packaging waste for each Member State of the European Union. It is between 50 and 65% by weight for recovery and, within that general target, between 25 and 45% by weight for recycling of the total packaging waste in each Member State. In addition, there is a minimum 15% recycling rate target set for each packaging material. Consequently, the percentage of waste plastic recycling in Europe is likely to increase.

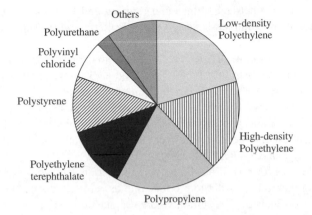

Figure 3.8   *The main plastic types found in municipal solid waste. Source: APME 2004.*

In many countries, plastic is collected from commercial and industrial sources as separate plastic fractions, much of which is recycled directly back into the plastic product manufacturing process. Although plastics make up between 5 and 15 wt% of municipal solid waste it comprises 20–30% of the volume. The plastics in municipal solid waste are mainly in the form of plastic film and rigid containers. Plastic film comprises about 3–4 wt% of the household waste stream and is almost impossible to recycle. However, plastic containers are more easily collected separately or segregated from the waste stream. There are six main plastics which arise in European municipal solid waste. These are high-density polyethylene (HDPE), low-density polyethylene (LDPE), polyethylene terephthalate (PET), polypropylene (PP), polystyrene (PS) and polyvinyl chloride (PVC). Figure 3.8 shows the proportion of each plastic found in municipal solid waste (APME 2004). Separation of the plastics from waste is mainly by hand, either by the householder prior to collection, or at a materials recycling facility. New developments attempt to automate the separation of plastics into different types. For example, separation schemes for segregating plastic types using X-ray analysis, flotation and colour sensors, have all been researched (Basta et al 1995; Warmer Bulletin 1992).

Plastic bottles represent the main plastic product which is separated from municipal solid waste. Bottles are of low density, for example, the number of plastic bottles required to make one tonne of polyethylene terephthalate is 20 000. The bottles are transferred to the materials recovery facility where they are separated from any unwanted contaminating materials which can range from 5 to 20%, and are then sorted, usually by hand, into the three main types of bottle, PET, PVC and HDPE. The bottles are compacted and baled to produce bales of approximately one cubic metre and weighing between 150 and 300 kg, and sent for further processing. Processing takes the form of shredding, washing, drying and bagging the plastic, which is then sold on to the end user of the recycled plastic (RECOUP 1995, 1996).

The separated plastic material is processed by the end user by being granulated or pelletised, melted or partially melted and extruded to form the end product. The recycled plastic may be added to virgin plastic during the process. Outlets for single types of recycled plastics include, for example, HDPE for dustbin sacks, pipes and garden furniture, PVC for sewer pipes, shoes, electrical fittings and flooring and PET for egg cartons, carpets, fibre filling material and audio cassettes (Warmer Bulletin 1992). Applications for plastic mixtures have included plastic fencing, industrial plastic pallets, traffic cones, playground equipment and garden furniture. There is resistance from the customer market for recycling plastics to produce film which may be used for food packaging, because of the perceived associated health hazard. There is concern that compounds acquired during use may survive the recycling process to contaminate foods held in containers made from recycled materials (Sadler 1995; Kuznesof and VanDerveer 1995). Other uses for recycled plastic products in the construction industry include pipes, damp-proof membranes, plastic lumber and plastic/ wood composites (WRAP 2003(a)).

The low-grade uses for mixed plastic recycled materials has led to research into alternative processing methods in order to produce higher value products. One example is via tertiary recycling or feedstock recycling, where the plastic waste materials are processed back to produce basic petrochemicals that can be used as feedstock to make virgin plastic (Meszaros 1995; Lee 1995). The process has the advantage that mixed plastics can be used, since all of the feedstock is reduced to petrochemicals. The plastic is identical to virgin plastic and can therefore be used in any application. Tertiary recycling can be via hydrogenation at high temperature and pressure or via pyrolysis in an inert atmosphere at atmospheric pressure, to thermally degrade the plastics. Pyrolysis of plastics at high temperature thermally degrades the plastic by breaking the bonds of the polymer to produce lower molecular weight oligomers and monomers. The vapours resulting from the process are condensed to produce a waxy hydrocarbon product which has a high degree of purity and can be refined at the petroleum refinery to produce a range of petrochemical products, including virgin plastic.

Table 3.6 shows the plastics content in municipal solid waste for various European countries, which ranges from 5 to 15% (APME 2004). Interestingly, there are significant variations in the quantities of plastic waste content in municipal solid waste when corrected for population statistics. The kg/capita quantities of plastics in municipal solid waste range from 20 kg/person in Greece to 56 kg/person in Switzerland. Table 3.7 shows the plastics waste management for the proportion of wastes collected for various countries throughout Europe (APME 2004). The recycling rate varies across the countries of Europe, with some countries such as Greece and Portugal having low rates of recycling at 2% and 3% respectively, whilst other countries such as Spain, the Netherlands and Finland have rates of 14–16%. Germany has a high rate of recycling since they have a targeted plastics recycling scheme for households.

3.3.2   Glass

Glass is made from relatively cheap raw materials: silica sand, limestone and sodium carbonate. However, glass making is energy intensive and glass recycling can reduce the energy used, since recycled glass melts at a lower temperature than the raw materials (WRAP 2003(b)). For example, increasing the amount of glass waste or 'cullet' in the furnace to 50% can result in a 15% saving in energy. In addition, when using virgin raw materials, 15% of the weight of input is lost as waste gases but, whilst using cullet, there are no waste gases, also water used in the processing is reduced by up to 50% (Warmer Bulletin 49, 1996). Glass recycling also reduces the emissions of carbon dioxide. The

Table 3.6 *Plastics content in municipal solid waste for various European countries*

Country	Plastics content (%)	Tonnage (1000 tonnes)	Plastics content kg/capita
Austria	12.5	228	28
Denmark	8.0	216	41
Finland	5.0	87	17
France	10.5	2108	36
Germany	5.5	1934	24
Greece	7.5	214	20
Ireland	10.5	118	31
Italy	11.5	2315	40
Portugal	10.0	263	26
Spain	10.5	1450	37
Sweden	7.5	216	24
The Netherlands	11.0	693	44
UK	10.0	2430	41
Norway	8.0	112	25
Switzerland	15.0	402	56

Source: APME 2004.

Table 3.7 *Total waste plastics management for the collected plastic waste fraction (1000 tonnes)*

Country	Collectable (1000 t)	Recycling (1000 t)	Recycling rate (%)	Recovered (1000 t)	Disposal (1000 t)
Austria	350	67	19	73	210
Denmark	351	36	10	242	73
Finland	162	22	14	29	111
France	3120	287	9	998	1835
Germany	3161	983	31	806	1372
Greece	317	6	2	0	311
Ireland	204	16	8	0	188
Italy	3396	438	13	428	2530
Portugal	453	13	3	110	330
Spain	2095	314	15	266	1515
Sweden	384	32	8	173	179
The Netherlands	1027	166	16	542	318
UK	3682	295	8	295	3093
Norway	181	19	10	77	85
Switzerland	545	40	7	378	128

Source: APME 2004.

disadvantages of using recycling glass cullet are mainly associated with contaminants. Some contaminants, such as metals, are easily separated. However, problematic contaminants such as ceramic glass material and heat resistant borosilicate-type pyrex glasses are more difficult to detect. These give rise to defects in the finished glass product and can block the liquid glass flow in the glass moulding machinery (WRAP 2003(b)).

The processing of waste glass consists of a number of stages. The glass from the bottle banks is delivered to the recovery facility, where it is sorted by colour. The glass is stored by colour in separate bunkers until required, when it is fed to conveyor belts, where ferrous materials such as bottle caps are removed by magnetic separation and hand sorting is used to remove unwanted contamination. The glass is then crushed, screened and lightweight non-ferrous contamination, such as aluminium caps, plastics and paper labels are removed by vacuum suction. A final stage of removal of opaque material, such as china pieces from the clear glass, may be made electronically. The crushed processed glass is then available for recycling into the glass-making process. Usually the glass-cullet-making process is operated by a separate company to the glass-making company. Glass manufacture is fully automated, three tonne batches of mixed recycled cullet and virgin raw materials are heated rapidly to 1540 °C. The melted materials are blown into glass bottles at the rate of about 150/minute. A hot end-coating process coats the bottles with tin oxide to smooth the surfaces. The cooled glass is reheated and slowly cooled to anneal them which reduces stresses in the glass. A coating of polyethylene may be given at the end of the process to reduce surface scratching. Up to 80% of glass cullet can be accepted into the glass-making furnace and there is no limit to the number of times that glass can be recycled (Warmer Bulletin 49, 1996; BGRC Fact File 1996).

Waste glass bottles, jars and other containers comprise the main glass waste in municipal solid waste. The main system for collecting waste glass or 'cullet' throughout Europe is through bottle banks, kerbside collection schemes and through materials recycling facilities, or to collect glass from individual households as part of a city recycling scheme, as in Milton Keynes and Adur (Poll 1995). In addition, flat glass for window and door glazing will end up in the 'construction and demolition' waste category. There will also be some in-house recycling of glass waste within the industrial process itself.

The majority of glass in municipal solid waste is in the form of 'packaging', that is bottles, jars etc., and is therefore covered by the EC Packaging and Packaging Waste Directive (1994). The Directive sets particular targets for 'recycling' and 'recovery' of packaging waste, including glass. Recycling targets for glass have to be met through recycling of the material back into the industry, or for other recycling applications, since glass comprises part of the non-combustible fraction of waste and therefore there is no option of energy recovery to meet overall recycling targets.

Table 3.8 shows the tonnages of waste glass arising in municipal solid waste and the recycling rate for several countries across Europe. Glass recycling in Europe is well established, with several countries achieving recycling rates of over 75%. Higher recycling rates achieved in some European countries are due to more effective collection methods such as higher densities of bottle banks. However, the data may also represent different methods of defining glass recycling in different countries.

Glass waste collected from the municipal solid waste stream consists of mainly clear, brown and green bottles and jars. The glass recycling industry rarely accepts glass waste in the form of mixed colour cullet and requires separation into colour categories. In addition, a high degree of purity is required for each load of cullet and rejection of a waste glass consignment load may occur if it is contaminated with the wrong colour glass or if there is any ceramic contamination (Manser and Keeling 1996). Consequently, glass bottle bank containers are usually segregated into clear, brown and green glass waste. Elimination of contamination from the colour-segregated containers

Table 3.8   *Glass recycling rates from municipal solid waste in Europe*

Country	Waste glass (1000 tonnes)	Recycled (1000 tonnes)	Recycling rate (%)
Belgium	556	420	76
Finland	58	37	64
France	3364	1691	50
Germany	3763	3146	84
Ireland	111	36	32
Portugal	315	138	44
Spain	1532	575	38
The Netherlands	507	396	78
United Kingdom	2155	583	27

Source: European Commission 2003.

by hand sorting is then easier. However, where the material recycling facility handles non-segregated coloured glass waste or even non-segregated municipal solid waste, then separation of the different colours of glass waste becomes difficult and results in lower glass recycling rates.

To overcome the problem of different coloured glass and the need for sorting, automatic optical colour separators have been developed to segregate glass into different colours, although the main process for colour separation of glass at material recycling facilities is still sorting by hand (Manser and Keeling 1996). In addition, a USA company has developed a process to cover clear glass with coloured organic coatings which, when the glass is being recycled, simply melt away (Warmer Bulletin 49, 1996). The result is that there would be no need for coloured glass to be manufactured and, consequently, no limit to the amount of cullet which could be recycled. Other developments include photographic quality ink-only labels applied directly to the containers, eliminating the need for labels made of foil, plastic and paper with the consequent need for removal during recycling (Warmer Bulletin 49, 1996).

Alternative uses for glass, other than as containers, has been investigated and potential applications of finely ground glass cullet in the production of cement, road surfacing material, building aggregates, bricks, tiles, pipes and insulation materials have been suggested (Poll 1995; Manser and Keeling 1996). However, in many such applications, the glass cullet is competing with low-cost products and hence the requirement for a competitive low-cost material incorporating recycled glass. In addition, in many of the alternative applications, the glass has to be ground to a fine particle size which involves further processing costs.

The influence of the EC Directive on Packaging and Packaging Waste in Europe has meant a strengthening of the glass recycling industry. However, the Directive calls for a minimum of a 15% recycling rate for each packaging material, and most countries already achieve this target for glass recycling (Warmer Bulletin 49, 1996). Glass makes up a significant proportion of household packaging (Table 3.8), however, commercial packaging contains virtually no glass. Because of the very low value of glass cullet, there is virtually no international trade, as the costs of transporting between countries makes the process uneconomic. Recycling of glass containers in the USA is approximately 19%

due to the growth in alternative containers in competition to glass (US EPA 2003). Japan has also seen a significant rise in the recycling of glass to reach a level of almost 80% (OECD 2004).

3.3.3 Paper

Table 3.9 shows a comparison of the quantities of paper, paperboard and paper products produced, and consumed and the collection, utilisation and recycling rates for various European countries (CEPI 2003). In most European countries the rate of recycling for paper is high. It should also be noted that some paper cannot be recycled, such as toilet tissue, paper used in construction materials, archival books and papers, paper products which are contaminated during use, etc. It is estimated that 19% of all paper is non-recyclable (CEPI 2003). Some countries, such as Germany, the United Kingdom and Belgium, export recycled paper. The major use for recycled paper is in the packaging sector, for such applications as packaging case materials, carton boards, wrappings, etc. About 25% of recycled paper is used in the production of newsprint for newspapers and magazines. However, the use of recycled paper for quality printing paper and writing paper is low, due to quality issues. The demand for recycled waste paper is very dependent on market conditions, particularly as waste paper is an internationally traded commodity (Making Waste Work 1995). There have been many fluctuations in the demand for recycled paper and board. For example, over-supply of waste paper in Europe and North America during the early 1990s led to a collapse in the market, but there was a recovery towards the end of the decade.

Waste paper is graded across Europe into different categories based on quality. Under the 2001 European List of Standard Grades of Recovered Paper and Board,

Table 3.9 *Consumption, collection and utilisation of paper, paperboard and paper products in Europe*

Country	Production (1000 tonnes)	Consumption (1000 tonnes)	Collection rate[1] (%)	Utilisation rate[2] (%)	Recycling rate[3] (%)
Austria	4419	2015	61.4	43.0	94.3
Czech Republic	882	1055	44.5	43.0	35.9
Finland	12 776	1067	71.7	5.5	65.8
France	9798	11 241	49.7	58.2	50.8
Germany	18 526	18 984	72.2	65.0	63.4
Hungary	518	797	44.9	67.4	43.8
Ireland	43	500	33.8	109.3	9.4
Italy	9273	10 995	44.9	56.0	47.2
Norway	2114	842	67.7	21.6	54.2
Portugal	1522	1047	45.3	22.4	32.6
Spain	5365	6948	52.1	81.5	62.9
Sweden	10 723	2155	68.8	17.4	86.4
The Netherlands	3338	3549	64.8	71.1	66.8
UK	6217	12 411	47.6	74.2	37.1

[1] Collection rate: Collection compared with consumption, including net imports minus exports.
[2] Utilisation rate: Percentage of recovered paper utilisation as raw material compared with paper consumption.
[3] Recycling rate: Percentage of recovered paper utilisation compared with paper consumption.
Source: CEPI 2003.

there are five main grades of paper: Group 1 – Ordinary; Group 2 – Medium; Group 3 – High; Group 4 – Kraft; and Group 5 – Special. Within each group the different categories of waste paper and board are designated using a standard numbering system. Table 3.10 shows examples of the standard European list of grades of recycled paper and board (CEPI 2002).

The degree of reprocessing of the recycled paper and board required, depends on the grade of paper collected as waste, and the end use. The higher quality grades collected such as paper-mill production scrap and office waste, require less processing and are used as a primary paper pulp substitute in applications such as paper printing and tissues. Intermediate grades of waste paper, such as newspapers, require further processing to de-ink the paper and can be recycled back into the newspaper industry for newsprint.

Table 3.10 *Examples from the European list of standard grades of recovered paper and board*

Group 1: Ordinary Grades
1.01 Mixed paper and board, unsorted, but unusable materials removed
1.02 Mixed papers and boards (sorted) (maximum 40% newspapers and magazines)
1.04 Supermarket corrugated paper and board (minimum 70% corrugated board)
1.06 Unsold magazines
1.07 Telephone books
1.08 Mixed newspapers and magazines 1 (minimum 50% newspapers)
1.09 Mixed newspapers and magazines 2 (minimum 60% newspapers)
1.10 Mixed newspapers and magazines 3 (minimum 60% magazines)

Group 2: Medium Grades
2.01 Newspapers (less than 5% coloured material)
2.02 Unsold newspapers (free from coloured material)
2.05 Sorted office paper
2.07 White wood-free books (wood-free paper)
2.08 Coloured wood-free magazines
2.09 Carbonless copy paper

Group 3: High Grades
3.01 Mixed lightly coloured printer shavings
3.06 White business forms
3.07 White wood-free computer print-out
3.08 Printed bleached sulphate board
3.11 White heavily printed multiply board
3.14 White newsprint
3.17 White shavings (shavings and sheets of white unprinted paper)

Group 4: Kraft Grades (chemical sulphate pulp)
4.01 New shavings of corrugated board
4.02 Used corrugated Kraft 1
4.03 Used Kraft sacks

Group 5: Special Grades
5.01 Mixed recovered paper and board
5.02 Mixed packaging

Source: CEPI 2002.

Lower quality waste paper is used mainly for packaging material and constitutes the main route for recycled paper and board. Paper and board, in municipal solid waste, consists mainly of newspapers and magazines which are readily recycled back into the paper pulping industry, provided they are collected as clean paper before it has been contaminated with organic waste (Manser and Keeling 1996). Collected paper from unsegregated waste materials recycling facilities is not generally suitable for use in the paper industry (CEPI 2002). A significant proportion of material in the 'paper and board' category, consists of packaging material and is also not easily recycled since it might be contaminated with organic waste, it might be plastic coated or could contain aluminium foil.

The recycling process involves an initial stage where the paper and board is pulped, followed by various stages of screening to remove contamination, cleaning to remove glues and other contaminants and then de-inking and further processing to clean and thicken the pulp. In the case of higher quality papers, a final bleaching stage may be included. Figure 3.9 shows the various stages involved in the reprocessing of waste paper to produce recycled fibre for the paper industry (Huston 1995). Waste paper is recycled to recover its cellulosic fibre content as the pulped fibres bind together under pressure to form paper. In addition to the cellulose, non-fibrous fillers are used to give added strength to the paper. Virgin paper pulp is made largely from wood derived from softwoods, such as spruce or pine, or hardwoods, such as poplar and eucalyptus, and are farmed as an industrial crop. It is estimated that between 10 and 17 trees are required to produce a tonne of paper. In addition, the production of one tonne of paper requires 130 kg of calcium carbonate, 85 kg sulphur, 40 kg of chlorine and 300 000 l of water (Warmer Bulletin 43, 1994). The wastewater is treated and returned to the water system via sewerage works. The paper manufacturing process also consumes large quantities of energy, with estimates of over 20 GJ per tonne of paper, being required.

Recycling waste paper has a number of environmental advantages, including the reduced need for wood pulp from trees. However, in most cases the wood is harvested as a commercial farming crop, coupled with tree replanting schemes. In addition, recycling can reduce the energy requirements by up to 40% and water consumption by 60%. Also, emissions to air and water and solid waste can be reduced when recycled paper is used in comparison to virgin paper (Warmer Bulletin 43, 1994). There is a practical limit to the number of times that paper can be recycled because the fibres eventually breakdown and become too small for the paper-making process. Estimates suggest that a maximum number of four recycles would be possible.

Across the world, paper recycling rates vary. For example, Western Europe averages a more than 50% recycling rate, whereas for Latin America the rate is less than 35%, and for Africa it is less than 30% (WRAP 2002). In the USA, many states have set mandatory recycling goals, which have resulted in a rapid recent rise in the recycling of paper and board (Warmer Bulletin 50, 1996). In particular, there have been State or Government initiatives to increase the quantities of recycled fibre in newsprint and in Federal Government used paper. The paper and board content of municipal solid waste in the USA is about 36% with a recycling rate of 45% (US EPA 2002). Japan has also seen a significant increase in the level of recycling during the last decade to reach a recycling rate of almost 60% (OECD 2004).

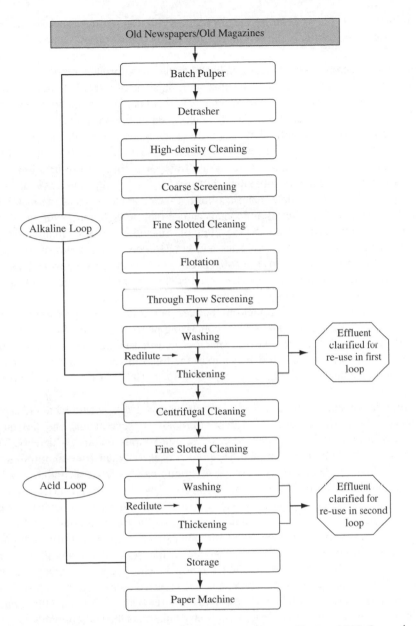

Figure 3.9 *Stages in the reprocessing of waste paper. Source: Huston 1995. Reproduced by kind permission of Apringer Science and Business Media.*

3.3.4 Metals

Metals produced via the ore mining, smelting and refining process, are referred to as primary metals. Scrap metals or secondary metals are derived either from industrial operations, such as the smelting or refining plant, and from the manufacture of metal shapes and products, or as post-consumer metal products collected from the waste stream. The industrial operations scrap metal is a purer form of the metal since it comes directly from the manufacturing industry. It is usually of known quality and composition and often uncoated and is therefore readily recycled back into the metal-production process. However, the scrap metal collected as post-consumer scrap is made up of discarded, used, or worn-out products and, as such, usually contain residues of other contaminant components. Whether or not metal scrap is recovered is generally determined by the economics of the collection and processing system, in particular, the purity of the recovered products, the market for the recycled products, and the monetary value of the metal (Campbell 1996). In addition, the costs of collection and transport, sorting and transformation into re-usable metal and the cost of disposing of any residual material, also have to be considered. Metals have the advantage over other recyclable materials, in that the characteristics of the metal are not changed by the recycling process and metals can be recycled over and over again.

The ferrous scrap derived from the recycling of iron and steel comes mainly in the form of bulky waste such as scrapped vehicles and kitchen goods such as washing machines, cookers, fridges, etc. These items are collected by scrap metal merchants and sorted into various types and grades of scrap metal, for transfer up the waste metal infrastructure to the metal refiners. The bulky scrap items are broken down by fragmentisers which fragment the waste metal in a hammer mill and are then sorted by air classifiers to produce a metal-rich stream and a residual stream containing plastics, wood, fabric, etc., which is eventually landfilled. The scrap ferrous metal is recycled in steelworks using either the basic oxygen furnace or an electric arc scrap re-melting furnace. Up to 25% scrap may be incorporated into the basic oxygen process and up to 100% in the electric arc process (Holt 1995).

The aluminium scrap collected from scrap metal merchants is shredded and separated from other metals, using magnetic separators to remove ferrous metals, and sink-and-float methods to separate the aluminium from other non-ferrous metals by density. Further processing may include, for example, the removal of the lacquer from aluminium drinks cans prior to re-melting. The major user of aluminium alloys from secondary smelters is the automotive industry, followed by the engineering and building industries.

Aluminium can be continuously recycled since there is no loss of quality during the recycling or re-melting process. In addition, the re-melting of aluminium from recycled scrap can save up to 95% of the energy required to extract and process primary aluminium. The recycled aluminium has a high value due to the demand for aluminium for end product use mainly for use in the automotive industry, where 75% of the recycled aluminium is used. Therefore, there is a high demand for recycled aluminium and most European countries have a comprehensive infrastructure to recycle it. The recycling rates for aluminium are high, for example, for aluminium drinks cans the recycling rate is 41%, for building and construction the rate is 85%, and for the automotive industry the rate is 95% (EAA 2004). Some individual countries within Europe have very high recycling rates for aluminium drinks cans. For example, Sweden has a recycling rate of 92% and Switzerland 88%.

The metals content of household waste is estimated at between 5 and 10% and is mainly in the form of tin plated steel cans used for drinks, and tinned foods and aluminium drinks cans. Table 3.11 shows the potential sources of non-ferrous metals in the household waste stream (Waste Management Paper 28, 1992). Recycling of metal cans, both steel

Table 3.11 Non-ferrous metals in the household waste stream

Metal	Form	Modes of occurrence
Copper	Metal Alloy Electroplate	Electrical fittings and wire, plumbing fittings, kitchen ware Constituent of brass, screws, plated products Decorative waste
Zinc	Metal Alloy Galvanised	Carbon–zinc batteries Components in diecastings, door fittings, domestic appliances, toys Domestic kitchen and garden ware
Tin	Electroplated Alloys	Cans, containers, toys, kitchen ware, electrical contacts Solder, constituent of bronze
Lead	Metal Alloys Chemical	Pipes, electric bulb contacts, wine bottle closures Lead–acid batteries, plumbing products As oxide and sulphate in lead–acid batteries, lead-based paints
Nickel	Metal Alloys	Plating Cutlery, components of kitchen goods

Source: Waste Management Paper 28, 1992.

and aluminium, represents a significant source of scrap metal. For example, it is estimated that almost 2 billion aluminium drinks cans are recycled in the UK and 62 billion aluminium drinks cans in the USA each year. Steel cans are separated from other metals and materials in the waste stream by magnetic separation. Re-processing of the cans involves a de tinning stage, using hydrometallurgical processing. A further significant source of ferrous metals is from magnetic separation from mass burn municipal waste incinerator ash. Alloys are mixtures of metals and are difficult to separate into their constituent metals. Consequently, alloys are usually recovered and recycled as alloys.

Other metals in the household waste stream include copper, zinc and lead. However, because of the diverse nature of the mode of occurrence of these metals, very little is recycled from household waste. In-house, industrial and commercial recycling of these metals is relatively high, since they occur as readily identifiable and easily sorted metals in significant concentrations.

Important EC Directives which will influence the level of metals recycling throughout Europe are the Packaging and Packaging Waste Directive (Council Directive 94/62/EC 1994), the End-of-Life Vehicle (ELV) Directive (Council Directive 2000/53/EC) and the proposed Waste Electrical and Electronic Equipment Directive (COM (2000-0158) 365, 2000). Each of the Directives is designed to increase the level of recycling of the various components of each waste stream, including the metals. Ferrous metals and aluminium already have high recycling rates. For example, metals recycling from scrap or end-of-life vehicles is of the order of 75%, and packaging in the form of food tins and aluminium drinks cans have similarly high recovery rates. However, for other non-ferrous metals and for metals contained in difficult-to-recycle waste streams, such as electrical and electronic goods, the recycling rates are significantly lower. The recycling of lead from automotive batteries is an exception. Because the waste is easily identifiable and can be sorted and refined by secondary lead smelting, high rates, of the order of 80%, of lead-acid batteries are recycled.

3.3.5 Tyres

Scrap tyres have been designated as a priority waste stream by the European Commission and, as such, are subject to recycling targets and recommendations regarding environmentally acceptable treatment and disposal methods. It is estimated that Europe produces about 250 million scrap car and truck tyres each year, representing about 3 million tonnes by weight of tyre. Figure 3.10 shows the generation of scrap tyres in selected European countries. The worldwide generation of scrap tyres is estimated at 1000 million tyres/year. The generation of waste tyres are linked to the generation of end-of-life vehicles. The European Union has attempted to control the management of end-of-life vehicles and consequently tyres through the 2000 End-of-Life Vehicle Directive (Council Directive 2000/53/EC). Amongst a series of measures in relation to scrap vehicles, the Directive stipulates the separate collection of tyres from vehicle dismantlers and encourages the recycling of tyres. In addition, the 1999 Waste Landfill Directive (Council Directive 1999/31/EC) bans the landfilling of tyres, both whole and shredded, by 2006. These two Directives will inevitably impact on the management of tyres throughout the European Union with a trend away from landfill and towards increased re-use, recycling and recovery. Figure 3.11 shows the recent trends in the management of scrap tyres for the Member States of the European Union listed in Figure 3.10 (ETRA 2004). The main routes for management of waste tyres (2002) are landfill, energy recovery, material recycling, re-treading and export (Figure 2.21, Archer et al 2004).

The decline in the use of landfill throughout the European Union has been prompted by the banning of scrap tyres going to landfill under the EC Waste Landfill Directive (1999). However, there has been growing resistance to landfilling of tyres, since tyres do not degrade easily in landfills as they are bulky and take up valuable landfill space while preventing waste compaction. They can also cause instability within the landfill and may 'float' to the surface of the landfill site. In addition, they can be a breeding ground for insects and a home for vermin (Lemieux and Ryan 1993; Bressi 1995). Many landfill sites refuse to take tyres and, consequently, open dumping and stockpiling have increased,

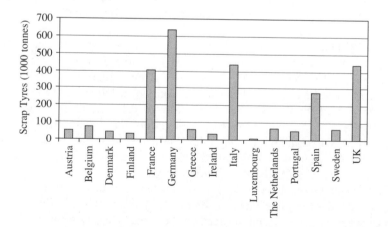

Figure 3.10 *Arisings of scrap tyres in European countries. Source: European Tyre Recyclers Association, Paris, France 2004.*

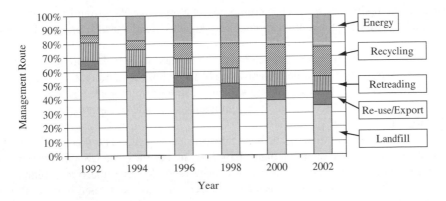

Figure 3.11 *Recent trends in management of scrap tyres for member states of the EU. Source: European Tyre Recycling Association 2004.*

with the potential for accidental fires or arson resulting in high pollution emissions to the atmosphere and water courses. The resultant fires are very difficult to extinguish since they burn at very high temperatures, producing large volumes of thick smoke. Organic compounds identified in the fumes from the combustion of scrap tyres includes a wide range of toxic chemicals such as benzene, chrysene, benzo[b]fluoranthene, benzo[a]pyrene and heavy metals (Lemieux and Ryan 1993; DeMarini et al 1994). Large tyre dump fires generate pyrolytic or semi-pyrolytic conditions near the base of the fire, generating large quantities of oil which can cause contamination of land and water courses. The fires can cause major pollution episodes. For example, the Hagersville tyre dump fire in the USA took 17 days to extinguish and involved 200 fire-fighters. Other tyre fires have been known to smoulder for several years with the constant risk of re-ignition. There has consequently been a trend towards an increase in re-use, re-treading, recycling and energy recovery from scrap tyres.

Re-treading of tyres involves the removal of the tread surface and replacement with a new tread surface. The process involves grinding of the tyre surface to produce a clean surface onto which the new rubber is bonded. The grinding produces a rubber crumb which can be used in other applications. About 80% of re-tread tyres are for passenger car use and 20% for truck and bus tyres, aeroplanes also use re-treaded tyres (ETRA 2004). There is a growing use of re-treaded tyres amongst municipal automotive fleets such as buses, because of their durability, safety and cost-effectiveness. However, there is competition for the expansion of re-treading tyres from the importation of low-cost tyres from abroad. Tyres are produced from petroleum oil and it has been estimated that the re-treading process achieves savings of 4.5 gallons of oil for a car tyre and 15.0 gallons of oil for a truck tyre, compared with the oil used in the production of a new tyre (Ogilvie 1995).

Crumb is fine-grained or granulated tyre material obtained from the shredding and grinding of tyres. The process also involves the removal of the tyre reinforcing fibre and steel. Granulate tyre rubber is produced via mechanical processes at ambient temperature or cryogenic cooling to very low temperatures, which make the rubber brittle, followed by crushing. The main uses for rubber crumb are as childrens' playground surfaces, sports surfaces, carpet backing and other applications, such as absorbents for oils and hazardous and chemical wastes. Grinding rubber to a powder particle size range of less than 1 mm, enables the use of the rubber in applications such as shoe soles and automotive parts.

Rubber reclaim consists of devulcanisation of the rubber by temperature and pressure, using various reclaiming chemicals and solvent treatments to produce a rubber which can be used in low-grade applications. Such applications include cycle tyres, conveyor belts and footwear.

Incineration of tyres utilises their high calorific value, which is typically 32 MJ/kg, to produce energy. Tyre incineration plants have operated in Italy, the USA and Japan. Such plants are dedicated tyre incinerators, but energy recovery from the tyre combustion is widely practised, using cement kilns where the tyres are co-incinerated with fossil fuels. Co-incineration in cement kilns utilises temperatures of up to 1500 °C, as found in cement kilns, to dispose of the tyres, whilst also substituting for the use of fossil fuels, with the consequent saving in costs. The metal and ash components of the tyre are incorporated into the cement clinker. The emissions from tyre waste incineration and the use of waste as scrap tyres in cement kilns are covered by the EC Waste Incineration Directive (2000). Other thermal treatment of scrap tyres occurs in the pulp and paper industry as well as some industrial boilers and furnaces.

An alternative technology for the thermal treatment of tyres is pyrolysis which is the thermal degradation of the tyre in an inert atmosphere. Pyrolysis of tyres has been established for many years but is currently receiving renewed attention. The process produces an oil, char and gas product, all of which have the potential for use. The recovery of oil can be as high as 58 wt% of tyre rubber and has a high calorific value, of the order of 42 MJ/kg. It can be used as a fuel, chemical feedstock or can be added to petroleum refinery feedstocks (Williams et al 1995). The char can be used as a solid fuel, or can be upgraded for use as activated carbon or carbon black. The yield of char from the process is of the order of 35–38 wt% of tyre rubber. The derived gas has a sufficiently high calorific value to provide the energy requirements for the pyrolysis process. In addition, the steel reinforcement can be easily separated from the friable char and recycled back into the steel industry. A number of commercial pyrolysis units have been developed. For example, in the UK, Coalite Tyres Services operate a batch pyrolysis system to process shredded scrap tyres with a planned capacity of 90 000 tonnes per year. The system uses a series of retorts previously used for the retorting of coal, to produce smokeless coal. The retorts are purged with nitrogen to eliminate any oxygen and to ensure pyrolysis conditions are maintained. The retorts are heated initially by fossil fuel burners, but as the pyrolysis of the tyres proceeds, the fossil fuel is replaced by the pyrolysis gas derived from the scrap tyre pyrolysis. The evolved pyrolysis oil vapours and gases pass through a heat exchanger to cool the vapours and then into an oil condensing tank, the non-condensed gases then pass to a scrubbing unit to clean the gases, before passing to the burner. The oil from the condensing tank is pumped to a local oil refinery for the production of chemical feedstocks (Goucher 2001). In addition, other commercial or demonstration units for tyre pyrolysis are available worldwide which are based on different designs. For example, vacuum pyrolysis of tyres has been used in Canada for the production of high-value chemicals, such as limonene, in the derived pyrolysis oil (Pakdel et al 1991). Other commercial plants in operation are located in South Korea and Taiwan (Archer et al 2004).

Other uses for scrap tyres include their use in the automotive industry. For example, battery casings, door facings, flexible tubing, arm rests, etc. They have been used extensively in civil engineering applications in landfill engineering, as lightweight fill material for road construction, as subgrade insulation for roads, highway embankments and as shock absorbers for boats and marinas (ETRA 2004; US Rubber Manufacturers Association 2002)

3.4   Economic Considerations

Waste recycling relies on several inter-related requirements, all of which must be in place for an economically successful scheme to be realised. These are (Waste Management Paper 28, 1992):

1. a secure and stable supply of waste materials;
2. a suitable collection system and transportation to the materials recovery facility;
3. a reliable materials separation and clean-up process to produce the end recycled materials and products;
4. secure and stable markets for the raw materials and products.

Secure and stable supplies of waste are required for the market to invest in the long-term development of recycling process facilities. Over-supply of waste or loss of markets for the end recycled products means that disposal costs for the treatment of the un-recycled waste become a factor in the assessment of the economic appraisal of the project. The generation of waste is a continuous process and treatment and disposal options in the case of fall-off in the demand for recycled products should be insured against by ready alternatives being available. The collection and transportation of the waste to the recycling facility should also be stable and able to undertake preliminary sorting of the waste. The 'bring' and 'collect' schemes have a number of advantages but also disadvantages. For example, 'bring' systems are cheaper, but rely on the public to supply the recycled materials and the recovery rates for such schemes tend to be lower than 'collect' systems and are very dependent on public attitude. 'Collect' systems generally produce higher recovery rates than 'bring' systems, but are more expensive. The materials recycling facility should be able to handle the collected recyclable material and produce a marketable endproduct from the process.

Contamination of the materials is also a factor in determining the economic viability of a recycling scheme. Placing non-recyclable waste into the recycling collection container can mean at best a significant increase in the time required for sorting and a consequent increase in costs, and at worst the scrapping of the whole container load. The level of contamination of dirt, grease, food waste, etc., on the recyclable materials means an increase in the level of clean-up of the materials and a further increase in costs. The level of tolerable contamination is interlinked with the collection system, the processing facility design and the end market recycled product. The price of recycled end products have generally to be competitive with products made from virgin materials, since they are usually of lower quality and therefore are often sold at lower prices. The presence of contaminants means that recycled materials are impure compared with virgin materials and therefore tend to be aimed at lower specification markets, which are also lower value. However, for some products such as recycled paper, which for some grades may be in short supply, higher prices can be obtained than for virgin paper. For example, the demand by consumers for quality recycled writing paper increased the cost of recycled paper above the equivalent paper made from virgin wood pulp. Recycled materials, like any other commodity traded in the market place, are subject to supply and demand with the additional proviso that there will be competition from virgin materials. In addition, some recycled materials are traded internationally and therefore subject to competition from recycling schemes in other countries, which may be subsidised or which produce recycled materials of higher quality. However, international trade is a two-way process and excess supply of recycled materials can be exported. Examples include scrap ferrous metal, aluminium and waste paper. The interlinking of all the stages of the recycling process means that at times of low demand collection systems, particularly where private companies are involved, can sometimes be curtailed.

Costs of recycling are difficult to assess and compare due to differences in the cost factors included in the assessments. For example, some recycling costs are not separated from the general costs of waste management, some schemes include the income from the sale of the recycled materials with the costs of collection while others do not, and some report the costs of collection only (Warmer Bulletin 45, 1995).

Difficulties in the comparison of the costs of recycling have led to the introduction of the terms 'diversion rate' and 'cost difference' to adequately compare costs (Warmer Bulletin 48, 1996). Costs, whether of recycling or waste management, are very dependent on local conditions such as the price of land, labour costs, equipment costs, local taxes and subsidies, treatment and disposal costs, etc. The use of the 'diversion rate' and the 'cost difference' involves the use of ratios to compare directly the costs before and after implementation of a recycling scheme. The 'diversion rate' uses the ratio of the amount of materials recovered as recyclable material to the total amount of waste generated. The 'cost difference' is the cost of waste management with recycling, minus the cost of waste management without recycling, ratioed to the cost of waste management without recycling (Figure 3.12, Warmer Bulletin 48, 1996; White et al 1995).

$$\text{Diversion Rate } (\%) = \frac{A}{B} \times 100$$

$$\text{Difference in Cost } (\%) = \frac{C-D}{D} \times 100$$

A = Amount of material recovered as recycled materials
B = Total amount of waste generated
C = Cost of waste management with recycling
D = Cost of waste management without recycling

Figure 3.12   *Calculation of 'diversion rate' and 'cost difference' for comparison of recycling schemes. Sources: Warmer Bulletin 48, 1996; White et al 1995.*

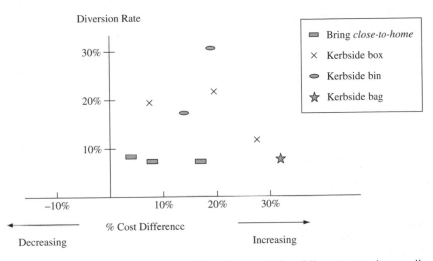

Figure 3.13 *'Cost difference' versus 'diversion rate' for different recycling collection systems. Source: Warmer Bulletin 48, 1996. Reproduced by permission of R.C. Strange.*

Therefore the cost of recycling becomes a percentage increase or decrease in overall waste management costs. Similarly, the effectiveness of the recycling system gives a percentage recovery of diversion of the recyclable materials out of the waste stream, instead of going to final disposal. Thereby, comparison of recycling schemes can be made without the distortion of local variations.

Figure 3.13 shows the use of the 'cost difference' and 'diversion rate' to compare different 'bring' and 'collect' collection systems (Warmer Bulletin 48, 1996). The results show that 'bring' schemes which are close to home achieve low diversion rates of the recyclable materials, but are relatively low in the cost difference compared with waste management schemes without recycling. Higher diversion rates are found for kerbside collection, but significant differences can be seen depending on the container for the recyclable materials, the kerbside bag in particular achieving low rates of recovery or diversion, but resulting in high comparative cost.

A number of factors can influence the diversion rate and cost difference, and these can be identified as internal and external factors. Internal factors include changes in the operation of the collection system or the re-processing facility or an expansion of the recycling scheme. External factors include the market prices paid for the recycled materials and the costs of the final disposal of the waste. Figure 3.14 shows the analysis of the impact of these internal and external factors on the diversion rate and cost difference for a number of recycling schemes throughout Europe (Warmer Bulletin 48, 1996). For a particular programme, two scenarios can be identified:

Scenario A. For a city with low waste management costs and where there is a small market for the recyclable end product, it would be difficult for a recycling programme to produce very much recycled material without drastically increasing the costs of waste management (line **A** in Figure 3.14).

Scenario B. For a city with high costs of waste management and where there is a strong market for the recycled end products, higher rates of recycling (diversion) can be

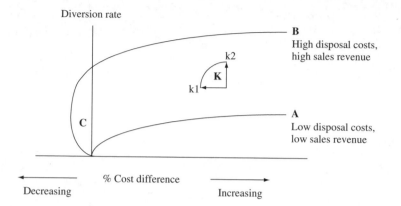

Figure 3.14   *The influence of internal and external factors on the 'cost difference' versus 'diversion rate'. Source: Warmer Bulletin 48, 1996. Reproduced by permission of R.C. Strange.*

obtained for relatively small increases in cost difference, until a point is reached where increasing the rate of diversion by a small percentage results in a large increase in the cost difference (line **B** in Figure 3.14).

Scenario C. Represents the scenario where implementation of a recycling programme results in a *decrease* in the cost difference of waste management for the city (Area **C** in Figure 3.14).

K. Represents the area of potential improvement in cost difference or diversion rate for a city recycling programme (**x**) which might be via decreasing the costs difference whilst maintaining the diversion rate (**k1**) or increasing the diversion rate whilst maintaining the cost difference (**k2**).

The initial capital investment required for a recycling scheme is high, with the added uncertainty as to the constancy of the income. The costs of materials recovery facilities are very high and significant costs are incurred in setting up the household recyclable sorting bins and collection vehicles. In addition, prices obtained for materials recovered from waste have shown fluctuations, influencing the overall economics of recycling. Price fluctuations result from fluctuations in the market price of the recovered material and also with the degree of contamination. To minimise such price fluctuations, the recycling industry is seeking to obtain longer contracts with agreed specifications of the recycled materials. The required specification may be either on the recycled materials sold on to the end user, or indeed placed on the waste collector as specifications for the quality of the incoming waste to the materials recycling facility.

3.5   Life Cycle Analysis of Materials Recycling

Life cycle analysis is the analysis of a product throughout its lifetime to assess its impact on the environment (Figure 3.15, Warmer Bulletin 46, 1995). The concept of life cycle analysis is a useful one in waste management and aids in the determination of whether

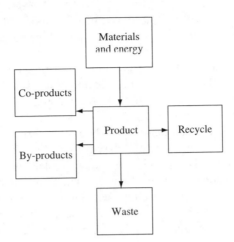

Figure 3.15   *Life-cycle analysis. Source: Warmer Bulletin 46, 1995.*

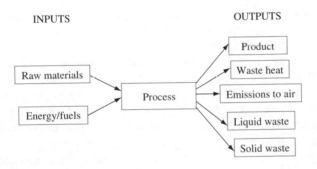

Figure 3.16   *The processes examined in conducting a life-cycle analysis of a product. Source: Warmer Bulletin 46, 1995.*

waste reduction, re-use, recovery or disposal is the best practicable environmental option. Life cycle analysis has been comprehensively applied in the area of solid waste management (White et al 1995; McDougall et al 2001). The analysis quantifies how much energy and raw materials are used and how much solid, liquid and gaseous waste is generated at each process stage of the manufacture of a product (Figure 3.16, Warmer Bulletin 46, 1995). It can be a particularly useful tool in the assessment of the full environmental impact of the production of a product via materials recycling versus product production from virgin materials.

The life cycle analysis of a product involves making detailed measurements during the manufacture of the product from the mining of the raw materials, including the energy inputs used in its production and distribution, through to its use, possible re-use or recycling, and its final disposal. Defining the boundaries of the life cycle analysis and the methodologies involved can vary from analysis to analysis. For example, some analyses have included the environmental impacts in terms of emissions to air, water and on to land when the final waste is disposed of in landfill compared with incineration. Others may include the life cycle analysis of the machinery used in the manufacture of the process. Most studies would normally ignore second-generation impacts, such as the energy required to fire the bricks used to build the kilns which are then used to manufacture the raw material.

The first stage of the analysis involves the collation of the data relevant to the processes involved in the manufacture of the product; the second stage of the life cycle analysis is the interpretation of the data. For example, the production of an aluminium drinks can involves production of aluminium oxide from the raw material bauxite, which involves various stages of reaction with inputs of calcium oxide and sodium hydroxide at high temperatures and pressure. The aluminium oxide is then electrolysed in a solution containing cryolite (calcium aluminium fluoride) and fluorspar (calcium fluoride) and aluminium is produced at the carbon cathode, with carbon dioxide given off at the anode. In addition, fluorine is also given off during the electrolysis process. The aluminium metal is then used in the manufacture of the aluminium can and may involve rolling, pressing and fusing of the metal. Therefore, a full life cycle analysis of the aluminium drinks can should include the mining of the bauxite, the mining of rock salt and limestone used to produce the sodium hydroxide and calcium oxide, respectively, and the energy used in their production. The production of aluminium oxide would involve temperature, pressure and emissions, such as the sludge by-product from the process. The electrolysis stage includes the production of the carbon anodes and cathodes from petroleum, the energy used in the electrolysis process and emissions of carbon dioxide, fluorine and other gases. Finally, the production process from the aluminium involves the production of aluminium sheeting, and rolling and pressing with the associated energy inputs and waste and emissions outputs. In addition, not included in this survey would be the environmental impact of transport at each stage. Recycling of the product can also influence the interpretation of the life cycle of the product. For example, the aluminium drinks can may be recycled, perhaps several times.

Once collected, the data can also be liable to different interpretations, for example, comparison of the environmental impact of high energy demand in one process with high water demand in another. Similarly, it is difficult to compare the environmental impact of, for example, emissions of the pollutant gases, sulphur dioxide, nitrogen dioxide or hydrogen fluoride. To overcome these problems some analyses aggregate the various environmental impacts into categories such as the impact on the ozone layer, the contribution to acid rain or the impact on global warming.

Life cycle analyses comparing recycling with manufacturing of the product from virgin materials have been used to highlight the benefits of recycling. Figure 3.17 shows the comparative life cycle analyses for products derived from recycled material and from virgin materials (White et al 1995; McDougall et al 2001). The life cycle for recycled products includes the assessment of the environmental impacts in terms of energy used and emissions at each of the processes involved in recycling. These would include the separation of the recyclable materials from the solid waste at the waste treatment plant, transportation of the recovered materials to the processing plant and the various processes involved in reprocessing the recovered materials into useable materials. The recycling of a product can be compared with the production based on virgin materials, using life cycle analysis to determine which process has the minimum environmental impact. Table 3.12 shows a comparison of a life cycle assessment for recycled paper versus virgin paper using

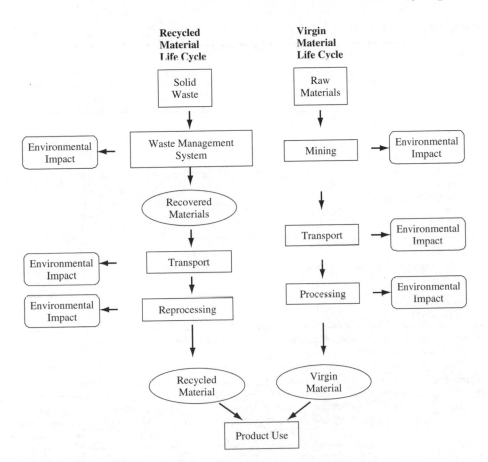

Figure 3.17 *Life-cycle assessment for recycled and virgin materials. Source: White et al 1995. Reproduced by kind permission of Apringer Science and Business Media.*

Table 3.12 *Energy consumption and emissions; comparison of recycled paper with virgin paper production*

Source	Recycled paper (per tonne produced)	Virgin paper (per tonne produced)	Savings (per tonne produced)
1. Energy consumption (GJ)	14.4	22.7	8.3
2. Air emissions (g)			
Particulate	357	4346	3989
CO	383	3165	2782
NO_x	2295	5114	2819
N_2O	280	345	65
SO_x	6054	10868	4814
HCl	0	4	4
HF	0.004	0.01	0.006

Table 3.12   *Continued*

Source	Recycled paper (per tonne produced)	Virgin paper (per tonne produced)	Savings (per tonne produced)
H_2S	0	15	15
HC	4195	6258	2063
NH_3	2.9	3.4	0.5
Hg	0	0.004	0.004
3. Water emissions (g)			
BOD	1	2921	2920
COD	3	25423	25420
Suspended solids	1	1	0
TOC	25	30	5
AOX	0	3	3
Ammonium	0.331	0.876	0.545
Chloride	9	22	13
Fluoride	0.714	1.89	1.18
Sulphide	0	7	7
4. Solid waste (kg)			
Total solid waste	70.6	150.2	79.6

Sources: White et al 1995. Reproduced by kind permission of Apringer Science and Business Media.

Table 3.13   *Energy and emissions savings of recycling versus virgin production*

Material	Process energy saved (GJ/tonne)	Air emissions for recycling (GJ/tonne)	Water emissions for recycling (GJ/tonne)	Solid waste for recycling (kg/tonne)
Paper[1]	8.3	generally lower	generally lower	80 reduction
Glass[2]	3.8	generally lower	generally lower	25 increase
Metal – Fe[3]	13.5	generally lower	generally lower	278 reduction
Metal – Al	156.0	generally lower (except HCl)	generally lower	639 reduction
Plastic (LDPE)[4]	15.4	generally lower (except for CO_2)	few data	93 increase
Plastic (HDPE)[5]	25.6	generally lower	poor data may be higher for recycled	184 increase
Textiles[6]	52–59	no data	no data	no data

[1] Pulp and paper making included.
[2] Process to finish container included. Data for 100% virgin extrapolated as all glass making uses some used material.
[3] Data for tinplate recycling up to production of new tinplate.
[4] Incomplete data for reprocessing of low-density polyethylene.
[5] Incomplete data for reprocessing of high-density polyethylene.
[6] Energy range for woven and knitted wool only.

Sources: White et al 1995. Reproduced by kind permission of Apringer Science and Business Media.

data from Swedish and Swiss sources (White et al 1995). The table covers the energy and materials inputs and the emissions outputs. However, the energy used and emissions from the collecting and sorting of the material and the transport to the reprocessing plant are not included. The estimates also depend on the types of process plant used and the grade of input paper and required grade of output paper. Consequently, energy and emissions data can vary from analysis to analysis. Table 3.13 shows the comparison of the energy and emissions for recycled versus virgin production of a variety of materials (White et al 1995).

References

ACCR. 2000. State of Selective Collection and Recycling in European Cities. Association of Cities and Regions for Recycling in Europe, Brussels, Belgium.

APME. 2004. Good practices guide on waste plastics recycling, a guide by and for local and regional authorities. Association of Plastics Manufacturers in Europe, Brussels, Belgium.

Archer E., Klein A. and Whiting K. 2004. The scrap tyre dilemma: can technology offer commercial solutions. Waste Management World, January, James & James Science Publishers Ltd, London.

Atkinson W., Barton J.R. and New R. 1993. Cost assessment of source separation schemes applied to household waste in the UK. Warren Spring Laboratory, Report No. LR 945, National Environmental Technology Centre, Harwell.

Atkinson W. and New R. 1992. Monitoring the impact of bring systems on household waste in the UK. Warren Spring Laboratory, Report No. LR 944, National Environmental Technology Centre, Harwell.

Atkinson W. and New R. 1993. Kerbside collection of recyclables from household waste in the UK. Warren Spring Laboratory, Report No. LR 946, National Environmental Technology Centre, Harwell.

Aumonier S. and Troni J. 2001. Research Study on International Recycling Experience. DEFRA, UK, HMSO.

Barton J.R. 1989. Processing of urban waste to provide feedstock for fuel/energy recovery. In, Ferrero G.L., Maniatis K., Buekens A. and Bridgwater A.V. (Eds.), Pyrolysis and Gasification, Elsevier Applied Science, London.

Basta N., Fouhy K. and Moore S. 1995. Prime Time for Post-consumer Recycling. Chemical Engineering, February, 31–35.

BGRC Fact File. 1996. British Glass Recycling Company Ltd, Sheffield.

Bressi G. (Ed.) 1995. Recovery of Materials and Energy from Waste Tyres, Working Group on Recycling and waste Minimisation. International Solid Waste Association, Copenhagen, Denmark.

Campbell M.C. 1996. Non-ferrous metals recycling, Recycling Technology Newsletter. Natural Resources Canada, Ottawa, Ontario, Canada.

CEPI. 2002. European List of Standard Grades of Recovered Paper and Board. Confederation of European Paper Industries, Brussels, Belgium.

CEPI. 2003. Special Recycling 2002 Statistics. Confederation of European Paper Industries, Brussels, Belgium.

Council Directive 94/62/EC. 1994. Packaging and Packaging Waste Directive, L 365, 31.12.1994. Official Journal of the European Communities, Brussels, Belgium.

Council Directive 1999/31/EC. 1999. Landfill of Waste. Official Journal of the European Communities, L182, Brussels, Belgium.

Council Directive 2000/53/EC. 2000. End-of-life Vehicles. Official Journal of the European Communities, L269, Brussels, Belgium.

COM (2000–0158) 365. 2000. Proposal for a Directive of the European Parliament and of the Council on waste electrical and electronic equipment. Official Journal of the European Communities, E365E, Brussels, Belgium.

Crittenden B. and Kolaczkowski S. 1995. Waste Minimisation: A Practical Guide. Institution of Chemical Engineers, Rugby.

DEFRA. 2004. Construction and Demolition Waste, 1999 and 2001, Key Facts about Waste and Recycling. Department of Environment, Food and Rural Affairs, UK.

DeMarini D.M., Lemieux P.M., Ryan J.V., Brooks L.R. and Williams R.W. 1994. Mutagenicity and chemical analysis of emissions from the open burning of scrap rubber tyres. Environmental Science and Technology, 28, 136–141.

EAA. 2004. European Aluminium Association, Brussels.

ERRA. 1994. Technology Manual, European Recovery and Recycling Association (ERRA). Reference Multi-material Recovery, Brussels, Belgium.

ETRA. 2004. European Tyre Recyclers Association, Paris, France.

Eunomia. 2003. Eunomia Research and Consulting, Costs for Municipal Waste Management in the EU, Final Report to Directorate General Environment. European Commission, Brussels, Belgium.

European Commission. 2000. Proposal for a Directive of the European Parliament and of the Council on Waste Electrical and Electronic Equipment, COM(2000) 347 final. Office for Official Publications of the European Communities, Luxembourg.

European Commission. 2003. Waste Generated and Treated in Europe, European Commission. Office for Official Publications of the European Communities, Luxembourg.

European Environment Agency. 2002. Review of selected waste streams: sewage sludge, construction and demolition waste, waste oils, waste from coal fired power plants and biodegradable municipal solid waste. European Environment Agency, Copenhagen.

European Environment Agency. 2003. Europe's Environment: the third assessment. European Environment Agency, Copenhagen.

Everett J.W. and Peirce J.J. 1993. Curbside Recycling in the USA: Convenience and Mandatory Participation. Waste Management and research, 11, 49–61.

Gendebien A., Leavens A., Blackmore K., Godley A., Lewin K., Whiting K.J., Davis R., Giegrich J., Fehrenbach H., Gromke U., del Bufalo N. and Hogg D. 2003. Refuse derived fuel, current practice and perspectives. European Commission – Directorate General Environment, Brussels.

Gladding T. 2002. Issues in Environmental Science and Technology, 18, 53–72.

Goucher D.J. 2001. Coalite Tyre Services Pyrolysis Process. Engineering for Profit from Waste, V1, Institution of Mechanical Engineers, London.

Hogg D. and Hummel J. 2002. The Legislative-driven Economic Framework Promoting MSW Recycling in the UK. Eunomia, Bristol.

Holt G. 1995. Opportunities and Barriers to Metals Recycling. Recycling Advisory Unit, National Environmental Technology Centre, AEA Technology, Harwell.

Huston J.K. 1995. Manufacture of newsprint using recycled fibres. In, Technology of Paper Recycling, McKinney R.W.J. (Ed.). Blackie Academic and Professional, London.

Jackson D.V. 1988. A review of the developments in the production and combustion of refuse derived fuel. In, Energy Recovery through Waste Combustion, Brown A., Evemy P. and Ferrero G.L. (Eds.), Elsevier Applied Science, London.

Kuznesof P.M. and VanDerveer M.C. 1995. In, Rader C.P., Baldwin S.D., Cornell D.D., Sadler G.D. and Stockel R.F. (Eds.), Plastics, Rubber and Paper Recycling. American Chemical Society, Symposium Series 609, American Chemical Society, Washington.

Lee M. 1995. Feedstock Recycling: New Plastic for Old. Chemistry in Britain, July, 515–516.

Lemieux P.M. and Ryan J.V. 1993. Characterisation of air pollutants emitted from a simulated scrap tyre fire. Journal of the Air and Waste Management Association, 43, 1106–1115.

McDougall F.R., White P.R., Franke M. and Hindle P. 2001. Integrated Solid Waste Management: A Lifecycle Inventory. Blackwell Science, London.

Making Waste Work. 1995. Department of the Environment and the Welsh Office, HMSO, London.

Manser A.G.R. and Keeling A.A. 1996. Practical Handbook of Recycling Municipal Solid Waste. CRC Lewis Publishers, Boca Raton, US.

Meszaros M.W. 1995. Advances in plastics recycling. In, Rader C.P., Baldwin S.D., Cornell D.D., Sadler G.D. and Stockel R.F. (Eds.), Plastics, Rubber and Paper Recycling. American Chemical Society, Symposium Series 609, American Chemical Society, Washington.

OECD. 2004. Addressing the Economics of Waste. Organisation of Economic Cooperation and Development, ECD, Paris.

Ogilvie S.M. 1995. Opportunities and Barriers to Scrap Tyre Recycling. Recycling Advisory Unit, National Environmental Technology Centre, AEA Technology, Harwell.

Pakdel H., Roy C., Aubin H., Jean G. and Coulombe S. 1991. Environ. Sci. Technol., 25, 1646–1649.

Poll A.J. 1995. Opportunities and Barriers to Glass Recycling. Recycling Advisory Unit, National Environmental Technology Centre, AEA Technology, Harwell.

Project Integra. 2004. Hampshire Waste Services, Otterbourne, Hampshire.

Rampling T.W. and Hickey T.J. 1988. The laboratory characterisation of refuse derived fuel, Warren Spring Laboratory (National Environmental Technology Centre), Report LR 643 (MR), HMSO, London.

RECOUP. 1995. News Sheets, Recycling of Used Plastics (Containers) Ltd (RECOUP), Peterborough.

RECOUP. 1996. News Sheets, Recycling of Used Plastics (Containers) Ltd (RECOUP), Peterborough.

Sadler G.D. 1995. Recycling of Polymers for food use. A current perspective. In, Rader C.P., Baldwin S.D., Cornell D.D., Sadler G.D. and Stockel R.F. (Eds.), Plastics, Rubber and Paper Recycling. American Chemical Society, Symposium Series 609, American Chemical Society, Washington.

US EPA. 2002. Municipal solid waste in the United States: 2000 Facts and Figures. Office of Solid Waste and Emergency Response, US Environmental Protection Agency, Washington.

US EPA. 2003. Municipal Solid Waste in the United States: 2001 Facts and Figures. Office of Solid Waste, US EPA.

US EPA. 2004. Municipal Solid Waste. United States Environmental Protection Agency, Washington, US.

US Rubber Manufacturers Association. 2002. US Scrap Tyre Markets: 2001. Rubber Manufacturers Association, Washington, US.

Warmer Bulletin 43. 1994. Warmer Information Sheet: Papermaking and Recycling. Journal of the World Resource Foundation, Tonbridge, Kent.

Warmer Bulletin. 1995. Technical Brief: Materials Reclamation Facilities. The World Resource Foundation, Tonbridge, Kent.

Warmer Bulletin 45. 1995. The Cost of Recycling: How to ask the Right Questions. White P., Hummel J. and Willmore J., Journal of the World Recycling Foundation, Tonbridge, Kent.

Warmer Bulletin 46. 1995. Warmer Information Sheet: Lifecycle Analysis and Assessment. Journal of the World Resource Foundation, Tonbridge, Kent.

Warmer Bulletin 48. 1996. Affordable Recycling – the Critical Factors. Hummel J., White P.R. and Willmore J., Journal of the World Resource Foundation, Tonbridge, Kent.

Warmer Bulletin 49. 1996. Journal of the World Resource Foundation, High Street, Tonbridge, Kent, UK, May.

Warmer Bulletin 50. 1996. Markets for Waste Paper. Cooper I., Journal of the World Resource Foundation, Tonbridge, Kent.

Warmer Bulletin 59. 1998. Materials Reclamation Facilities. Journal of the World Resource Foundation, March.

Warmer Bulletin 91. 2003. Materials Reclamation Facilities. Journal for Sustainable Waste Management, July.

Warmer Bulletin, 1992 Information Sheet. Plastics Recycling. Warmer Bulletin, 32 February 1992. Journal of the World resource Foundation, Tonbridge, Kent.

Waste Management Paper 28. 1992. Recycling. Department of the Environment, HMSO, London.

White P.R., Franke M. and Hindle P. 1995. Integrated Solid Waste Management: A Lifecycle Inventory. Blackie Academic and Professional, London.

Williams P.T., Besler S., Taylor D.T. and Bottrill R.P. 1995. Pyrolysis of Automotive Tyre Waste. Journal of the Institute of Energy, 68, 11–21.

WRAP. 2002. UK Paper Mills: Review of Current Recycled Paper Usage. Waste and Resources Action Programme (WRAP), Banbury, UK.

WRAP. 2003(a). Survey of Applications, Markets and Growth Opportunities for Recycled Plastics in the UK. Waste and Resources Action Programme (WRAP), Banbury, UK.

WRAP. 2003(b). Recycled Glass Market Study and Standards Review. Waste and Resources Action Programme (WRAP), Banbury, UK.

4

Waste Landfill

Summary

Introduction. Site selection and assessment. Landfill design and engineering, considerations for landfills, operational practice. Types of waste landfilled, inert wastes, bioreactive wastes. Landfill design types, attenuate and disperse landfills, containment landfills, landfill liner materials, landfill liner systems, co-disposal landfills, entombment landfills, sustainable landfills (the controlled flushing bioreactor landfill). Landfill gas. landfill gas migration, management and monitoring of landfill gas. Landfill leachate, leachate management and treatment. Landfill capping. Landfill site completion and restoration. Energy recovery. Old landfill sites.

4.1 Introduction

For many wastes landfill is the largest route for disposal throughout the countries of Europe. Figure 4.1 shows that in many countries across Europe, waste landfill is the dominant disposal route for municipal solid waste. For some countries, including, the UK, Italy, Spain, Finland, Portugal, Italy, Ireland, Turkey, Romania, Poland, Hungary, Bulgaria, Lithuania and Latvia, more than 60% of municipal solid waste is disposed of to landfill (European Commission 2003). In countries where waste landfill is important as a major route for waste disposal, for example, Greece, Ireland, Italy, Portugal, Spain and the UK, there is likely to be a trend towards smaller numbers of larger landfills (Eunomia 2003). This is attributed to the higher costs of landfill resulting from the imposition of the EC Waste Landfill Directive (1999), leading to higher costs. This would tend to force the

Waste Treatment and Disposal, Second Edition   Paul T. Williams
© 2005 John Wiley & Sons, Ltd   ISBNs: 0-470-84912-6 (HB); 0-470-84913-4 (PB)

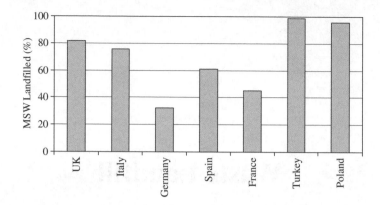

Figure 4.1 *Municipal solid waste landfilled in selected European countries. Source: European Commission 2003.*

smaller sites to close because of the advantages of economies of scale in reducing costs for the larger landfills (Eunomia 2003). For other wastes, including hazardous waste (Figure 4.2), industrial waste (Figure 4.3) and sewage sludge (Figure 4.4), waste landfill is a significant, if not a dominant disposal route (European Commission 2003; European Environment Agency 2002(a); DEFRA 2004).

The major advantage associated with landfilling of wastes is the low cost of landfill compared with other disposal options and the fact that a wide variety of wastes are suitable for landfill. It should also be remembered that, ultimately, many other waste treatment and disposal options require the final disposal route for the residues to be landfill. For example, incineration bottom and flyashes are disposed of in landfill sites. Also, in many cases, there is a strong interaction of the minerals extraction process with the infilling of the void space with waste. Extraction of rock materials has produced and continues to produce large holes in the ground which at some stage require infilling. Mineral extraction plans usually include the restoration of the site after completion of the extraction. Infilling

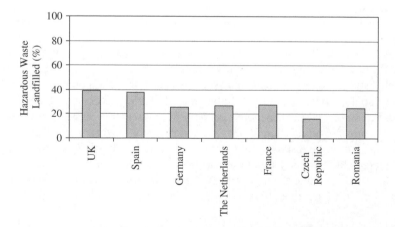

Figure 4.2 *Hazardous waste landfilled in selected European countries. Source: European Commission 2003.*

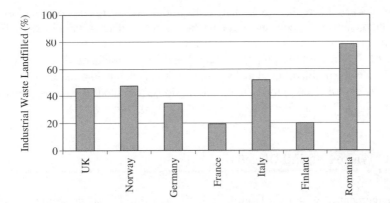

Figure 4.3   *Industrial waste landfilled in selected European countries. Source: European Commission 2003.*

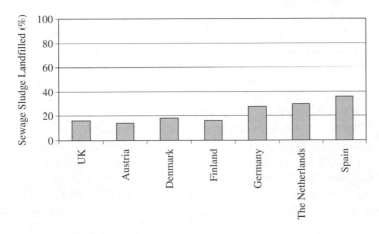

Figure 4.4   *Sewage sludge landfilled in selected European countries. Sources: European Environment Agency 2002(a); DEFRA 2004 (UK).*

of the mineral workings by waste is therefore an economical advantage for the site developer. The collection and utilisation of landfill gas as a fuel for energy generation is also an advantage. However, landfill achieves a lower conversion of the wastes into energy with about one-third less energy recovery per tonne from landfill gas than incineration. This is mainly due to the conversion of the organic materials in the waste into non-combustible gases and leachate and general losses from the system. Increasingly, there is an emphasis on regarding the modern landfill as a fully designed and engineered process with high standards of management.

There are however, some disadvantages with landfill. Older sites which are, in some cases, still under current use or have long been disused, were constructed before the environmental impacts of leachate and landfill gas were realised. Many of these sites are now sources of pollution with uncontrolled leakages. Landfill gas in particular can be hazardous since the largest component, methane, can reach explosive concentrations. This problem is emphasised when it is realised that many of the older sites were constructed close to areas of housing, or sometimes housing sites have been built on disused landfill sites. Therefore, all landfill sites are required to be monitored for landfill gas and the gas from operational sites must be controlled via proper venting. Landfill gas methane is also a 'greenhouse gas', leading to the problems of global warming, but with about 30 times the effect of carbon dioxide. The contribution to total methane emissions from landfill gas has been estimated by the European Commission to be 32% (COM(97) 105 Final 1997).

4.2   EC Waste Landfill Directive

The management of waste via landfill is dominated throughout the European Union by the EC Waste Landfill Directive (1999) and the implementation of its measures and targets. The background to the implementation of the Waste Landfill Directive is related to concerns over the environmental impact of waste landfills and the desire of the European Commission to increase the level of waste re-use, recycling and recovery, throughout the European Union. Earlier reviews of the management of waste landfills in the European community showed it to be very variable in terms of siting policies, lining requirements, leachate control requirements and approaches to waste input (Hjelmar et al 1995). Consequently, one of the aims of the Waste Landfill Directive was to harmonise standards throughout the EU for waste landfill facilities on the basis of a high level of environmental protection. In addition, the European Commission regards landfilling of waste as the least favourable option, due to the fact that landfilling does not make use of waste as a resource and may result in substantial negative impacts on the environment. The European Commission have identified 'emissions of hazardous substances to soil and groundwater, emissions of methane into the atmosphere, dust, noise, explosion risks and deterioration of land' as potential significant environmental impacts from the landfilling of waste (COM(97) 105 Final 1997).

Therefore, the Waste Landfill Directive (1999) introduced stringent operational and technical requirements for waste landfills to reduce pollution of surface water, soil and air, as well as the risk to human health from the practice of landfilling of waste. The Directive contains significant measures, procedures and operational detail to ensure the minimal environmental impact of waste landfill on the environment. Initially, an environmental impact assessment is required under the provisions of another EC Directive: Assessment of the Effects of Certain Public and Private Projects on the Environment (85/337/EEC 1985). The characteristics of the waste accepted by the landfill site are documented and include visual inspection of the waste and the types, origin and producer and collector of the waste. In the case of hazardous waste, the precise location where the waste was deposited on the site should be documented.

To contain the landfill gas and leachate, a barrier system which lines the landfill and acts as a barrier to the outside environment is required, with minimum stipulations for the minimum permeability and thickness of the landfill liner system, depending on the category of waste permitted to be deposited in the landfill site. The landfill gas must be collected via a system of porous pipework within the landfill site and then treated and used or flared. The leachate is collected and treated to remove pollutants to environmentally acceptable levels. Throughout the lifetime of the landfill site, which can be several decades, sampling and analysis of the landfill gas and leachate and gases are required, including monitoring outside the landfill site to assess any impact on the surrounding environment.

Table 4.1 shows the monitoring requirements for waste landfill sites in Member States of the European Union for the operating and post-closure period of a landfill site (Waste Landfill Directive 1999). The samples of leachate must be collected at representative points throughout the landfill site and analysed separately for volume and composition. The detailed compositional analysis of the leachate will be determined by the regulatory authority, based on the type of wastes landfilled. The landfill gas emanating from the biodegradation of the organic waste in the landfill site must be analysed for the main landfill gas components of methane, carbon dioxide and oxygen, together with other gases, at representative sampling points throughout the landfill. Table 4.1 also shows that groundwater monitoring is required to determine any impact from the possible contamination of leachate. Consequently, sampling and analysis of groundwater is required at a measuring point in the groundwater inflow region and at two points in the outflow region. In some cases meteorological data may also be required, for example, where the potential for build-up of leachate in the landfill site is monitored by the calculation of the balance of water into and out of the site. In such cases, the volume of rainwater falling on the site, the ambient temperature, the force of the prevailing wind, the evaporation rate and the atmospheric humidity may also be required on a daily basis (Waste Landfill Directive 1999).

One particular area of environmental impact that the European Commission addressed through the Waste Landfill Directive, was the issue of landfill gas emissions from the biodegradation of biodegradable municipal solid waste in landfills. Biodegradable wastes are defined as wastes, such as food and garden waste, paper, cardboard, textiles wood, etc., that are degraded over long periods of time by various aerobic and anaerobic bacteria to produce a liquid leachate and landfill gas. The organic proportion of municipal solid waste comprising the food and garden waste represents about 25–55% of European municipal solid waste. However, when other potentially biodegradable wastes such as paper and board, wood, textiles, etc., are included, then the biodegradable fraction increases to much higher figures (Table 4.2, European Environment Agency 2002(b)). The landfill gas is

Table 4.1   *Monitoring requirements for waste landfills in the European Union*

Monitoring	Operating phase frequency	After-care phase frequency
Emission data		
Leachate volume	Monthly	Six-monthly
Leachate composition[1]	Quarterly	Six-monthly
Volume and composition of surface water	Quarterly	Six-monthly
Potential gas emissions and atmospheric pressure for; CH_4, CO_2, H_2, O_2, H_2S etc.	Monthly	Six-monthly
Groundwater data		
Level of groundwater	Six-monthly	Six-monthly
Groundwater composition	Site-specific frequency	Six-monthly

[1] The compositional data are dependent on the types of waste landfilled.

Source: Waste Landfill Directive 1999.

Table 4.2 *Baseline data for biodegradable municipal solid waste in selected European Union countries*

Country	Year	MSW (1000 tonnes)	Biodegradable MSW (%)	Biodegradable MSW landfilled (%)
Austria	1995	2644	66	30
Belgium	1995	5014	86	56
Denmark	1995	2591	99	18
Finland	1994	2100	90	97
France	1995	34 700	80	62
Germany	1993	40 017	72	70
Greece	1990	3000	90	100
Ireland	1995	1550	69	94
Italy	1996	24 524	88	93
Portugal	1995	3884	85	99
Spain	1995	14 914	78	76
Sweden	1994	3200	83	36
The Netherlands	1994	8161	89	35
UK	1995	29 000	74	90

Source: European Environment Agency 2002(b).

composed of mainly methane and carbon dioxide, which are greenhouse gases contributing to global warming. The European Community has a strategy, in relation to international climate change agreements, to reduce the emissions of greenhouse gases. Consequently, in order to reduce the production of methane and carbon dioxide emissions from landfill sites, the Waste Landfill Directive sets targets to reduce the amount of biodegradable waste sent to landfill to 75% of the 1995 levels by 2006, 50% of 1995 levels by 2009 and 35% of 1995 levels by 2016 (Waste Landfill Directive 1999). Countries that landfilled more than 80% of their municipal solid waste at the 1995 target date are allowed to extend the deadlines by four years (Figure 4.5, Waste Landfill Directive 1999; European Environment Agency 2002(b)). Some difficulties arise in measuring the attainment of targets for individual countries since, in some cases, municipal solid waste data for the 1995 baseline is not available, particularly that fraction of the waste designated as bio-degradable. In addition, the definition of municipal solid waste, the different types of waste collected by 'municipalities' across Europe and also which categories of waste are included in the statistical data, vary between European countries (European Environment Agency 2002(b)). However, the proportion of municipal solid waste which is regarded as biodegrad-able has been estimated by the European Statistical Agency, Eurostat (Table 4.2, European Environment Agency 2002(b)). It is clear that a high proportion of the waste is regarded as biodegradable and that, for most countries, the main route for disposal of biodegradable municipal solid waste in 1995 was via landfill. Other countries, such as Austria, Denmark and the Netherlands have, in effect, already met the Waste Landfill Directive targets on the diversion of biodegradable municipal solid waste away from landfill. Several European countries, including Germany, Finland and France have already introduced limits or guidelines for biodegradable wastes going to landfill (COM(97) 105 Final 1997). The Directive also states that the diversion of biodegradable waste away from landfill should

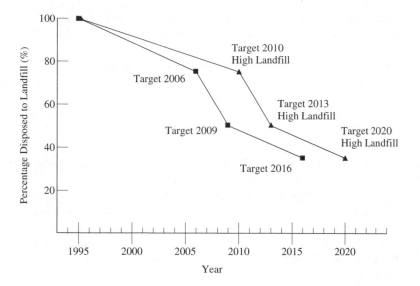

Figure 4.5 *Percentage of biodegradable municipal waste produced in 1995 that can be consigned to landfill. Sources: Waste Landfill Directive 1999; European Environment Agency 2002(b).*

aim to encourage the separate collection of biodegradable waste and to increase recovery and recycling.

Waste landfills in Member States of the European Union are categorised into three different types of landfill: landfills for hazardous waste; landfills for non-hazardous wastes; and landfills for inert waste. Each type of designated landfill can only accept the particular waste for which it is designated. Hazardous waste landfills can therefore only accept hazardous waste as defined in the European Waste Catalogue (Commission Decision 2000). Similarly, non-hazardous waste landfill sites are only permitted to accept non-hazardous waste which, again, are defined as those waste categories listed in the 2000 European Waste Catalogue, which are not hazardous. Within the many categories of waste listed in the Waste Catalogue, those which are regarded as hazardous are clearly noted, and consequently, those that are not highlighted are regarded as non-hazardous. Inert landfill sites are only permitted to accept inert waste.

To control the management of waste landfills throughout the European Union, a system of permits for each landfill site is required under the provisions of the Waste Landfill Directive. Since landfilling of waste is included in the 1996 EU Directive on Integrated Pollution Prevention and Control (IPPC) (Council Directive 96/61/EC), it is a designated IPPC process and consequently, is covered by the IPPC permitting process. The permit is issued and monitored by the 'competent authorities' of each Member State of the EU, such as the UK Environment Agency, the German Federal Environment Agency (Umweltbundesamt), the Danish Environmental Protection Agency, etc. These Member State agencies are supported by the European Environment Agency.

Since landfills are only permitted to take certain categories of waste, namely, hazardous, non-hazardous and inert wastes, the permits issued by the regulatory authority in each

Member State must contain a description of the types and total quantity of waste to be deposited. Consequently, a waste acceptance procedure is required which classifies the wastes acceptable to the different categories of landfill, including standard limit values and a system of sampling and analysis of wastes. Full documentation of the waste input is required, including a register of the quantities and characteristics of the waste deposited indicating origin, date of delivery and identity of the producer or collector, in the case of municipal solid waste. In the case of hazardous waste, a register of the precise location of the waste on the landfill site is required. If the waste is turned away from the site for any reason, the site operator is required to inform the relevant regulatory authority.

The regulatory authority in each Member State of the EU issuing a permit to operate the landfill site, will also require information on the capacity of the site, a description of the site, including its hydrogeological and geological characteristics, and the proposed methods of pollution prevention and control. In addition, the permit must state the proposed operation, monitoring and control plan and the plan for closure of the site and aftercare procedures. The site operator must report on the types and quantities of waste landfilled, and the results of the environmental monitoring programme, to the relevant regulatory authority, at least once per year. Where significant variations in the environmental affects of leachate or landfill gas occur, the operator must inform the regulatory authority. Any costs involved in remediation designated by the regulatory authority will fall on the site operator.

After the site is closed, the obligations of the landfill site operator do not end. The monitoring of the site is carried out after closure, for example, at six-monthly intervals, until the regulating authority agrees that stabilisation of the site has occurred and the landfill poses no hazard to the environment. Waste landfills represent a long-term process, with potential financial implications for the operator in terms of waste stabilisation in the landfill and aftercare monitoring and control of emissions. Consequently, the site operator is required to demonstrate financial security, to ensure that commitments to safeguard the environment are in place for the future.

A key area which will impact on the waste landfill industry is that the Waste Landfill Directive states that waste must be treated before it is landfilled. Treatment is defined in the Directive as the physical, thermal, chemical or biological processes, including sorting, that change the characteristics of the waste in order to reduce its volume or hazardous nature, to facilitate its handling or enhance recovery. Several countries throughout Europe, particularly, Austria and Germany, but also Italy, France, Belgium and the Netherlands, use Mechanical–Biological Treatment (MBT) or Biological–Mechanical Treatment (BMT) processes as a pre-treatment of municipal solid waste (Juniper 2004). The aim of such treatments is to minimise the environmental impact of the end disposal process through the recovery of metals and, in some cases, energy. The 'biological' part of the MBT or BMT process obviously can only apply to the organic fraction or biodegradable fraction of the waste and usually takes place through composting or anaerobic digestion in closed vessels. MBT has a mechanical sorting first step where the incoming municipal solid waste is crushed and screened to recover the recyclable material, followed by biological degradation through anaerobic digestion or closed-vessel composting to produce a humus-like material, which is usually landfilled, or may also be used as a refuse-derived fuel. BMT, on the other hand, is where the biological part of the process is the first step, involving drying/composting followed by mechanical sorting of the biodegraded waste to produce recyclable material and refuse-derived fuel. The organic residue from MBT and BMT processes requires a maturation stage to stabilise the residue. Figure 4.6 shows a schematic diagram of typical MBT and BMT processes (Smith et al 2001; Juniper 2004). Since there are a number of process configurations which combine biological and mechanical steps in the processing of the waste, a general term of 'MBT' is used to refer to all such integrated mechanical and biological processes.

MBT is not regarded as a disposal method, or a single technology, but a mixture of integrated processing operations. The biodegradation process of the organic waste, either through MBT or BMT processes, produces an inert stabilised compost-like organic residue. The residue formed from the MBT processing is approximately 25% of the original waste. A further advantage of the residue when it is landfilled is that it can be compacted to very high densities in the landfill site, increasing the potential void space in the landfill

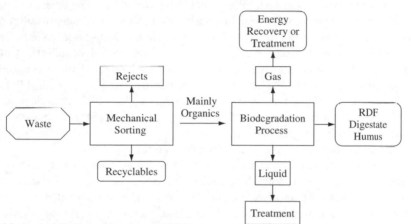

Mechanical-Biological Treatment (MBT)

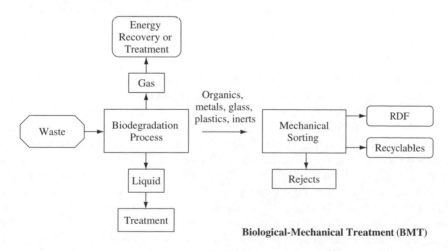

Biological-Mechanical Treatment (BMT)

Figure 4.6 *Schematic diagrams of the MBT and BMT processes. Sources: Juniper 2004; Smith et al 2001. Reproduced with the permission of Juniper Consultancy Services Ltd, Uley, UK.*

and reducing settlement. However, the residue is of poor quality and cannot be used in agricultural or horticultural applications (Smith et al 2001). If compost for agricultural or horticultural use is required, then the municipal solid waste would be source segregated to isolate a clean suitable putrescible material. MBT processes produce a poor quality organic residue since it is derived from a non-source segregated waste. Consequently, the residue is either landfilled or incinerated as refuse-derived fuel. There are some uses for the residue in applications such as landfill cover material and in land restoration (Waste Not Want Not 2002). The processes of MBT do, however, reduce the potential of the residue to produce landfill gas in a waste landfill since the residue has already been stabilised through the biodegradation processes which took place in the controlled MBT process. The biodegradability of the residue is not completely removed, but it is drastically reduced, in that the stabilised residue from MBT processes has been shown to produce only 10% of the landfill gas and 10% of the leachate which would have been generated by the untreated waste (Smith et al 2001; Waste Not Want Not 2002). The level of reduction being dependent on the amount of time the waste is subject to the biodegradation of the MBT processes. The time required for the biodegradation and maturation stage before the residue is stabilised may be from 40 days to several months (Smith et al 2001). Coupled with pre-treatment is the need to reduce the quantities of biodegradable waste going to landfill under the targets set by the European Commission in the Waste Landfill Directive (1999). The fact that the MBT residue is still biodegradable and capable of producing landfill gas, means that the waste would still be regarded as 'biodegradable' under the Waste Landfill Directive targets of reducing the amount of biodegradable waste going to landfill. The residue has, however, been reduced in volume.

If the residue is required for use as a refuse-derived fuel, then further processing is usually required (Waste Not Want Not 2002). The refuse-derived fuel would have a higher calorific value than unprocessed municipal solid waste, since the metals, glass and inert objects would have been removed and the liquid fraction of the residue would have been reduced.

4.3   Site Selection and Assessment

The selection of a site for a waste landfill depends on a wide range of criteria, including the proximity of the site to the source of waste generation, the suitability of access roads, the impact on the local environment of site operations and the geological and hydrogeological stability of the site. Before a permit to operate the landfill site is issued by the regulatory or 'competent authority' of the Member States of the European Union, the site operator has to satisfy the authority that the location chosen for the landfill site will have no significant impact on the environment. The EC Waste Landfill Directive (1999) sets out the location requirements on which the landfill operator has to report, to the competent regulatory authority, to ensure that the proposed landfill does not pose a serious environmental risk. These include the distances from the boundary of the site to, for example, residential areas, waterways and other agricultural or urban sites. The operator should report on the geological and hydrogeological conditions of the area and the risks of hazards such as flooding, landslides and even avalanches.

The main aim of the landfill site assessment investigation is the identification of the possible pathways and receptors of landfill gas and leachate in the surrounding environment and the environmental impact of site operations (Waste Management Paper 26B, 1995; Tchobanoglous and O'Leary 1994). Site assessment involves appraisal of geological and hydrogeological conditions at the site. This may include the use of existing surveys, aerial photography, boreholes, geophysical investigations, geological mapping and sampling, etc. The information allows the assessment of soil and bedrock grain sizes, mineralogy and permeabilities and ground water levels. In addition, the previous use of the site, meteorological data, transport infra-structure and planning use designations as well as the planning strategy of the area, would also be assessed. A topographical survey is undertaken to calculate the available void space and therefore the waste capacity of the site. Since daily, intermediate and final covering materials will be required extensively in the operation of the site, the availability of these materials in natural form should be assessed. At an early stage, background levels of water and air quality may be taken to assess the impact of the site in the long-term.

For large landfill sites an Environmental Assessment is also required to determine the impact on the environment. An Environmental Assessment is required under the EC Directive on the Assessment of the Effects of Certain Public and Private Projects on the Environment (EC Directive 85/337/EEC). The Directive requires environmental assessments to be made for certain prescribed processes, which includes large landfill sites and sites in environmentally sensitive areas. Environmental Assessment involves a description and an assessment of the direct and indirect effects of the project on human beings, fauna, flora, soil, water, air, climate and landscape, material assets and the cultural heritage (Petts and Eduljee 1994; Energy from Waste: Best Practice Guide 1996).

4.4   Considerations for Landfills

A waste landfill is a major design and engineering project and requires a number of considerations to be made as part of the process:

Final landform profile   The profile of the final landform is a key factor in design in that it dictates the after-use of the site, the waste capacity of the site and settlement of the site after completion and landscaping. Final landform gradients after emplacement of the capping material would normally be between 1 in 4 and 1 in 40, depending on the final use for the site to, ensure adequate safety of the steep slopes and a minimum gradient for suitable drainage (Waste Management Paper 26E, 1996). However, steep slopes greater than 1 in 10 may require control to offset erosion of the site.

Site capacity   The capacity of the site is clearly a key factor in site design and the determination of how much waste can be accommodated in the site depends on waste density, the amount of intermediate and daily cover, the amount of settlement of the waste during the operation of the site, and the thickness of the capping system (Waste Management Paper 26B, 1995).

Waste density   The density of the waste within the landfill will vary depending on the degree of pre-compaction of the waste before emplacement, the variation in the components within the waste, the progression of biodegradation, the amount of daily and intermediate cover and the mass of overlying waste. The degree of pre-compaction of the waste influences the amount of waste that can be accepted into the landfill, and also influences to a marked degree the amount of settlement of the landfill (Tchobanoglous and O'Leary 1994). Typical waste densities range from 0.65 to 0.85 tonnes/m^3, although different types of waste may reach densities as low as 0.4 tonnes/m^3 or up to 1.23 tonnes/m^3, depending on the amount of biodegradable and inert waste present. Inert wastes have higher densities, typically about 1.5 tonnes/m^3 (Waste Management Paper 26B, 1995).

Settlement   Settlement of the waste in the landfill occurs initially due to physical rearrangement of the waste soon after emplacement. As the biological, physical and chemical degradation processes proceed, further settlement occurs from overburden pressure due to compaction by its own weight. Typical long-term settlement values for municipal solid waste are 15–20% reduction, although values of up to 40% have been reported where there is a high organic content in the waste (Waste Management Paper 26B, 1995; Tchobanoglous and O'Leary 1994). Settlement can take place over periods of time up to 50 years, but the major settlement period (up to 90%) occurs within the first five years of the final emplacement of the waste (inert wastes, which do not significantly biodegrade and tend to be more dense than municipal solid waste, have low values of settlement (Waste Management Paper 26B, 1995; Sharma and Lewis 1994)).

Three main stages of settlement can be identified and these are initial compression, primary compression and secondary compression (El-Fadel and Khoury 2000). Initial compression is virtually instantaneous and is due to compaction of the void space and particles caused by the compression of the overlying waste and the compaction vehicles. Primary compression is due to the dissipation of pore water and gases from the void space and usually takes around 30 days. The secondary compression may take many years and is due to waste creep and to the biodegradation processes of the waste which produce leachate and landfill gas and thereby reduce mass and volume. The amount and rate of landfill settlement is related to a range of different factors including the initial void volume of the waste, the composition of the waste and thickness of the waste in the landfill, where deeper landfills exhibit faster rates of settlement (El-Fadel and Khoury 2000).

Materials requirements   The containment landfill requires various materials for site development, operation and restoration. Included in these requirements are the natural fill materials such as clay, sand, gravel and soil, which are used in various applications such as sand for lining the site to protect the liner materials, clay to provide an additional low permeability layer to the site, gravel for drainage for leachate collection, clay for capping material and restoration soils. The availability of such materials on-site increases the ease of operation and also reduces costs (Waste Management Paper 26B, 1995). Daily cover is also required which controls wind-blown litter, scavengers, fires and odours. The daily cover material may be soil, garden waste or alternative materials such as re-useable blanket material and inert wastes (Haughey 2001). These materials are required to be imported into the site and the close location of the source is a consideration to reduce costs. In addition, the daily cover uses valuable void space, reducing the potential income and the operating lifetime of the site.

Drainage   Drainage of the rainwater falling on the site is required to ensure that excessive water does not infiltrate the waste directly or from run-off from surrounding areas.

Cut-off drains both around and inside the site will keep the waste from becoming too wet and increasing the production of leachate (Waste Management Paper 26B, 1995; Tchobanoglous and O'Leary 1994). The effective capping of finished sections of the landfill site at the end of certain phases of the landfill operation, will also be necessary to reduce the precipitation falling on the site from entering the mass of waste. The capping of the landfill involves the use of a barrier system of natural and synthetic materials to control the inlet and outlet of water and gases. The capping acts in a similar way to the barrier system at the base and sides of the landfill.

Operational Practice   The daily operation of the landfill, including traffic movement, cover material storage and laying, security, etc. are major considerations for the design of a landfill site. The typical modern landfill site consists of a secure, fenced, landscaped site with access routes for waste transport vehicles. The sequence of operations for an incoming waste vehicle may include the weighing of the lorry on a weighbridge, document inspection and waste inspection. Once cleared, the lorry would move to the waste disposal area where the waste is tipped, the wheels of the lorry are cleaned and the lorry is then weighed out of the site to determine the weight of waste deposited. The driver collects any documentation. The tipped waste is compacted using specially designed compacting vehicles and a daily cover of soil or alternative material is added at the end of each working day (Pescod 1991–93).

The landfill site is normally developed, operated and restored in a series of phases to allow the most beneficial use of the site area, a method used throughout Europe and North America (Waste Management Paper 26B, 1995; Tchobanoglous and O'Leary 1994; Sarsby 1995). Each phase may last from 12 to 18 months (Waste Management Paper 26B, 1995). This serves to concentrate the waste disposal operation in specific areas and spreads the costs of leachate and landfill gas-control systems throughout the lifetime of the operation. Therefore, within a typical site, one part of the site might be being prepared with liner material and leachate and landfill gas collection systems in order to accept waste, while another part might be being filled with waste, yet another part could be being capped while part may be fully capped and restored. The phases are separated by the use of separation bunds. The phase would be a sub-area of the site, whereas cells (described below) are sub-divisions of the phase. Each phase is generally completed from preparation through to capping and restoration but with a temporary unrestored open face (Waste Management Paper 26B, 1995).

Within each phase are cells, which are sub-divisions of the phase. The size of each cell depends on the volume of waste being deposited but they are typically between 3 and 5 m, whilst other influencing factors include rainfall and the absorptive capacity of the waste, and thereby the minimisation of leachate production, and additionally the number of vehicles bringing waste to the site. The cell delineates the tipping area for a week or month. The waste is deposited, usually on shallow working faces, and compacted using a variety of specially designed compacting vehicles which break up and compact the waste. The daily cover is added at the end of the working day and is used to prevent windblown litter and odours, to prevent fires, deter scavengers, birds and vermin and to improve the visual impact of the site. The cover material usually consist of about 15 cm of soil; alternative cover materials such as shredded green waste and re-useable plastic sheeting and geotextiles have been used. As the number of cells in a phase increases, intermediate cover material is deposited, this is about 1 m thick and is used to minimise the ingress of rain water and thereby increase the production of leachate.

4.5   Types of Waste Landfilled

The Waste landfill Directive stipulates the types of waste allowed in each category of landfill site as hazardous, non-hazardous and inert. The lists of wastes acceptable at each designated and permitted landfill site are part of the permitting process of the regulatory authorities for each country. The general principle underlying the acceptance of a particular waste at a landfill site are that the composition, leachability, long-term behaviour and general properties of the waste to be landfilled should be known as precisely as possible. The acceptance criteria for a landfill would consequently be based on standardised waste analysis and characteristic and properties limit values. All wastes are listed in the European Waste Catalogue (Commission Decision 2000/532/EC 2000) which replaced earlier lists of non-hazardous and hazardous wastes in one unified document (described in Chapter 1). The Catalogue contains categories or 'chapters' of waste each with a two-digit code, each chapter contains sub-chapters and each sub-chapter lists the specific waste categories. Consequently, each category of waste has a specific six-digit code.

Hazardous waste categories within each sub-chapter are highlighted by an asterisk, some of which are shown in Table 2.16. Hazardous waste is defined as waste which is dangerous or difficult to keep, treat or dispose of and may contain substances which are corrosive, toxic, reactive, carcinogenic, infectious, irritant, harmful to human health or which may be toxic to the environment. The 2000 European Waste Catalogue replaced earlier lists and properties of hazardous waste, including a 1991 European Commission amendment (91/689/EEC) to the 1975 Waste Framework Directive (Council Directive 75/442/EEC; Waste Framework Directive 1975) and by a 1994 Council Decision which listed over 200 different types of hazardous waste (Council Decision 94/904/EC 1994). All such wastes are now included in the European Waste Catalogue. Some types of hazardous waste are not permitted to go to landfill sites, even hazardous waste landfill sites, and these include liquid waste, flammable waste, explosive or oxidising wastes, infectious clinical or hospital waste and used tyres. Hazardous wastes are deposited into designated and permitted 'hazardous waste landfill sites' with a high specification containment barrier liner system to contain the derived leachate and landfill gas and to allow for their collection and treatment.

Non-hazardous wastes are also listed in the European Waste Catalogue and make up the vast majority of wastes listed in the catalogue. Non-hazardous waste includes municipal solid waste and a wide range of industrial wastes, such as organic and inorganic wastes, provided that they are non-hazardous. A significant feature of non-hazardous wastes are that many, including most significantly municipal solid waste, are 'bioreactive wastes' which undergo biodegradation within the landfill environment. Non-hazardous wastes are permitted to be deposited into 'non-hazardous waste landfill' sites. However, stable, non-reactive hazardous wastes, for example, those that are solidified or vitrified, are also permitted to be deposited into non-hazardous waste landfills, provided that their leaching behaviour is equivalent to the general category of non-hazardous waste. There is also the requirement that such non-reactive hazardous wastes are deposited in cells within the landfill that do not contain biodegradable wastes. The site requires a containment barrier liner system to control, contain and collect and then treat the produced leachate and landfill gas.

Inert waste is defined in the Waste Landfill Directive (1999) as waste that does not undergo any significant physical, chemical or biological transformations. In addition, inert waste will not dissolve, burn or otherwise physically or chemically react, biodegrade or adversely affect other matter with which it comes into contact in a way likely to give rise to environmental pollution or harm human health. The total leachability and pollutant content of the inert waste and the ecotoxicity of the leachate must be insignificant and, in particular, not endanger the quality of surface water and/or groundwater (Waste Landfill Directive 1999). Inert wastes are therefore deemed not to pose a significant environmental risk either now or in the future since, as their name suggests, they are wastes of no or low reactivity. As such, inert wastes do not undergo significant chemical, biological or physical degradation to yield polluting materials and consequently, only the minimum barrier containment system is required. Typical inert wastes include bricks, glass, tiles and ceramic materials, concrete, stones, etc.

It should also be noted that, under the Waste Landfill Directive, only waste that has been subject to treatment is allowed to be landfilled. Treatment applies to hazardous waste and non-hazardous waste, but not always to inert waste, and has the aim of reducing the quantity of waste landfilled and therefore reducing the hazards to human health and the environment.

4.6   Landfill Design and Engineering

Landfill disposal is seen in many respects as the bottom rung of the hierarchy of waste disposal options, when considering the concept of sustainable waste management. However, the modern landfill site has developed from a site used for merely dumping of waste with little or no thought, to a site which is an advanced treatment and disposal option, designed and managed as an engineering project.

Key to the design of a waste landfill site is the containment barrier system at the base, sides and eventually the cap of the waste landfill, which contains the products of the biodegradation and physical and chemical degradation processes of the waste. The minimum requirements of the containment barrier system used are described in the European Commission Waste Landfill Directive (1999). The type and requirements of the barrier system are dependent on the type of waste permitted to be deposited in the landfill site which, in turn, is dictated by the permitted classification of the landfill. Under the EC Waste Landfill Directive waste landfills are categorised into:

- waste landfills for hazardous waste;
- waste landfills for non-hazardous waste;
- waste landfills for inert waste.

The EC Waste Landfill Directive (1999) sets out the minimum requirements for the barrier liner system for each type of waste landfill. The requirements are underpinned by the need to prevent pollution of the soil, groundwater or surface water and to ensure efficient collection of the leachate. In addition, the landfill liner system is designed to control the accumulation and migration of landfill gas. Protection of the environment from the impact of waste landfilling is through the combination of a geological barrier and a liner system to the bottom, sides and cap of the landfill. The system is designed for the operational, active and post-closure phases of the landfill.

The waste and the leachate and landfill gas produced by the site is contained in the landfill by a combination of a geological barrier and an artificial or synthetic sealing system. The geological barrier required for containment is determined by the geological and hydrogeological conditions below the landfill and in the vicinity of the landfill site. The characteristics of the geological barrier may be achieved by the use of either natural or synthetic materials. The landfill base and sides should consist of a natural mineral layer or an equivalent artificial synthetic layer to produce a sufficient environmental protection of the surrounding soil and groundwater. This is achieved by using liner materials which provide sufficient permeability and thickness requirements at least equivalent to the following:

- For hazardous waste landfill sites, an hydraulic conductivity of $\leq 1.0 \times 10^{-9}$ m/s and liner material thickness of ≥ 5 m.
- For non-hazardous waste landfill sites, an hydraulic conductivity of $\leq 1.0 \times 10^{-9}$ m/s and liner material thickness of ≥ 1 m.
- For inert waste landfill sites, an hydraulic conductivity of $\leq 1.0 \times 10^{-7}$ m/s and liner material thickness of ≥ 1 m.

Hydraulic conductivity is defined in Box 4.1. Where an artificial or synthetic liner system is used to replicate the hydraulic conductivity or permeability of a geological liner, the EC Waste Landfill Directive sets a minimum thickness of 0.5 m.

Box 4.1

Hydraulic Conductivity (Permeability)

Darcy's Law is an empirical law describing the flow of a fluid through a porous material. The law relates the flow rate of the fluid to a cross-sectional area of the porous material and the hydraulic gradient by way of a constant, the coefficient of permeability.

$$Q = kiA$$

Q = Flow rate
k = coefficient of permeability, permeability or hydraulic conductivity
i = hydraulic gradient (the pressure difference between the top and bottom of the layer of material)
A = cross-sectional area

Hydraulic conductivity or permeability therefore represents the ease with which a fluid such as leachate will flow through the liner material. The units of measurement are typically cm/s or m/s. Typical hydraulic conductivities of natural and synthetic or processed materials and waste are shown below.

Material	Hydraulic conductivity (permeability) (m/s)
Natural Materials	
Well-graded, clean gravels, gravel–sand mixture	2×10^4
Poorly graded, clean sands, gravely sands	5×10^{-4}
Silty sands, poorly graded, sand–silt mixture	5×10^{-5}
Inorganic silts and clayey silts	5×10^{-8}

Continued on page 187

Continued from page 186

Mixture of inorganic silt and clay	2×10^{-9}
Inorganic clays of high plasticity	5×10^{-10}
Synthetic or Processed Materials	
Compacted clay liner	$1 \times 10^{-8}-10^{-10}$
Bentonite-enhanced soil	5×10^{-10}
Geosynthetic clay liner	$1 \times 10^{-10}-10^{-12}$
Flexible membranes	1×10^{-13}
Geotextile	$1 \times 10^{-4}-10^{-5}$
Geonet	2×10^{-1}
Waste	
Municipal solid waste as placed	1×10^{-5}
Shredded Municipal solid waste	$1 \times 10^{-4}-10^{-6}$
Baled municipal solid waste	7×10^{-6}

Sources: Sharma and Lewis 1994; McBean et al 1995.

In addition to the geological barrier liner system, a leachate sealing and collection system must be added to ensure that the accumulation of the leachate is kept to a minimum. For both hazardous and non-hazardous categories of waste landfill site an artificial or synthetic sealing barrier is required, and also a drainage layer which should be ≥0.5 m. Inert landfill sites may require a leachate sealing and collection system, depending on the recommendations of the local regulatory competent authorities. In some cases, the regulatory or competent authorities may require a top liner system to act as a barrier to prevent the formation of leachate. In such cases a non-hazardous waste landfill is required to have a surface-sealing, barrier-liner system consisting of a gas drainage layer, an impermeable mineral layer, a drainage layer of greater than 0.5 m thickness and a top soil cover layer of greater than 1 m thickness. For a hazardous waste landfill site, a surface-sealing, barrier-liner system consisting of an artificial sealing liner, an impermeable mineral layer, a drainage layer of greater than 0.5 m thickness and a top soil cover layer of greater than 1 m thickness is required. Where landfill gas control is required, the capping liner layer will also contain landfill gas collection immediately beneath the capping material (Allen 2001).

4.7   Landfill Liner Materials

There are a large variety of natural and synthetic mineral materials and synthetic polymeric materials which are used in the construction of hazardous, non-hazardous and inert waste landfill sites (McBean et al 1995; Sharma and Lewis 1994; Bagchi 1994; Christensen et al 1996(a); Waste Landfill Directive 1999). The choice of liner material and barrier liner system will depend on the type of waste, the geological and hydrogeological conditions in the surrounding environment, the prediction of the properties of the derived leachate, and the resistance of the liner to the leachate. The minimum requirements for the geological barrier and the leachate collection and drainage system have been discussed in Section 4.6. A critical factor in the type of liner material used for containment types of landfill is the permeability which is measured as 'hydraulic conductivity'. Hydraulic conductivity is discussed in Box 4.1.

Natural clay   Clays are unconsolidated rocks composed of clay minerals formed as breakdown products from the weathering of pre-existing rocks. The clay minerals include, for example, montmorillanite, illite and kaolinite and are of extremely fine grain size. The very fine grain size means that porosity is also extremely low and consequently permeability is very low. Clay minerals are formed from sheets of alumino-silicates stacked in layers bonded by cations. Water molecules are absorbed between the layers which causes swelling and hence the observed low permeability. In addition, in situ clay may be utilised as the underlying material of the landfill if the local geological environment lends itself to this choice of site selection, the clay acting as a further low permeability barrier beneath the liner system. However, when used as a design barrier, the properties of the clay should be carefully evaluated due to potential inhomogeneities in the clay strata. In such cases, the in situ clay would be excavated and re-laid. Normal practice for natural clay liners in landfill sites is to use local or imported clay which is compacted into layers of between 0.6 and 1.0 m thick to form a homogeneous, low-permeability layer (McBean et al 1995; Sharma and Lewis 1994).

The clay liner consists of a mixture of clay minerals and fine silt particles which blend to form a clay or clay soil with suitable low permeabilities. The factors which can affect the suitability and performance of a particular clay or clay soil in its use in a landfill liner system include porosity and permeability, which in turn depend on clay mineralogy, particle size distribution, plasticity, strength, moisture content and compaction. Moisture is an important factor in determining permeability, and adjustment of the moisture content of the clay material to obtain suitable permeability may be required before use. Optimum moisture content is determined by a standard test which produces a maximum dry density when a clay soil is compacted in a standard mould. In many cases, lower permeabilities may be obtained by clay liners which are slightly wetter than the optimum moisture content, in such cases care must be taken to stop this water from migrating away. Clay mineralogy also determines the permeability of a clay, for example, clays with a higher proportion of montmorillanite have lower permeabilities than clays with a high proportion of illite, which in turn have lower permeabilities than clays with a high proportion of kaolinite (Bagchi 1994).

The clay material is excavated from the source site and blended to form a homogeneous material. It may also be necessary to sieve and remove large rocks. The clay material is prepared by adjusting the moisture content to achieve the lowest permeability. The clay is then transported and spread by bulldozers or scrapers at the site. The clay liner is then compacted on site by large roller vehicles to form a more homogeneous layer by breaking up large pieces of clay and thereby greatly reducing the void space between the pieces and serving to also increase the density.

Bentonite-enhanced soils   Bentonite is a mixture of clay minerals, principally of the montmorillanite type. Sodium bentonites and calcium bentonites exist with the sodium form having lower permeabilities. Where the naturally occurring clay soil does not have a high enough level of clay minerals to produce a suitably low permeability, bentonite clay is added to form a bentonite-enhanced soil. The bentonite-type clay minerals have a high swelling characteristic on absorbing moisture, which forces the hydrated bentonite clay around the soil particles to form a synthetic clay. In addition, the bentonite-enhanced soil further swells under pressure and consequently lower permeabilities are obtained as the mass of waste builds up in the overlying landfill. Sodium bentonite is added at between 5 and 15%, depending on the type of original soil. Where the calcium form of bentonite is used, larger quantities are required due to its lower swelling properties and therefore higher permeability. To ensure the formation of a homogeneous low permeability clay the bentonite–soil mixture should be thoroughly mixed and the moisture content adjusted to produce the lowest permeability-enhanced soil. Mixing is usually carried out before application to the site location. Suitable host soils for bentonite application include sands and silty sands, whereas more cohesive materials may be difficult to mix, producing an inhomogeneous variable permeability mixture (Waste Management Paper 26B, 1995; McBean et al 1995).

Geosynthetic clay liners   Geosynthetic clay liners are a mixture of bentonite clay mechanically or chemically adhered to a geotextile fabric. Alternatively, the bentonite layer may be sandwiched between two layers of the geotextile fabric where the layers are joined by adhesives, needlepunching or stitching. Typically the geosynthetic clay liners are approximately 1 cm thick and are available in 5 m × 30 m rolls (Sharma and Lewis 1994). The rolls are laid out on natural clay soils or polymeric flexible membranes and joined by bentonite cement. Geosynthetic clay liners are often used as alternatives to natural compacted clay liners. However, since they are a relatively new material in landfill applications, their long-term stability in landfills has not been tested over very long periods. Table 4.3 shows a comparison of the natural and geosynthetic types of clay liner (Sharma and Lewis 1994).

Flexible membrane liners   Flexible membrane liners are synthetic, polymeric plastic materials with extremely low permeabilities. Whilst permeabilities are extremely low, there is some diffusion of leachate through the membrane. Also, if the liner became punctured

Table 4.3   *Comparison of geosynthetic clay liners and natural clay liners*

Characteristic	Geosynthetic clay liner	Natural clay liner (compacted)
Materials	Bentonite clay, adhesives geotextiles and geomembranes	Native soils or blend of soil and bentonite
Construction	Manufactured and then installed in the field	Construction in the field
Ambient temperature	Installation at low temperature permissible	Installation at or below freezing temperature not permissible
Thickness	Approximately 10 mm	Approximately 0.5–1 m
Clay hydraulic conductivity	10^{-8}–10^{-10} cm/s typical	10^{-7}–10^{-8} cm/s typical
Speed and ease of construction	Rapid, simple installation	Slow complicated installation
Water content	Essentially dry, cannot desiccate	Nearly saturated
Settlement	Adjusts to differential settlement	Performance is poor in case of differential settlement
Leachate	Cannot be used in direct contact with most leachate	Can be used in direct contact with most leachate

Sources: Sharma and Lewis 1994; Bagchi 1994.

Table 4.4   *Types of synthetic flexible membrane liners*

Thermoplastic polymers	Thermoset polymers	Combinations
PVC	Butyl or isoprene-isobutylene (IIR)	PVC-nitrile rubber
Polyethylene e.g. LDPE[1], HDPE[2]		
Chlorinated polyethylene (CPE)	Ethylene propylene diene monomer (EPDM)	PVC-ethyl vinyl acetate
Elasticized polyolefin (3110)	Polychloroprene (neoprene)	Cross-linked CPE
Ethylene interpolymer alloy (EIA or XR-5)	Ethylene polypropylene terpolymer (EPT)	Chlorosulphonated polyethylene
Polyamide	Ethylene vinyl acetate	(CSPE or Hypalon)

[1] LDPE = Low-density polyethylene.
[2] HDPE = High-density polyethylene.

Source: Sharma and Lewis 1994. Copyright © 1994. Reprinted with permission of John Wiley & Sons, Inc.

with small holes either during manufacture or emplacement, leakage of leachate would also occur. There are a variety of membrane liners available and the major types used in landfill applications are shown in Table 4.4 (Sharma and Lewis 1994). The commonest types used are high-density polyethylene and PVC. The membranes come in sheets or rolls ranging from 5 to 15 m wide and up to 500 m in length and range in thickness typically from 0.75 mm to 3.00 mm. There are a range of properties which define the suitability for use of the various membranes in landfill applications, including density, tensile strength, puncture resistance, tear resistance, resistance to ultraviolet (UV) light and ozone and chemical resistance (Sharma and Lewis 1994; Waste Management Paper 26B, 1995; Tchobanoglous and O'Leary 1994). Membrane chemical resistance is very important since the leachate may contain a range of organic and inorganic acids and alkalis and organic hydrocarbons. There are a range of standard tests available to determine the properties of the membranes. The membrane sheets or rolls are seamed together using heat sealing or liquid solvents, usually carried out on site. Fast rates of seaming can be achieved ranging from 30 m/h to 100 m/h depending on the system of seaming (Sharma and Lewis 1994). Table 4.5 compares the advantages and disadvantages of commonly used synthetic flexible membrane liners (Bagchi 1994).

Table 4.5   *Advantages and disadvantages of commonly used synthetic flexible membranes*

Synthetic flexible membrane	Advantages/disadvantages
Butyl rubber	Good resistance to UV, ozone and weathering elements Good performance at high and low temperatures Low swelling in water Low strength characteristics Low resistance to hydrocarbons Difficult to seam
Polychlorinated polyethylene (CPE)	Good resistance to UV, ozone and weather elements Good performance at low temperatures

	Good strength characteristics Easy to seam Poor resistance to chemicals, acids and oils Poor seam quality
Chlorosulphonated polyethylene	Good resistance to UV, ozone and weather elements Good performance at low temperatures Good resistance to chemicals, acids and oils Good resistance to bacteria Low strength characteristics Problem during seaming
Ethylene propylene rubber (EPDM)	Good resistance to UV, ozone and weather elements High strength characteristics Good performance at low temperatures Low water absorbance Poor resistance to oils, hydrocarbons and solvents Poor seam quality
LDPE and HDPE	Good resistance to most chemicals Good strength and seam characteristics Good performance at low temperatures Poor puncture resistance
PVC	Good workability High strength characteristics Easy to seam Poor resistance to UV, ozone, sulphide and weather elements Poor performance at high and low temperatures

Source: Bagchi 1994. Copyright © 1994. Reprinted with permission of John Wiley & Sons, Inc.

Geotextiles   Geotextiles are fabric materials used as protection for polymeric plastic membranes and filtration material to filter out fine-grained particles from the leachate so that drainage layers do not become blocked. Geotextiles are composed of polypropylene or polyester fibres which are manufactured to form a fabric-type material. Manufacture consists of either weaving, using traditional weaving techniques, or needlepunching through the fibre web to produce an interlocked fabric. Geotextiles are not used for containment and therefore have relatively high permeabilities. As with polymeric plastic membranes, there are a wide variety of tests which may be carried out on the geotextile material to determine, for example, the tensile strength, tear resistance, burst strength, chemical resistance, etc. Seaming of the geotextile may be by overlapping or simple stitching of the fabric (Sharma and Lewis 1994; McBean et al 1995; Christensen et al 1996(a)).

Geonets   Geonets are porous sheets of plastic netting used as drainage layers to carry leachate or landfill gas. The nets are typically about 5 mm thick and are usually composed of polyethylene (Sharma and Lewis 1994). To prevent the net from clogging due to particles in the leachate, there is usually a geotextile fabric material bonded to the geonet. The main role of geonets is drainage and they are used as alternatives to naturally well-drained materials such as coarse sands or gravels, but require less thickness to achieve the same effectiveness. For example, a 4.5 mm thick geonet layer would have a similar drainage function to 300 mm of sand.

4.8   Landfill Liner Systems

The containment landfill, where the waste and product leachate and landfill gas are contained by the multiple use of liner materials in order to build-up a barrier liner system, represents the current type of landfill design predominantly used throughout Europe and North America (Waste Management Paper 26B, 1995; Waste Management in Japan 1995; Sarsby 1995; Hickman 1996; Bagchi 1994; Hjelmar et al 1995; Westlake 1995; Allen 2001). The modern landfill is often described as an engineering project involving construction engineering and process engineering. The project involves design details and drawings, construction of the base, installation of the lining, operation of the site through its lifetime (involving emplacement of the waste and reacting to biological, chemical and physical changes of the waste), monitoring and control of gas and leachate, and finally aftercare.

The minimum requirements for the containment of barrier liner systems for hazardous waste, non-hazardous waste and inert waste landfill sites are laid down in the EC Waste Landfill Directive (1999). However, due to leakage problems of landfill systems and the requirement to contain all leachate within the landfill, the design of waste landfills has resulted in more and more complex liner systems (Allen 2001). Consequently, waste landfills usually install elaborate multi-layered barrier systems. There are a variety of landfill liners available which may be designed in different combinations to produce a large number of different liner systems (McBean et al 1995; Sharma and Lewis 1994; Tchobanoglous and O'Leary 1994; Waste Management paper 26B, 1995; Allen 2001). Increasing complexity of liner design and the use of multilayers of materials inevitably increases costs and a balance has to be achieved between cost and protection of the environment. To this end, risk assessment is used as the criteria for selection of the most appropriate landfill liner system. The assessment of risk is on a site-specific basis in relation to the surrounding geology and hydrogeology and the type, composition and amount of waste. Risk assessment is used in the determination of a suitable landfill system in defining the acceptable seepage rate of leachate into the surrounding environment, such that the pollutants do not cause an unacceptable level of contamination. This would take into account the processes of leachate formation and the dilution, dispersion and attenuation in the surrounding environment. The pathways for potential leakage of landfill gas and impacts on the environment should also be assessed.

All liner materials allow a certain low level of seepage through the liner to an extent determined by their permeability or hydraulic conductivity. Similarly, the system should not allow groundwater to seep into the landfill and increase leachate levels. The liner system also minimises the release of landfill gas, coupled with a collection system of porous pipework throughout the waste to collect the gas. The design of liner system should also be resistant to the variety of chemical properties of the leachate throughout the lifetime of the site, which may be over a 50-year period, particularly for the chemicals that may be present in hazardous waste. The selection of the liner system will also be influenced by local geological conditions, for example, if clay is the local environment, then clays will be used as a component of the system.

The liner systems rely on combinations of liner materials and liquid collection layers to contain and collect the leachate and landfill gas. There are several different types of liner systems for example (Waste Management Paper 26B, 1995): single liner system; composite liner system; double liner system; multiple liner system.

Single liner system In most cases it is not sufficient to use only one primary liner, since failure of the liner results in leachate escape. However, the single liner system may be appropriate for certain low-risk wastes in sites where escape of leachate poses negligible risk of contamination. Figure 4.7 shows an example of the single liner system (Waste Management Paper 26B, 1995). The single liner of the system is comprised of a primary barrier consisting of a layer of clay, bentonite-enhanced soil or hydraulic asphalt. Above and below the primary liner or barrier would be a separation/protection layer of geotextile material, for example, non-woven, needlepunch fabrics composed of polyester or polypropylene fibre. The material acts as a protective layer and filter for fine suspended solids. Between the waste and the separation/protection layer would be a leachate collection system consisting of a series of drainpipes or a drainage layer. Similarly, beneath the liner, if necessary, may be a groundwater collection system, such as a drainage layer of gravel or a synthetic, polymeric net material (geonet).

Composite liner system Figure 4.8 shows a schematic diagram of a composite liner system (Waste Management Paper 26B, 1995). The use of two different types of liner material, a clay-based mineral layer and a polymeric membrane layer provides the composite liner system with a more secure containment than the single liner system and more protection

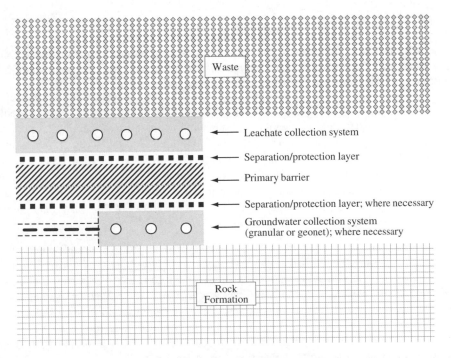

Figure 4.7 *Schematic diagram of a typical single liner system. Source: Waste Management Paper 26B, 1995. Crown copyright material is reproduced with the permission of the Controller of HMSO and the Queen's Printer for Scotland.*

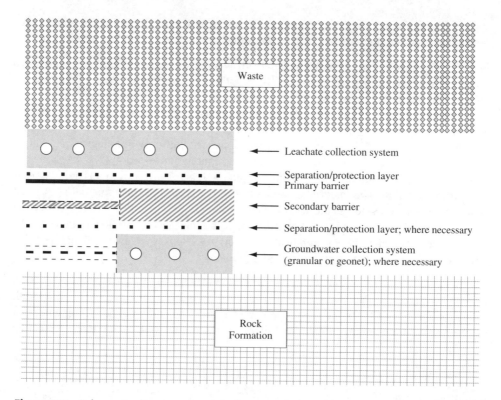

Figure 4.8 *Schematic diagram of a typical composite liner system. Source: Waste Management Paper 26B, 1995. Crown copyright material is reproduced with the permission of the Controller of HMSO and the Queen's Printer for Scotland.*

for the environment. The primary liner or barrier, in the case of the composite liner system, is a synthetic polymeric flexible membrane liner with a separation/protective layer of thin polymeric or geotextile material. The polymeric membrane liner consists of, for example, high-density polyethylene (HDPE), low-density polyethylene (LDPE), PVC, etc. The secondary barrier would be a layer of clay, bentonite-enhanced soil or geosynthetic clay liner. Below the secondary barrier would be a separation/protection layer of geotextile material, followed by a groundwater collection system. Between the waste mass and the liner system is the leachate collection system.

Double liner system   Figure 4.9 shows a schematic diagram of a double liner system (Waste Management Paper 26B, 1995). The system incorporates an intermediate high-permeability drainage layer between the primary and secondary liner barriers. The intermediate drainage system is in addition to the leachate drainage collection system and the groundwater drainage collection system and is used to monitor and remove leachate and landfill gas from between the barriers. Where a multi-barrier system is used there is the possibility of build-up of leachate or landfill gas between the low permeability layers which may cause problems of lateral migration of leachate or gas, hence the need for an intermediate drainage layer. The primary and secondary barriers may be composed of

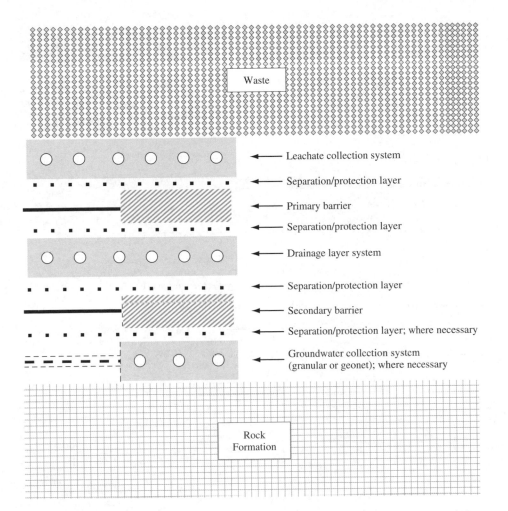

Figure 4.9 *Schematic diagram of a typical double liner system. Source: Waste Management Paper 26B, 1995. Crown copyright material is reproduced with the permission of the Controller of HMSO and the Queen's Printer for Scotland.*

mineral materials such as, clay, bentonite-enhanced soil or a geosynthetic clay liner such as bentonite matting, or alternatively a synthetic, polymeric membrane liner is used. As before, each layer is separated by a separation/protection layer of geotextile fabric filter material.

Multiple liner system   Figure 4.10 shows a schematic diagram of the multiple liner system (Waste Management Paper 26B, 1995). The multiple liner system combines some of the attributes of the composite and double liner systems. Primary, secondary and tertiary liner barriers are incorporated, with intermediate drainage to remove leachate and landfill gas from between the barriers. The primary barrier is usually composed of synthetic polymeric membrane in intimate contact with the secondary liner barrier

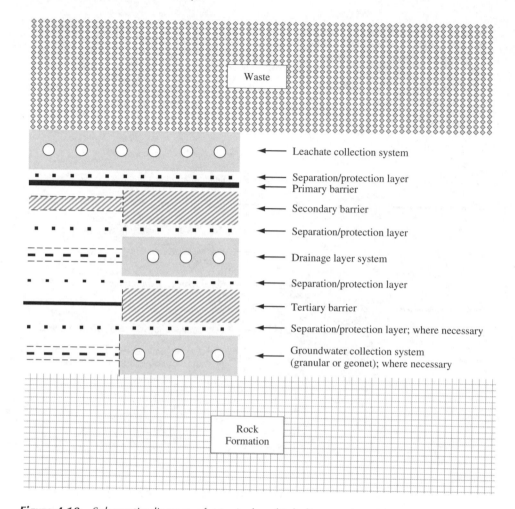

Figure 4.10   Schematic diagram of a typical multiple liner system. Source: Waste Management Paper 26B, 1995. Crown copyright material is reproduced with the permission of the Controller of HMSO and the Queen's Printer for Scotland.

composed of mineral materials such as clay, bentonite-enhanced soils or bentonite matting. Intimate contact between primary and secondary barriers is required, to prevent lateral movement of leachate or gas. As before, separation/protection layers of geotextile fabric filters and leachate drainage collection and groundwater drainage collection layers are also incorporated.

Table 4.6 summarises the typical liner materials used for the various liner barriers in the four systems discussed (Waste Management Paper 26B, 1995). Several other liner systems have been described for different countries and a range of different waste types (see for example, Sharma and Lewis 1994; McBean et al 1995; Tchobanoglous and O'Leary 1994; Bagchi 1994; Ham 1993; Bishop and Carter 1995; LaGrega et al 1994; Allen 2001).

Table 4.6 *Typical liner materials used in landfill liner systems*

Liner type	Primary barrier	Secondary barrier	Tertiary barrier
Single liner system (Figure 4.7)	Clay Bentonite enhanced soil or hydraulic asphalt	Not applicable	Not applicable
Composite liner system (Figure 4.8)	Flexible membrane liner	Clay Bentonite enhanced soil or geosynthetic clay liner	Not applicable
Double liner system (Figure 4.9)	Clay Bentonite enhanced soil Geosynthetic clay liner or flexible membrane liner	Clay Bentonite enhanced soil Geosynthetic clay liner or flexible membrane liner	Not applicable
Multiple liner system (Figure 4.10)	Flexible membrane liner	Clay Bentonite enhanced soil or geosynthetic clay liner	Clay Bentonite enhanced soil, geosynthetic clay liner or flexible membrane liner

Source: Waste Management Paper 26B, 1995. Crown copyright material is reproduced with the permission of the Controller of HMSO and the Queen's Printer for Scotland.

4.9 Processes Operating in Waste Landfills

4.9.1 Processes Operating in Hazardous Waste Landfills

Hazardous waste is derived from a large number of industrial sources. However, the main sources are from organic and inorganic industrial processes. Other important sources include inorganic wastes from thermal processes, inorganic waste from metal treatment facilities and wastes from waste treatment facilities. Since the landfilling of liquid wastes is specifically banned under the EC Waste Landfill Directive, only solid hazardous wastes are allowed to be landfilled in hazardous waste landfill sites. Additionally, under the EC Directive the waste must be 'treated' before it is landfilled, to reduce the hazardous nature of the waste in terms of impact on human health and the environment.

Once the treated hazardous waste is placed in the landfill it will be subject to a range of biological, physical and chemical processes which will degrade the components of the waste. These processes include biodegradation, filtration, redox reactions, complexation, ion exchange, adsorption, precipitation, neutralisation, etc. The migration of leachate through the hazardous waste mass in the landfill will disperse and dilute the pollutants. In addition, leachate may be absorbed into or adsorbed onto the components of the waste. Chemical reactions of the leachate derived from the hazardous waste will be attenuated or reduced by interaction between the leachate and other components of the surrounding waste and other material, including daily and intermediate cover material to chemically alter or fix the leachate. Such reactions include interaction of cations and anions via ion-exchange, removal of leachate pollutants by precipitation reaction, or the formation of large ion complexes, which effectively remove pollutants from the environment by fixation in a large complex molecule, and oxidation–reduction reactions. Some of the wastes may be biodegradable and undergo the decomposition reactions described in Section 4.9.2.1. Pavelka et al (1993) in the USA have examined the leachate from a number of hazardous waste only, landfill sites. They showed that significantly high concentrations of a number of analysed species were found in the leachate from this type of landfill (Table 4.7, Pavelka et al 1993).

4.9.2   Processes Operating in Non-hazardous Waste Landfills

Municipal solid waste is the most significant category of waste that is permitted to be deposited in a non-hazardous waste landfill. Municipal solid waste contains a high proportion of organic material which can be degraded by the range of micro-organisms found in waste landfills including food and garden waste, paper and board, wood and some textiles. The proportion of biodegradable waste has been estimated at more than 66% for a range of countries across Europe (Table 4.2, European Environment Agency 2002(b)). Industrial and commercial wastes may contain over 60% of dry weight biodegradable organic material (Waste Management Paper 26B, 1995). Provided that they are non-hazardous, under the definitions of the Waste Landfill Directive (1999), they would be acceptable in a non-hazardous waste landfill and be subject to biodegradation processes.

The processes of degradation of organic bioreactive waste in landfills involves not only biological processes but also inter-related physical and chemical processes (Waste Management Paper 26B, 1995; McBean et al 1995; Tchobanoglous and O'Leary 1994; Westlake 1995; Pescod 1991–93; Waste Management Paper 26, 1986; Christensen et al 1996(b); Gendebien et al 1992). The processes operate on any organic waste, consequently, such biodegradation processes may also occur, not only in non-hazardous waste sites accepting municipal solid waste, but also in hazardous waste landfills where biodegradable hazardous wastes are accepted. The stages involved in the degradation of bioreactive solid wastes can take many decades to complete.

Table 4.7 *Leachate composition from hazardous waste only landfill sites*

Constituent	Hazardous waste-only leachate (mean concentration µg/l)
Methyl ethyl ketone	19800
Methyl isobutylketone	19700
Acetone	17400
Phthalic acid	19300
Phenol	21700
Arsenic	17000
Nickel	2160
Zinc	950

Source: Pavelka et al 1993.

4.9.2.1 Decomposition Processes of Bioreactive Wastes in Landfills

The organic components of the waste are degraded by micro-organisms in the landfill. The organic materials occurring in waste can be classified into broad biological groups represented by proteins, carbohydrates and lipids or fats. Carbohydrates are by far the major component of biodegradable wastes and include cellulose, starch and sugars. Proteins are large complex organic materials composed of hundreds or thousands of amino acids groups. Lipids or fats are materials containing fatty acids. Five main stages of degradation of biodegradable wastes have been identified (Kjeldsen et al 2002; Waste Management Paper 26B, 1995; McBean et al 1995). Figure 4.11 shows the decomposition pathways of the major organic and inorganic components of biodegradable wastes, and Figure 4.12 shows the process in more detail (Waste Management Paper 26B, 1995). Throughout the process of degradation, because of the heterogeneous nature of waste, all the different stages may be progressing simultaneously until all the waste has reached stage five and stabilisation of the landfill has been reached.

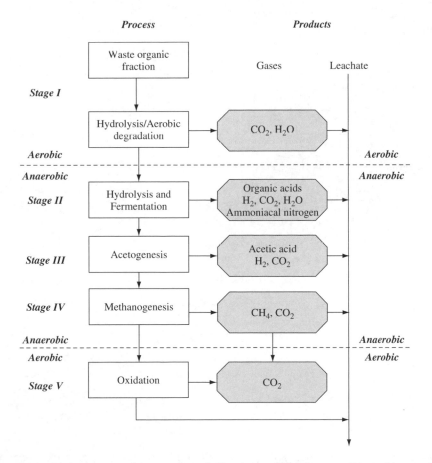

Figure 4.11 *Major stages of waste degradation in landfills. Source: Waste Management Paper 26B, 1995. Crown copyright material is reproduced with the permission of the Controller of HMSO and the Queen's Printer for Scotland.*

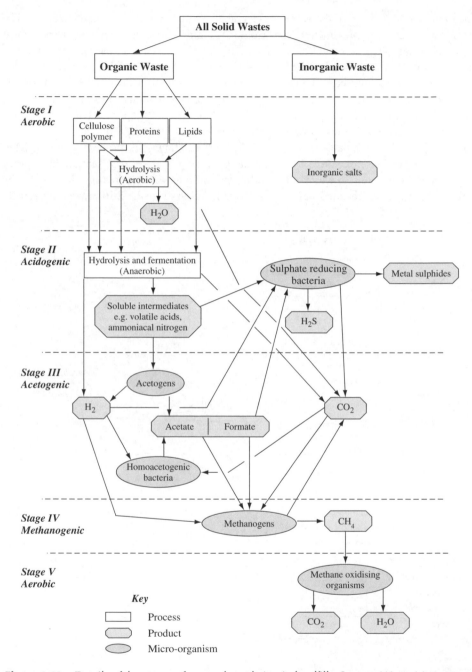

Figure 4.12 *Details of the stages of waste degradation in landfills. Source: Waste Management Paper 26B, 1995. Crown copyright material is reproduced with the permission of the Controller of HMSO and the Queen's Printer for Scotland.*

Stage I. Hydrolysis/aerobic degradation   The hydrolysis/aerobic degradation stage occurs under aerobic (in the presence of oxygen) conditions. This occurs during the emplacement of the waste and for a period thereafter which depends on the availability of oxygen in the trapped air within the waste. The micro-organisms are of the aerobic type, that is, they require oxygen and they metabolise the available oxygen and a proportion of the organic fraction of the waste to produce simpler hydrocarbons, carbon dioxide, water and heat. The heat generated from the exothermic degradation reaction can raise the temperature of the waste to up to 70–90 °C (McBean et al 1995; Waste Management Paper 26B, 1995). However, compacted waste achieves lower temperatures due to the lower availability of oxygen. Water and carbon dioxide are the main products, with carbon dioxide released as gas or absorbed into water to form carbonic acid, which gives acidity to the leachate.

The aerobic stage lasts for only a matter of days or weeks depending on the availability of oxygen for the process, which in turn depends on the amount of air trapped in the waste, the degree of waste compaction and how quickly the waste is covered.

Stage II. Hydrolysis and Fermentation   Stage I processes result in a depletion of oxygen in the mass of waste and a change to anaerobic (absence of oxygen) conditions. Different micro-organisms, the facultative anaerobes, which can tolerate reduced oxygen conditions become dominant. Carbohydrates, proteins and lipids are hydrolysed to sugars which are then further decomposed to carbon dioxide, hydrogen, ammonia and organic acids. Proteins decompose via deaminisation to form ammonia and also carboxylic acids and carbon dioxide. The ammonia is derived largely from the deaminisation of proteins, which also form carboxylic acids and carbon dioxide. The derived leachate contains ammoniacal nitrogen in high concentration. The organic acids are mainly acetic acid, but also propionic, butyric, lactic and formic acids and acid derivative products, and their formation depends on the composition of the initial waste material. The temperatures in the landfill drop to between 30 and 50 °C during this stage. Gas concentrations in the waste undergoing stage II decomposition may rise to levels of up to 80% carbon dioxide and 20% hydrogen (Waste Management Paper 26, 1986; Waste Management Paper 26B, 1995).

Stage III. Acetogenesis   The organic acids formed in Stage II are converted by acetogen micro-organisms to acetic acid, acetic acid derivatives, carbon dioxide and hydrogen under anaerobic conditions. Other organisms convert carbohydrates directly to acetic acid in the presence of carbon dioxide and hydrogen. Hydrogen and carbon dioxide levels begin to decrease throughout Stage III. Low hydrogen levels promote the methane-generating micro-organisms, the methanogens, which generate methane and carbon dioxide from the organic acids and their derivatives generated in the earlier stages. The acidic conditions of the acetogenic stage increase the solubility of metal ions and thus increase their concentration in the leachate. In addition, organic acids, chloride ions, ammonium ions and phosphate ions, all in high concentration in the leachate, readily form complexes with metal ions, causing further increases in solubilisation of metal ions. Hydrogen sulphide may also be produced throughout the anaerobic stages as the sulphate compounds in the waste are reduced to hydrogen sulphide by sulphate-reducing micro-organisms (Christensen et al 1996(b)). Metal sulphides may be a reaction product of the hydrogen sulphide and metal ions in solution. The presence of the organic acids generate a very acidic solution which can have a pH level of 4 or even less (Moss 1997).

Stage IV. Methanogenesis   The methanogenesis stage is the main landfill gas generation stage, with the gas composition of typical landfill gas generated at approximately 60% methane and 40% carbon dioxide. The reactions are relatively slow and take many years for completion. The conditions maintain the anaerobic, oxygen-depleted environment of Stages II and III. Low levels of hydrogen are required to promote organisms, the methanogens, which generate carbon dioxide and methane from the organic acids, and their derivatives such as acetates and formates, generated in the earlier stages. Methane may also form from the direct micro-organism conversion of hydrogen and carbon dioxide to form methane and water. Hydrogen concentrations produced during Stages II and III therefore fall to low levels during this fourth stage. There are two classes of micro-organisms which are active in the methanogenic stage, the mesophilic bacteria which are active in the temperature range 30–35 °C and the thermophilic bacteria active in the range 45–65 °C. Therefore, landfill gas can be generated during the methanogenic stage over a temperature range of 30–65 °C, with an optimum temperature range of gas generation between 30 and 45 °C. In fact, most landfill sites fall within this temperature range with an average range for UK landfill sites of between 30 and 35 °C. Where temperatures in the mass of waste drop significantly, for example, to below 15 °C in cold weather in shallow sites, then the rate of biological degradation falls off. The organic acids formed during Stages II and III are degraded by the methanogenic micro-organisms, and as the acid concentration becomes depleted, the pH rises to about pH 7–8 during the methanogenesis stage. Ideal conditions for the methanogenic micro-organisms are a pH range from 6.8 to 7.5, but there is some activity between pH 5 and pH 9. Stage IV is the longest stage of waste degradation, but may not commence until 6 months to several years after the waste is placed in the landfill, depending on the level of water content and water circulation. Significant concentrations of methane are generated after between 3 and 12 months, depending on the development of the anaerobic micro-organisms and waste degradation products. Landfill gas will continue to be generated for periods of between 15 years and 30 years after final deposition of the waste, depending on waste and site characteristics (Landfill Gas Development Guidelines 1996). However, low levels of landfill gas may be generated up to 100 years after waste emplacement.

Stage V. Oxidation   The final stage of waste degradation results from the end of the degradation reactions, as the acids are used up in the production of the landfill gas methane and carbon dioxide. New aerobic micro-organisms slowly replace the anaerobic forms and re-establish aerobic conditions. Aerobic micro-organisms which convert residual methane to carbon dioxide and water may become established.

Figure 4.13 shows the changes in composition of landfill gas and leachate as the five stages of waste degradation progress with time (Waste Management Paper 26B, 1995; 26A, 1995). Initial formation of hydrogen and carbon dioxide in the hydrolysis/aerobic degradation, hydrolysis and fermentation and acetogenesis stages is followed by the main landfill gas generation stage, the methanogenesis stage. The characteristic landfill gas composition is methane and carbon dioxide with other minor components and water vapour. The final stages mark the end of the reaction and a return to aerobic conditions. Hydrogen sulphide gas may also form, derived from sulphate-reducing micro-organisms, in wastes with a high concentration of sulphate.

Changes in leachate composition throughout the five-stage degradation period are also shown in Figure 4.13. Throughout the five stages, cellulose becomes depleted by

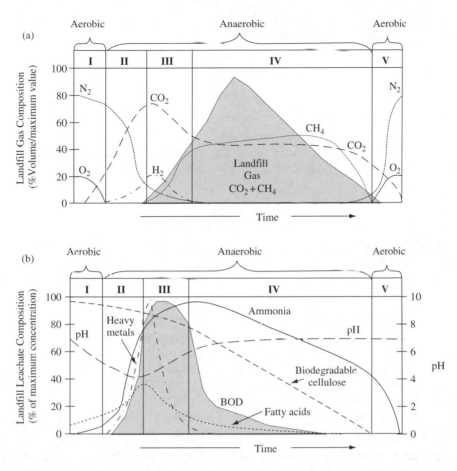

Figure 4.13 (a) Landfill gas composition and (b) leachate composition in relation to the degradation of biodegradable solid wastes. Source: Waste Management Paper 26B, 1995.

reaction with the various micro-organisms. Fatty acids in the leachate initially formed from the action of anaerobic micro-organisms in Stages II and III, become depleted in the methanogenic stage as they are converted to methane and carbon dioxide. The leachate becomes acidic in Stage II due to the derived organic acids. The presence of the organic acids results in solubilisation of heavy metals such as chromium, iron and manganese, into the acidic leachate. The organic acids become depleted in the methanogenic phase as the micro-organisms convert the acids to methane and carbon dioxide. As the pH begins to rise again, the heavy metals come out of solution as sulphide, hydroxide and carbonate precipitates. The chemical oxygen demand (COD) increases through the second and third stages but decreases in the methanogenic Stage IV (McBean et al 1995; Waste Management Paper 26B, 1995; Waste Management Paper 26, 1986; Westlake 1995). Table 4.8 compares leachate composition in the early and later stages of waste degradation. COD and biochemical oxygen demand (BOD) are standard parameters which measure the quality of water and are discussed in Box 4.1 (Tchobanoglous and Burton 1991; Fifield and Haines 1995).

Table 4.8   *Comparison of early and late stage leachate composition*

Stage II/III leachate	Stage IV leachate
High content of fatty acids	Low content of fatty acids
Acidic (Low pH)	Neutral/alkaline (pH 7–8)
High BOD/COD ratio	Low BOD/COD ratio
High ammoniacal nitrogen	Lower levels of ammoniacal nitrogen
High organic nitrogen	
Heavy metals, e.g., Cr, Fe and Mn in solution	

Source: Waste Management Paper 26B, 1995.

4.9.2.2   *Factors Influencing Waste Degradation in Landfills*

There are numerous factors influencing the degradation of the waste, and these have been reviewed by Westlake (1995), Christensen et al (1996(b)) McBean et al (1995), Waste Management Paper 26 (1986), Waste Management Paper 26B (1995), Waste Management Paper 27 (1994) and Gendebien et al (1992);

Site characteristics   Landfill sites with waste depths exceeding 5 m tend to develop anaerobic conditions and greater quantities of landfill gas. Shallower sites allow air interchange and lower anaerobic activity, and consequently lower landfill gas production. However, if the site is well capped, anaerobic conditions will be created. Similarly, rapid covering of the waste will reduce the aerobic phase, and since this is the increasing temperature phase, this will tend to keep waste temperatures down. Also, rapid covering of the waste will reduce the chance of rainfall increasing the moisture content of the waste, which in turn reduces the initial rate of biodegradation.

Waste characteristics   The major components of municipal solid waste include the biodegradable fraction, that is, the paper and board, food and garden waste and textiles, and non-biodegradable components, plastics, glass and textiles. The amount of biodegradation will vary depending on the proportion of biodegradable components in the waste. This fraction has been shown to vary depending on a number of factors, for example, higher concentrations of garden waste are produced in spring and autumn, and more industrially developed countries produce more paper. Also, older landfill sites have been shown to contain lower proportions of biodegradable waste than modern sites due to the changing nature of waste over the last few decades. In addition, the composition of the organic components, that is, the proportion of cellulose, proteins and lipids, will similarly influence the degradation pathway.

Shredding or pulverisation of the waste prior to landfilling results in increased available surface area and consequent increased homogeneity and increased rates of biological degradation. The density or degree of compaction of the waste in the landfill will increase the amount of biodegradable material available for degradation and therefore increase the production of landfill gas per unit volume of void space in the landfill. Too high a degree of compaction, however, may limit the percolation of water through the site, which is necessary for the free flow of nutrients for the micro-organisms.

Moisture content of the waste　The waste biodegradation process requires moisture and is in fact a major factor in determining the production of landfill gas and leachate. Even in certain types of containment landfill where dry conditions are a design requirement, the degradation processes will continue, albeit at much slower rates of reaction. Increased rates of gas production are found with high moisture content landfill sites. The moisture content within the site will depend on the inherent moisture content of the waste, the level of rainfall in the area, percolation of surface and groundwaters into the site and, since water is a degradation product, the rate of biodegradation of the waste. The range of moisture contents for typical municipal solid waste ranges from 15 to 40% with a typical average value of 30% (Moss 1997). However, it is not only the moisture content of the waste that is important, but the movement of the moisture to distribute the micro-organisms and nutrients and flush away the degradation products. Artificially induced continuous flushing of the site with leachate or water increases the rate of degradation by flushing out degradation products and replenishing nutrients for the micro-organisms. Whilst this flushing may increase the rate of gas production, leachate production will also increase.

Temperature　The temperature range indicates the type of micro-organisms which are active. Initially aerobic bacteria may increase the temperature up to levels of 80 °C if the waste is left well aerated as the micro-organisms break down the waste to produce methane and carbon dioxide. However, compacted waste achieves lower temperatures due to the lower availability of oxygen. As aerobic conditions are replaced by anaerobic conditions throughout Stages II–IV, and the aerobic micro-organisms are replaced by anaerobic micro-organisms, the temperature drops to between 30 and 50 °C (Waste Management Paper 26B, 1995). The active temperature phase for the methanogenic micro-organisms fall into two ranges, 30–35 °C for the mesophilic-type bacteria and 45–65 °C for the thermophilic bacteria. The majority of landfill sites have temperatures between 30 and 35 °C during the main landfill gas generation phase. If the site is cold then significantly less gas is produced than at higher ambient temperatures. Chaiampo et al (1996) have monitored the temperature changes with depth throughout a 20 m deep municipal solid waste landfill in Italy. They showed that the first 1–2 m were in the temperature range of 10–15 °C, but the temperature increased to 35–40 °C at the 3–5 m depth and to 45–65 °C in the 5–20 m depth region. They equated the temperature regions with the mesophilic bacteria in the 1–5 m range and thermophilic bacteria in the deeper layers.

Acidity　The acidity of the landfill site influences the activity of the various micro-organisms and therefore determines the rate of biodegradation. The pH of a typical landfill site would initially be neutral, followed by acidic phases, Stages II and III, where organic acids are produced from waste degradation by the acetogenic micro-organisms, and the pH falls to as low as 4. The resultant organic acids provide the nutrients for the methanogenic bacteria and as the acids are consumed, the pH rises. The methanogenic bacteria are most active in the pH range 6.8–7.5 and, if the pH rises or falls outside this optimum range, then gas production is significantly reduced. The formation of organic acids and a drop in pH is an essential step in the waste biodegradation process, in that the organic acids provide the nutrients for the main gas generation phase IV micro-organisms, the methanogens.

4.9.3   Processes Operating in Inert Waste Landfills

By definition, inert waste landfills may only accept waste that does not undergo any significant physical, chemical or biological transformations. In addition, the inert waste should not be reactive to produce leachate or landfill gas in any significant quantities that may give rise to environmental pollution or harm human health. The major source of inert wastes are from the construction and demolition industries and so inert waste mostly consists of bricks, glass, tiles, ceramic materials, concrete, stones, etc. The pollutant content of the waste should be insignificant and should not contain any waste that would significantly biodegrade or be physically or chemically transformed in the landfill over time, to produce leachate which contains any pollutants or is ecotoxic. In particular, the properties of the leachate should not endanger the quality of surface water and/or groundwater (Waste Landfill Directive 1999). Consequently, the biological, physical and chemical processes that occur in an inert waste landfill should be negligible.

However, it has been shown that, in the past, some landfill sites designated as 'inert' and only accepting inert non-domestic waste have produced some leachate containing pollutants in significant concentrations, and also significant production of landfill gas. Table 4.9 shows the production of carbon dioxide and methane from designated inert landfill sites. In the majority of cases, insignificant quantities of landfill gas are generated, but in one particular case, high and potentially hazardous levels of gas were generated (Environmental impacts from landfills accepting non-domestic wastes 1995). Table 4.10

Table 4.9   *Landfill gas composition at six inert waste landfill sites (%)*

Gas	Inert Landfill Site					
	1	2	3	4	5	6
Methane	0	0	0	2	4	55
Carbon dioxide	3	1	0	10	5	35

Source: Environmental impacts from landfills accepting non-domestic wastes 1995.

Table 4.10   *Leachate composition from inert waste landfills (mg/l)*

Parameter	Inert landfill site					
	7	8	9a	9b	9c	10
pH	8.81	7.83	7.7	8.5	7.92	7.82
COD	600	85	100	95	300	—
TOC	290	31	43	32	20	140
Acetic acid	0	0	<20	<20	<20	0
Butyric acid	0	0	0	0	0	0
Propionic acid	0	0	0	0	0	0
Valeric acid	0	0	0	0	0	0
Iso-valeric acid	0	0	0	0	0	0
Phosphate	13	0.01	0.2	10.3	0.3	0.7
Chloride	1700	130	94	32	99	180

Sulphate	220	51	330	250	300	120
Nitrate	0.3	52	<0.1	0.5	<0.1	1.9
Ammonia	95	26	5.2	3.6	0.4	39
Calcium	110	150	460	340	380	570
Copper	<0.1	<0.1	0.5	<0.1	<0.1	<0.1
Iron	1.2	1.5	380	1.8	5.4	30
Potassium	180	38	25	16	12	26
Magnesium	110	38	45	20	20	47
Manganese	0.3	0.3	3	1.2	2.6	2.1
Sodium	3000	150	65	45	60	200
Lead	<0.1	<0.2	0.4	<0.2	<0.2	<0.2
Zinc	0.3	0.2	2.8	<0.1	<0.1	0.3

Source: Environmental impacts from landfills accepting non-domestic wastes 1995.

shows the variation in leachate composition for designated inert landfill sites (Environmental impacts from landfills accepting non-domestic wastes 1995). The data shown represent a different set of landfill sites than given in Table 4.9. Whilst the concentrations in the leachate are generally low, significant contaminants are found, indicating that biodegradation is occurring to some extent in the so-called 'inert' wastes.

4.10   Other Landfill Design Types

Apart from the main design types of hazardous waste, non-hazardous waste and inert landfills, there are variations in the design and operation of the landfill. These include, the 'sustainable' or 'controlled flushing bioreactor', or 'sustainable' type of landfill and the 'entombment' or 'dry' type of landfill. In addition, the well designed and managed landfill site can be compared with those of the developing world.

4.10.1   Controlled Flushing Bioreactor Landfill

The controlled flushing bioreactor, or sustainable type of landfill, is designed and operated to achieve stabilisation of the waste within a 30–50 year time span. The landfill is designed for the non-hazardous type of landfill and, in particular, for biodegradable wastes. The landfill operates as a controlled bioreactor, to accelerate the biodegradation processes operating on the waste by continuously recirculating water and/or leachate through the waste.

This type of waste landfill operation is referred to as sustainable through a strategy that the present generation should deal with the waste it produces and not leave it to future generations. In this context a generation is regarded as between 30 and 50 years (Waste Management Paper 26B, 1995). Consequently, a sustainable type of landfill is designed and operated to produce stabilisation of each phase of the site 30–50 years after completion of landfill operations. The number of controlled flushing bioreactors throughout the world is relatively small, but they have received worldwide attention as a means of accelerating the stabilisation of biodegradable wastes, for example, in the USA, EU, Canada, Australia,

South America, South Africa, Japan and New Zealand (Reinhart et al 2002; Maier et al 1995; Reinhart 1996; Townsend et al 1996; Lee and Jones-Lee 1993; Hjelmar et al 1995; Westlake 1995; Waste Management Paper 26B, 1995; Yuen et al 1995; Van den Brook et al 1995). Many non-hazardous landfills containing biodegradable wastes achieve recirculation of leachate by surface spraying it, but more with the aim of leachate management and treatment than to accelerate biodegradation, environmental sustainability or to increase the rate of landfill gas production. The recirculation of leachate by surface spraying onto the waste landfill results in enhanced evaporative losses of the leachate.

Continuous flushing of the site with leachate increases the rate of degradation by flushing out degradation products and replenishing nutrients for the micro-organisms and thereby achieves faster stabilisation of the waste. Increased biodegradation results in an increase in the rate of gas generation, but inevitably also an increase in the generation of leachate. Table 4.11 shows the advantages and disadvantages of leachate recirculation. A leachate collection and recirculation system is therefore required which evenly distributes leachate throughout the mass of waste. Table 4.12 shows the hydrogeological requirements for the sustainable type of landfill (Beaven 1996). Depending on the size of landfill, infiltration rates of 3000–10 000 mm/year of leachate are required. This compares with infiltration

Table 4.11   *Advantages and disadvantages of leachate recirculation*

Advantages

- Encourages early waste establishment and maintenance of methanogenesis. A high moisture content and the movement of moisture have both been shown to promote methanogenesis
- Develops a more uniform quality of leachate (measured as COD), so that the design and operation of treatment and disposal facilities is easier
- Optimises removal of hazardous organic contaminants by, for example, optimising conditions for biodegradation and stripping volatile organic material by increased gas production
- Minimises dry zones in the wastes, which could otherwise remain largely undegraded for many years
- Takes up the absorptive capacity of the biodegradable waste and reduces fluctuations in leachate flow rate
- Promotes enhanced evaporative losses of leachate by surface spraying
- Provides temporary storage of short-lived peak flow rates, allowing treatment facilities to be designed for flows closer to average values

Potential problems and concerns

- Surface flooding may be caused either by irrigation rates being locally too high or by the formation of inorganic solid layers
- Spray drift from leachate recirculation may result in health concerns and increased smells, particularly during the acetogenic phase
- Break-outs of leachate accumulated as perched water from the side slopes of landfills may occur, increased by the presence of compacted or low-permeability layers within the waste
- Clogging of sub-surface recirculation systems may occur
- Extremely high concentrations of dissolved salts may occur in sites accepting predominantly inorganic waste

Sources: Reinhart et al, 2002; Waste Management Paper 26F, 1994.

Table 4.12 *Hydrogeological requirements for sustainable landfill design*

Parameter	Design Requirement
Infiltration rate of water/leachate	3000–10 000 mm/year (dependent on landfill depth and flushing techniques)
Leachate control system	Enables the handling of 4000–11 000 l/tonne leachate
Bulk waste hydraulic conductivity	Between 10^{-5} and 10^{-6} m/s
Homogeneity of waste	Pulverisation or shredding of the waste to increase homogeneity and reactive surface area
Depth of waste	Limited to 20–30 m to maintain hydraulic conductivity between 10^{-5} and 10^{-6} m/s
Water/leachate distribution	Even distribution to wet all parts of the waste spaced approximately 10 m apart using injection pipes, trenches or wells
Daily cover	Restricted to high permeability or biodegradable materials, or removal of previous day's cover before emplacement of waste
Waste density	Low density, e.g., <0.8 tonnes/m^3 (dry density) to maintain hydraulic conductivity between 10^{-5} and 10^{-6}

Source: Beaven 1996.

rates from non-flushing landfills of between 50 and 250 mm/year. To obtain such high levels of liquid throughput, the hydraulic conductivity of the waste should be greater than 10^{-7} m/s and ideally in the range 10^{-5}–10^{-6}. Such high hydraulic conductivities are not normally found in waste landfills. Therefore the implication is that waste processing and operational parameters may have to be changed to achieve stabilisation within a generation. Such waste processing techniques may include pulverising or shredding the waste to produce a homogenous, high-surface-area material, which is more readily biodegraded. In addition, compaction of the waste to densities of greater than 0.8 tonnes/m^3 may reduce the bulk hydraulic conductivity to less than the minimum 10^{-7} m/s required. Similarly, pulverisation of waste causes more dense compaction at depth, and the requirement for a minimum hydraulic requirement may limit the depth of a sustainable landfill site to between 20 and 30 m. Much higher levels of leachate are generated from the flushing bioreactor, and generation rates of leachate for the lifetime of the waste are between 4000 and 11 000 l/tonne of waste. This compares with the containment type of landfill, without leachate flushing, of about 276 l/tonne of waste for the lifetime of the site. Consequently, the leachate management system for the flushing bioreactor is substantially greater than conventional systems (Beaven 1996).

Accelerated stabilisation using the flushing bioreactor implies accelerated biodegradation of the waste and a consequent increase in the rate of landfill gas generation. Consequently, the gas collection and control system is required to have a higher capacity, since the potential volume of gas is evolved in a shorter period of time.

Biodegradation to produce landfill gas is largely due to methanogenic degradation over a temperature range of 30–65 °C, with an optimum temperature range of gas generation between 30 and 45 °C. Accelerated biodegradation can therefore be enhanced by maintaining temperature control throughout the mass in the range 30–45 °C by the use of insulating capping and base liner systems. The system for the even distribution of leachate throughout the waste requires a horizontal arrangement of closely spaced injection pipework, trenches or wells. Even and uniform liquid distribution is dependent on the spacing of the injection system and the permeability of the waste. The recirculation system may be surface or sub-surface. Surface systems use rain-guns, or perforated pipes to spray the water/leachate onto the surface of the waste. Sub-surface systems use perforated pipes and have the advantage that they may be placed below the final cap.

A number of case studies have been reported involving leachate recirculation at landfill sites, to achieve increased waste stabilisation. For example, in the UK at the Brogborough landfill site in Bedfordshire and Lower Spen Valley landfill, West Yorkshire (Blakey et al 1995), in the USA at the Alachua County Southwest landfill in Florida USA (Townsend et al 1996), at eight sites throughout the USA including the landfills at Central Facility, Maryland, Winfield, Florida, Pecan Row, Georgia, Lemons, Missouri, Mill Seat, New York State, etc. (Reinhart 1996, Reinhart and Al-Yousfi 1996) and in Melbourne Australia (Yuen et al 1995). In summary, most case studies report that accelerated stabilisation of waste and increased gas production occurs at large-scale landfill sites where leachate recirculation is employed. Further points arising from the case studies suggest that, whilst operating parameters were variable from case to case, leachate quality may not be significantly affected, and permeability of the waste is of major importance to allow efficient recirculation of leachate. On-site storage of leachate is also recommended to reduce off-site charges for water and sewerage.

4.10.2   Entombment (Dry) Landfills

The entombment or dry landfill type of landfill site aims to contain the waste in a relatively dry form for long periods of time by preventing biodegradation or physical and chemical reaction of the waste, and thereby the formation of leachate or landfill gas. The entombment type of landfill is a containment landfill that is designed and operated on the principle that the landfill is contained indefinitely. This type of landfill is common in the USA and France and has been used for hazardous and non-hazardous types of waste (Kjeldsen et al 2002; LaGrega et al 1994; Freeman 1998; Moss 1997).

For biodegradable wastes, the entombment type landfill in theory generates minimal leachate and therefore should not pose an environmental problem. The typical features of an entombment type landfill are shown in Table 4.13 (Lee and Jones-Lee 1993). The dry waste approach prevents the infiltration of rain water, ground water and surface water. Consequently, the entombment waste landfill acts as a long-term storage site. Whilst preventing the infiltration of water is the aim of the design, in practice, particularly for biodegradable, municipal solid waste, water will inevitably be present with the waste and some degradation of the waste will occur, with the consequent production of leachate and landfill gas. Therefore, monitoring, collection and control of leachate and landfill gas is also required. To this end the liner materials and liner system are designed to ensure that no percolation of water into or out of the site boundary occurs. Any leachate that forms is quickly removed and treated to prevent increased biodegradation and further formation of leachate and landfill gas (Westlake 1995).

Waste storage has the advantages that the environmental hazard is 'contained' and that at some time in the future new technologies may be used to treat the waste. In addition, because the dry tomb approach produces low levels of leachate and landfill gas, the leachate

Table 4.13   *Typical features of the entombment-type (dry) landfill*

Liner – The liner is typically a composite liner composed of a compacted clay underlayer with an overlying flexible plastic membrane sheeting liner. The liner is designed to prevent escape of leachate from the landfill to the surrounding environment and also serves as a foundation for the leachate collection and removal system

Leachate collection and removal system – The leachate and collection removal system is a drainage system placed between the mass of waste and the liner system and is designed to collect and transport the leachate to where it can be removed by pumping or gravity flow

Cover – A low-permeability covering is placed over the landfill once it has been filled and is designed to keep rainwater and run-on water away from the landfill

Groundwater monitoring wells – A groundwater monitoring program is relied upon to signal the failure of the liner system to control the containment of the leachate

Other measures – Other systems may be incorporated in a 'dry tomb' landfill design to enhance the ability of the system to repel moisture or to manage leachate

Source: Lee and Jones-Lee 1993.

and gas collection, control and treatment requirements are much reduced and consequently cheaper. The principle of the entombment landfill, that formation of leachate and landfill gas is minimised by limiting the amount of water penetrating the site, emphasises the importance of the liner system. Developments in entombment design have therefore concentrated on increasingly sophisticated liner and capping systems. In most cases thick polymeric plastic liner materials are the basis of the liner system and are used to line the base, sides and cap of the site. In addition, the liner system uses layers of soil, clay, sand, gravel and geotextiles to protect the liner material and aid the containment of the waste and thus prevent ingress of water. In addition, the siting of the landfill is increasingly towards local climates and topographies which are naturally drier (Kjeldsen et al 2002). The conflict of landfilling biodegradable wastes but trying to prevent or limit the biodegradation of the wastes has been recognised, particularly as the integrity of the landfill liner system is required to be maintained for decades or even centuries into the future (Joseph and Mather 1993). The obligations and liabilities for the monitoring, collection and treatment of leachate and landfill gas will therefore also extend far into the future, leading to environmental and economic implications for the operator.

Whilst increasing sophistication of the landfill liner system gives theoretical containment lifetimes of leachate of the order of thousands of years, in practice the liner will eventually fail at some time in the future. The entombment-type landfill liner systems may therefore not be expected to contain the leachate forever. Routes for the liner system to eventually fail and emit water may be, for example, through burrowing animals breaking through the capping material, or the plastic liner may contain flaws or become damaged during emplacement, resulting in small holes in the liner, and freeze–thaw of water and long-term corrosion from the chemical and physical properties of the leachate may cause damage to the liner material. Whilst the time period may be many decades or even centuries, the liner material will fail, and release the leachate to the environment. In addition, the long-term ability of the leachate control and removal system and the monitoring system to continue operation for very long periods of time are questioned, particularly as such systems are prone to clogging and their maintenance would be difficult or impossible to carry out due to accessibility problems. Further problems associated with the long-term containment of waste are that the conditions within the landfill may change, resulting in environmental damage. Consequently, the entombment type of landfill is increasingly being criticised as not an environmentally sustainable route to waste management (Lee and Jones-Lee 1993).

4.10.3   Landfills in the Developing World

The great majority of waste generated in developing countries is mainly of domestic origin consisting of mostly food waste, green waste and relatively low concentrations of toxic materials (Diaz and Savage 2002). Consequently, there is less of a need for a sophisticated landfill liner system of the type required for the waste generated from industrialised countries. The sophistication of waste landfill design will clearly vary across the world. Comparison can be made with the well structured, monitored and highly legislated landfill designs in operation in Europe, with those in developing countries. In many of the developing countries in Africa, Latin America and Asia, waste landfill is better categorised in the majority of cases as 'open dumping' (Diaz and Savage 2002; Johannessen and Boyer 1999). Containment-type landfills which utilise a liner, barrier system are less common. Whilst the environmental impacts of poor waste management may be known, the management of waste remains a low priority for such economies. On-site waste pickers sort through the deposited waste to recover materials for recycling and to provide a source of income. Such waste picking involves whole families and in some cases villages are dependent on waste scavenging for income. As towns and cities become larger and more developed, the environmental nuisance from the uncontrolled open dumping leads to the development of a more controlled system of landfill involving natural and synthetic liner materials, some form of leachate management and passive venting of landfill gas (Johannessen and Boyer 1999).

4.11   Landfill Gas

Gases arising from the biodegradation of biodegradable wastes in landfills consist of hydrogen and carbon dioxide in the early stages, followed by mainly methane and carbon dioxide in the later stages. What is known as 'landfill gas' is a product mainly of the methanogenic stage of degradation of biodegradable wastes. Landfill gas is produced from municipal solid waste which contain a significant proportion of biodegradable materials (Table 4.2). Municipal solid waste is permitted to be deposited into non-hazardous waste landfills under the EC Waste Landfill Directive (1999). In addition, wastes permitted to be deposited into hazardous waste landfill may also contain biodegradable components which will degrade to produce landfill gas. The main gases are methane and carbon dioxide, but a wide range of other gases can potentially be formed. In addition, the gas is usually saturated with moisture. Table 4.14 shows the composition of the major constituents of landfill gas and Table 4.15 shows the range of trace components (Waste Management Paper 27, 1994; Waste Management Paper 26, 1986). The main chemical compounds found in landfill gas can be broadly categorised into saturated and unsaturated hydrocarbons, acidic hydrocarbons, organic alcohols, aromatic hydrocarbons, halogenated compounds,

Table 4.14 *Typical landfill gas composition*

Component	Typical value (% by volume)	Observed maximum (% by volume)
Methane	63.8	88.0
Carbon dioxide	33.6	89.3
Oxygen	0.16	20.9
Nitrogen	2.4	87.0
Hydrogen	0.05	21.1
Carbon monoxide	0.001	0.09
Ethane	0.005	0.0139
Ethene	0.018	—
Acetaldehyde	0.005	—
Propane	0.002	0.0171
Butanes	0.003	0.023
Helium	0.00005	—
Higher alkanes	<0.05	0.07
Unsaturated hydrocarbons	0.009	0.048
Halogenated compounds	0.00002	0.032
Hydrogen sulphide	0.00002	35.0
Organosulphur compounds	0.00001	0.028
Alcohols	0.00001	0.127
Others	0.00005	0.023

Source: Waste Management Paper 27, 1994. Crown copyright material is reproduced with the permission of the Controller of HMSO and the Queen's Printer for Scotland.

Table 4.15 *Trace components found in landfill gas*

Component	Concentration range (mg/m³)	Component	Concentration range (mg/m³)
Alkanes		*Alkenes*	
Propane	<0.1–1.0	Butadiene	<0.1–20
Butanes	<0.1–90	Butenes	<0.1–90
Pentanes	1.8–105	Pentadienes	<0.1–0.4
Hexanes	1.3–628	Pentenes	<0.5–2
Heptanes	4–1054	Hexenes	<0.5–136
Octanes	8.5–675	Heptadienes	<0.1–1.9
Nonanes	31–226	Heptenes	0.3–103
Decanes	81–335	Octenes	<1–144
Undecanes	12–164	Nonadienes	<0.1–9
		Nonenes	5.2–7.5
		Decenes	13–188
		Undecenes	<2–54
Cycloalkanes		*Cycloalkenes*	
Cyclopentane	<0.2–6.7	Limonene	2.1–240
Cyclohexane	<0.5–103	Other terpenes	14.3–311
Methylcyclopentane	<0.1–79	Methene	<0.1–29
Dimethylcyclopentanes	0.1–330		
Ethylcyclopentane	<0.1–<2		

Table 4.15   *Continued*

Component	Concentration range (mg/m³)	Component	Concentration range (mg/m³)
		Aromatic Hydrocarbons	
Methylcyclohexane	1.5–290	Benzene	0.4–114
Trimethylcyclopentanes	<0.1–58	Toluene	8–>460
Dimethylcyclohexanes	<2–54	Styrene	<0.1–7
Trimethylcyclohexanes	<0.1–27	Xylenes	34–470
Propylcyclohexanes	<0.5–8	Ethylbenzene	17–330
Butylcyclohexanes	<0.1–4	Methylstyrene	<0.1–15
		Propylbenzenes	36–292
		Butylbenzenes	5.8–138
		Pentylbenzenes	0.4–17.5
Halogenated compounds		*Organosulphur compounds*	
Chloromethane	<0.1–1	Carbonyl sulphide	<0.1–1
Chlorofluoromethane	<0.1–10	Carbon disulphide	<0.1–2
Dichloromethane	<0.1–190	Methanethiol	<0.1–87
Chlorodifluoromethane	<0.1–16	Ethanethiol	<0.1–<2
Dichlorofluoromethane	<0.1–93	Dimethyl sulphide	<0.2–60
Chloroform	<0.1–0.8	Dimethyl disulphide	0.1–40
Dichlorodifluoromethane	<0.1–48	Diethyl disulphide	<0.1–0.6
Trichlorofluoromethane	<0.1–20	Butanethiols	<0.1–2.4
Chloroethane	<0.1–46	Pentanethiols	<0.1–1.2
1,1-Dichloroethane	<0.1–130		
1,2-Dichloroethane	<0.1–8	*Alcohols*	
Vinylchloride	<0.1–32	Methanol	<0.1–210
1,1,1-Trichloroethane	<0.1–177	Ethanol	<0.1–>810
1,2-Dichloroethylenes	<0.1–302	Propan-1-ol	<0.1–110
Trichloroethylene	<0.1–170	Propan-2-ol	<0.1–>46
Tetrachloroethylene	<0.1–350	Butan-1-ol	<0.1–>19
1,1-Dichlorotetrafluoroethane	<0.1–1	Iso-butan-1-ol	<0.1–>5.3
1,2-Dichlorotetrafluoroethane	<0.1–10	Butan-2-ol	<0.1–210
1,1,1-Trichlorotrifluoroethane	<0.1–70		
Bromoethane	<0.1–<2	*Ethers*	
Chloropropanes	<0.1–<2	Dimethylether	0.02–<2
Dichlorobutanes	<0.1–<2	Methylethylether	<0.1–<2
Chlorobenzene	<0.1–2.1	Diethylether	<0.1–12
Dichlorobenzenes	<2–16	Dipropylethers	<0.1–220
Esters		*Other oxygenated compounds*	
Ethyl acetate	<0.1–64	Acetone	<0.1–3.4
Methyl butanoate	<0.1–15	1,3-Dioxolane	<0.1–5
Ethyl propionate	<0.1–136	Butan-2-one	0.4–38
Propyl acetate	<0.1–50	Tetrahydrofuran	<0.1–<2
Isopropyl acetate	<0.1–6	Pentan-2-one	<0.1–4.2
Methyl petanoate	<0.1–22	Methyl furans	<0.1–0.8
Ethyl butanoate	<0.1–350	Dimethyl furans	<0.1–12
Propyl propionate	<0.1–200	Camphor/fenchone	<0.1–13
Butyl acetate	<0.1–60	Carboxylic acids	<0.1–<2
Ethyl pentanoate	<0.1–27		
Propyl butanoate	<0.1–100		

Source: Waste Management paper 26, 1986. Crown copyright material is reproduced with the permission of the Controller of HMSO and the Queen's Printer for Scotland.

sulphur compounds and inorganic compounds (Allen et al 1997). The major constituents of landfill gas, methane and carbon dioxide are odourless, and it is the minor components such as hydrogen sulphide, organic esters and the organosulphur compounds which give landfill gas a malodorous smell. Landfill gas contains components which are flammable and when mixed with air can reach explosive concentrations in confined spaces. There have been problems associated with uncontrolled leakages of landfill gas into houses, shafts, culverts, pipework, etc., with potentially devastating effects (Williams and Aitkenhead 1991). The lower flammable limit, where ignition of the gas mixture can occur, is 4% for hydrogen and 5% for methane. In addition, the gas can cause asphyxiation where levels accumulate in such areas as manholes and culverts (Waste Management Paper 26, 1986; Waste Management Paper 27, 1994). This is particularly a problem where certain mixtures of landfill gas components result in the gas having a higher or lower density than air thus causing stratification of the air and gas. An asphyxiation hazard can occur in a confined space where the oxygen level has fallen from 21 to 18%. Some of the trace components of landfill gas have a toxic effect and may be hazardous if high enough concentrations are reached, for example, hydrogen sulphide (Waste Management Paper 27, 1994; Rettenberger and Stegmann 1996). Aromatic hydrocarbons are in low concentration but may potentially have an adverse effect on the workforce of the landfill site. A wide range of chlorinated hydrocarbons have been identified in landfill gas (Allen et al 1997). Chlorinated hydrocarbons are important because of their potential harm to the environment, but also when landfill gas is used as a fuel in landfill gas utilisation schemes there is the potential to form hydrogen chloride (Allen et al 1997; Rettenberger and Stegmann 1996). Young and Blakey (1996) also report a wide range of trace components in landfill gas, including trace organosulphur compounds.

A major review of a large number of epidemiological studies and exposure-risk studies associated with waste landfill sites has been carried out (Redfearn and Roberts 2002). The main route linking the source of pollution to the receptor (source–receptor pathway) was deemed to be via inhalation of landfill gas. Redfearn and Roberts (2002) concluded that the emission levels of trace constituents in landfill gas were too low to be sufficient to result in adverse health effects in the surrounding population, based on calculated exposure-risk assessments using air-dispersion-type models. The amount of dilution of the trace constituents in the atmosphere would render the concentrations, to which the population was exposed, several orders of magnitude below health-based criteria levels. However, some human health epidemiological studies on the populations around European landfill sites have suggested certain elevated risks, including birth defects and low birth weights. A larger number of other epidemiological health studies have shown no health risk associated with landfill sites. Redfearn and Roberts (2002) emphasise that the cause and effect of landfill sites and ill health is very difficult to assess due to the compounding influence of other factors, such as other industrial sources of pollution.

The major components of landfill gas, methane and carbon dioxide, are 'greenhouse gases'. The greenhouse effect is produced by certain gases in the atmosphere which allow transmission of short-wave radiation from the sun, but are opaque to long-wave radiation reflected from the earth's surface, thereby causing warming of the earth's atmosphere. A molecule of methane has approximately 30 times the greenhouse effect of a molecule of carbon dioxide (Porteous 1992).

The quantities of gas produced from waste depend on the biodegradable fraction of the waste, the presence of micro-organisms and suitable aerobic and anaerobic conditions

Table 4.16 *Theoretical production of carbon dioxide and methane from the major representative components of waste*

Component	Total Carbon dioxide + methane (m³/tonne)	Gas composition	
		Carbon dioxide (%)	Methane (%)
Cellulose (carbohydrate)	829	50.0	50.0
Protein	988	48.5	51.5
Fat	1430	28.6	71.4
Typical waste	300–500 (estimated)		
Typical waste	39–390 (measured)		

Source: McBean et al 1995.

and moisture. Theoretical work on the production of carbon dioxide and methane from the major fractions of biodegradable wastes, i.e., degradation of representative carbohydrates (cellulose), proteins and fats in the methanogenic stage of biodegradation of wastes in landfills are shown in Table 4.16 (McBean et al 1995). The table shows that the composition of the waste influences the composition and production of landfill gas. Estimates of the theoretical production of landfill gas from municipal solid waste, with a typical biodegradable composition, indicate that between 300 and 500 m³/tonne of gas would be generated throughout the lifetime of the site. Actual measurements of landfill gas generation rates, however, are highly variable (between 39 and 390 m³/tonne) due to the range of waste compositions, the fact that all the waste may not decompose and that gas may not always be the endproduct of degradation (McBean et al 1995).

4.11.1   Landfill Gas Migration

Gases generated in the landfill will move throughout the mass of waste in addition to movement or migration out of the site. The mechanism of gas movement is via gaseous diffusion and advection or pressure gradient. That is, the gas moves from high to low gas concentration regions or from high to low gas pressure regions (Kjeldsen et al 2002). Movement of gas within the mass of waste is governed by the permeability of the waste, overlying daily or intermittent cover, and the degree of compaction of the waste. Lateral movement of the gases is caused by overlying low permeability layers such as the daily cover and surface and sub-surface accumulations of water. Vertical movement of gas may occur through natural settlement of the waste, between bales of waste if a baling system is used to compact and bale the waste, or through layers of low permeability inert wastes such as construction waste rubble. Where landfill gas extraction is practised to recover the gas for energy use, the gas is collected in gas wells, and piped to the surface (Waste Management Paper 27, 1994).

Fully contained landfill sites where, after completion, the landfill is capped with an impermeable synthetic and natural containment system to prevent migration of landfill gas out of the site and which have gas recovery systems in place, have low gas emission levels (Mosher et al 1999). The capping liner system is also designed to prevent ingress of precipitation. For landfill sites where landfilling is still in operation and where the waste is only partially covered by an impermeable layer, then there are higher emissions of landfill gas. Tables 4.14 and 4.15 show that waste landfills are a source of volatile organic hydrocarbons, both to the site workers and to the surrounding neighbourhood. The contribution of a range of chemicals identified in landfill gas have been shown to be significant contributors to the toxic air pollutants in local neighbourhoods adjacent to landfill sites (Scheff et al 2001). Certain chemicals, including chlorinated hydrocarbons, have been identified as being derived from landfill as the major source. Additionally, other work has shown that chlorinated hydrocarbons are found in landfill gas at concentrations which exceed occupational exposure levels (Allen et al 1997). However, it is unlikely that long exposure to such levels would be experienced by landfill site workers and even less likely for members of the public (Allen et al 1997).

Sub-surface gas migration out of the mass of waste into the surrounding environment may occur from older sites, where containment was not practised, or through containment sites, where significant leakage has occurred. In addition, leachate movement out of such sites may cause later degradation to landfill gas. Migration of gas outside the site requires migration pathways such as high-permeability geological strata, through caves, cavities, cracks in the overlying capping layer and through man-made shafts, such as mine shafts and service ducts, etc. Gas may migrate considerable distances from the boundaries of the site through these possible pathways. It has been reported that changes in the major and trace components of landfill gas occur during subsurface migration (Ward et al 1996). For example, reduction in methane concentration occurs due to oxidation, and some alteration of trace landfill gases occurs due to adsorption onto soil particles, oxidation, degradation, condensation and dissolution.

4.11.2   Management and Monitoring of Landfill Gas

With the recognition of the formation of landfill gas and its associated hazards, and the potential to utilise the energy content of the gas, the modern landfill site is designed to trap the gases for flaring or use in energy recovery systems, particularly for the landfilling of biodegradable municipal solid waste in non-hazardous waste landfills. The priority for control of the gases is to protect the environment and prevent unacceptable risk to human health, and a landfill gas control system is therefore required. In addition, control mechanisms are required to minimise the risk of migration of the gases out of the site.

There are three types of system used to control landfill gas migration (Tchobanoglous and O'Leary 1994; McBean et al 1995; Pescod 1991–93; Waste Management Paper 26B, 1995; Waste Management Paper 27, 1994): passive venting; physical barriers; pumping extraction systems.

Passive venting   Passive venting systems are only recommended for old sites in the late stages of gas generation where gas generation rates are low, or where inert wastes are landfilled and similarly low or negligible rates of gas generation are found. The passive venting pit consists of a highly permeable vent of gravel material encased in a geotextile fabric to prevent ingress of fine material and reduction of permeability. The gases flow up the highly permeable layer and vent passively into the atmosphere through a permeable capping layer of sand and granular soil or crushed stone. The vent may also be constructed of granular material but with a central perforated plastic pipe, the pipe venting directly to the atmosphere. Construction of the passive venting system may be as emplacement of the waste proceeds or afterwards by drilling or excavation into the mass of waste. Typically the vents are placed at intervals of between 20 and 50 m. Other designs of passive gas venting systems include trenches, which are excavated into or at the boundary of the waste. The trench is lined at the outer edge with a low-permeability barrier and the trench is filled with a high-permeability gravel, or perforated pipes are used to vent the migrating gas to the surface (Waste Management Paper 26B, 1995; Waste Management Paper 27, 1994).

Physical barriers   Physical barriers use low-permeability barriers of, for example, flexible polymeric geomembranes, bentonite cement or clay, to contain and restrict the gas migration. Whilst these barriers might form part of a leachate containment system, they are less effective in containing gas. Coefficients of permeability for gas containment are required to be lower than 10^{-9} m/s. Efficiencies of barriers are improved if they are combined with a means of removing the gas by either passive venting or pumped extraction.

Pumping extraction systems   Pumping extraction systems pump the gas out of the landfill. The gas migrates to gas pits or wells within the waste, which consist of highly permeable gravel, stones or rubble with a central perforated plastic pipe. The gases pass through the high-permeability vent to a plain unperforated pipe which draws the gases through to the pump. Leachate vapour may also be pumped out with the gas, which has a high moisture content, and therefore a leachate condensation trap is required. Figure 4.14 shows a typical pumping extraction well. The gas pumped to the surface is either flared by self-sustaining combustion or the use of a support fuel, utilised in an energy recovery system, or if the gas concentrations are sufficiently low, discharged to the atmosphere.

Monitoring of landfill gas   The monitoring programme for landfill gas at waste landfill sites is recommended to determine whether landfill gas is causing a hazard to human health or the environment. Monitoring takes place throughout the operation of the plant and for many years during the post-closure period, until emission levels of methane and carbon dioxide are at environmentally insignificant levels, typically below 1.0% by volume of methane and 1.5% by volume of carbon dioxide. Monitoring takes place within the landfill and outside the site boundary. The monitoring programme, including the frequency of monitoring, will be dependent on the age of the site, the type of waste and the gas collection and control measures installed. Frequency between measurements will vary from weekly, monthly or even quarterly, depending on site-specific characteristics. More frequent monitoring may be required where migration of gas is suspected. Table 4.1 shows the monitoring requirements for landfill gas and leachate for waste landfill sites in Member States of the European Union for the operating and post-closure period of a land-fill site, as required by the EC Waste Landfill Directive (1999). Monitoring techniques for landfill gas include, for example, surface monitoring, sub-surface probes, gas monitoring wells and boreholes. Surface monitoring with portable instruments is mainly used to detect the presence of gas leaks throughout the site. Sub-surface monitoring using gas probes is used to monitor gas production and migration at depths of between 1 and 10 m in the mass of waste and in the surrounding environment. The probes may be left for long periods of time to monitor and map the production of gas from the site throughout site operation and post-closure. The probes are constructed of steel and plastic pipe, consisting of a porous lower section and a gas transfer pipe which transfers the gas to the surface,

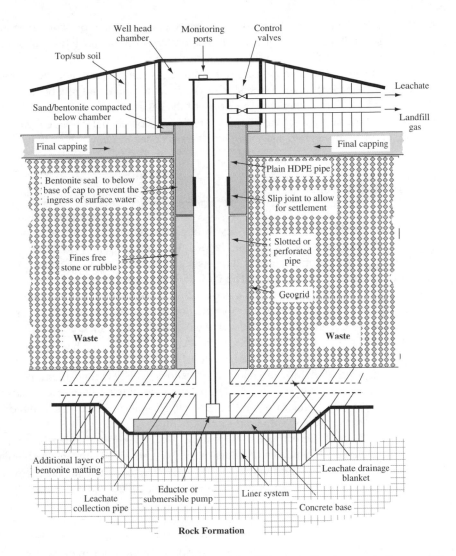

Figure 4.14 *Typical combined leachate and landfill gas collection well. Source: Waste Management Paper 26B, 1995. Crown copyright material is reproduced with the permission of the Controller of HMSO and the Queen's Printer for Scotland.*

where the gas sample is taken for analysis. Gas monitoring well and boreholes consist of a porous plastic casing in direct contact with the waste or geological strata. Probes or tubes may be permanently installed. They are installed within the mass of waste and in the surrounding environment.

Gas sample analysis may take the form of portable instruments for gas analysis or laboratory-based analysis. Portable analysers may be simple devices, such as gas indicator tubes, which produce a colour change to indicate a concentration of a particular gas in a sample. The gas is drawn through the tube on-site and an immediate indication of gas concentration is obtained. The method is, however, subject to error. More sophisticated instruments are available, such as infra-red gas analysers and flame ionisation detectors, which would normally be housed in a portable laboratory or at the analytical laboratory. The gas sample is piped directly to the analyser or else a sample of the gas is taken in suitable sealable containers, such as 'Teflon' bags or glass sample tubes and the sample is then transferred to the instrument for analysis. The most accurate and reliable technique for gas analysis is gas chromatography. A sample of the gas is taken in a suitable container to the laboratory for analysis. The gas chromatograph can separate out individual gas components and provide an accurate analysis, even at trace concentrations.

4.12   Landfill Leachate

Leachate represents the water which passes through the waste from precipitation, and water generated from the waste within the landfill site, resulting in a liquid containing suspended solids, soluble components of the waste and products from the degradation of the waste by various micro-organisms. The composition of the leachate will depend on the heterogeneity and composition of the waste and, for biodegradable wastes, the stage of biodegradation reached by the waste, the moisture content and the operational procedures (Kjeldsen et al 2002). The characteristics of the leachate are influenced by the waste material deposited in the site. For example, inert wastes will produce a leachate with low concentrations of components, whereas a hazardous waste leachate tends to have a wide range of components with highly variable concentrations. The decomposition rate of the waste also depends on aspects such as pH, temperature, aerobic or anaerobic conditions and the associated types of micro-organism. Associated with leachate is a malodorous smell, due mainly to the presence of organic acids.

The EC Waste Landfill Directive (1999) requires the containment of leachate within the landfill site by the implementation of a liner barrier system which contains the leachate. The landfill base and sides are lined with a natural mineral layer or an equivalent artificial synthetic layer to produce a sufficient environmental protection for the surrounding soil and groundwater. For hazardous waste landfill sites, a liner barrier which produces an hydraulic conductivity of $\leq 1.0 \times 10^{-9}$ m/s, coupled with a liner material thickness of ≥ 5 m, is required. For non-hazardous waste landfill sites the liner barrier system should have an hydraulic conductivity of $\leq 1.0 \times 10^{-9}$ m/s and liner material thickness of ≥ 1 m. Even inert landfill sites are required to have a liner barrier system to protect the surrounding environment and groundwater from the impact of leachate, and an hydraulic conductivity of $\leq 1.0 \times 10^{-7}$ m/s and liner material thickness of ≥ 1 m is required (EC Waste Landfill Directive 1999). The leachate is collected and treated to remove pollutants to environmentally acceptable levels. Throughout the lifetime of the landfill site, including during the operational and post-closure phases, sampling and analysis of the landfill leachate is required, this includes monitoring outside the landfill site to assess any impact on the surrounding environment.

Leachate from hazardous waste landfill sites tends to have highly variable concentrations of a wide range of components such as salts, halogenated organic compounds and trace metals and organic compounds (The World Resource Foundation 1996). Municipal solid waste deposited into non-hazardous waste landfills generates a leachate high in organic matter (COD>20 000 mg/l) falling to lower levels (COD~2000 mg/l) after several years.

They are also characterised by high concentrations of ammonia. Where the landfill contains lower levels of biodegradable waste, the leachate has lower levels of COD (COD < 4000 mg/l) throughout the lifetime of the site. Leachate from inert landfill sites tend to have low concentrations of most organic and inorganic species.

The production of leachate from the decomposition of municipal solid waste in non-hazardous waste landfill sites, changes with time as the waste degrades through the various five stages of biodegradation (Section 4.9.2.1). Table 4.17 compares the typical leachate of the acetogenic Stage III with the methanogenic Stage IV (Waste Management Paper 26B, 1995). The table shows that the pH of the early formed leachate is acidic/neutral with a pH range between 5.12 and 7.8, equating with the formation of acetic acid and other organic acids by the acetogenic micro-organisms under anaerobic conditions. The organic material of Stage III is very high, in the range 1010–29000 mg/l for the TOC.

Table 4.17 *Composition of acetogenic and methanogenic leachate from large landfill sites with high waste input rate and relatively dry environments (mg/l)*

Parameter	Acetogenic		Methanogenic	
	Range	Mean	Range	Mean
pH value	5.12–7.8	6.73	6.8–8.2	7.52
COD	2740–152000	36817	622–8000	2307
$BOD_{5\,day}$	2000–68000	18632	97–1770	374
Ammoniacal-N	194–3610	922	283–2040	889
Chloride	659–4670	1805	570–4710	2074
$BOD_{20\,day}$	2000–125000	25108	110–1900	544
TOC	1010–29000	12217	184–2270	733
Fatty acids (as C)	963–22414	8197	<5–146	18
Alkalinity (as $CaCO_3$)	2720–15870	7251	3000–9130	5376
Conductivity (μS/cm)	5800–52000	16921	5990–19300	11502
Nitrate-N	<0.2–18.0	1.80	0.2–2.1	0.86
Nitrite-N	0.01–1.4	0.20	<0.01–1.3	0.17
Sulphate (as SO_4)	<5–1560	676	<5–322	67
Phosphate (as P)	0.6–22.6	5.0	0.3–18.4	4.3
Sodium	474–2400	1371	474–3650	1480
Magnesium	25–820	384	40–1580	250
Potassium	350–3100	1143	100–1580	854
Calcium	270–6240	2241	23–501	151
Chromium	0.03–0.3	0.13	<0.03–0.56	0.09
Manganese	1.40–164.0	32.94	0.04–3.59	0.46
Iron	48.3–2300	653.8	1.6–160	27.4
Nickel	<0.03–1.87	0.42	<0.03–0.6	0.17
Copper	0.020–1.10	0.130	<0.02–0.62	0.17
Zinc	0.09–140	17.37	0.03–6.7	1.14
Cadmium	<0.01–0.10	0.02	<0.01–0.08	0.015
Lead	<0.04–0.65	0.28	<0.04–1.9	0.20
Arsenic	<0.001–0.148	0.024	<0.001–0.485	0.034
Mercury	<0.0001–0.0015	0.0004	<0.0001–0.0008	0.0002

N.B. Between 13 and 35 samples of acetogenic leachate and between 16 and 29 samples of methanogenic leachate were analysed to obtain range and mean results.
Units mg/l except for pH and conductivity.
Source: Waste Management Paper 26B, 1995.

Ammoniacal nitrogen levels tend to be higher in Stage III, due to the biodegradation of the amino acids of proteins and other nitrogenous compounds in the waste. The presence of organic acids of the acetogenic stage increases the solubility of metal ions into the leachate. BOD and COD levels are high, with high ratios of BOD:COD, indicating that a high proportion of the organic materials in solution are readily biodegradable. Methanogenic leachate has a neutral/alkaline pH reflecting the degradation of the organic acids of Stage III to methane and carbon dioxide by the methanogenic micro-organisms. As a consequence, the TOC in the leachate decreases compared with the acetogenic stage. Metal ions continue to be leached from the waste but as the pH of the leachate increases, the metal ions become less soluble and decrease in concentration in the leachate. The concentration of ammoniacal nitrogen decreases slightly, but remains high in the leachate. BOD and COD levels decrease compared with acetogenic leachates.

Box 4.2
BOD and COD

Biochemical oxygen demand is a standard test for the presence of organic matter in water. High levels of organic matter in water cause pollution problems since micro-organisms in the water biodegrade the organic material and thereby use up the dissolved oxygen in the water, leaving insufficient for fish and other aquatic life. The requirement for oxygen is the BOD. The standard BOD test involves a sample of water in a sealed bottle which is completely filled with the water sample and which is left in the dark for 5 days at 20 °C. The dissolved oxygen at the start and end of the test is measured using either a dissolved oxygen electrode or a titrimetric method, the Winkler method. The BOD is expressed in g/m^3, representing the amount of oxygen used up by the micro-organisms in biodegrading the organic materials in the water.

   Chemical oxygen demand, as its name suggests, is a chemical method of determining the total organic material in a water sample which can be oxidised chemically rather than biologically. It is a more rapid test, taking only a few hours rather than the 5 days of the BOD test. A strong oxidising agent such as potassium dichromate is used to react with the oxygen in the organic material and titrimetric analysis indicates the amount of oxidation which has taken place. COD is generally higher than the BOD since more compounds can be chemically oxidised than can be biologically oxidised.

Sources: Tchobanoglous and Burton 1991; Fifield and Haines 1995.

   In addition to the components listed in Table 4.17, a wide range of minor components have been detected in leachate from municipal solid waste. Table 4.18 shows the concentrations of trace organic compounds found in leachate from a municipal solid waste landfill site (Rugge et al 1995; White et al 1995). More than 200 trace xenobiotic organic compounds (XOC) have been identified in landfill leachate from municipal solid waste (Kjeldsen et al 2002).

   The pollutants found in leachate from municipal solid waste landfill sites can be broadly categorised into four main groups (Kjeldsen et al 2002).

Table 4.18   *Trace organic components found in municipal solid waste leachate*

Component	Organic component (mg/l)
1,1,1-Trichloroethane	0.086
1,2-Dichloroethane	0.01
2,4-Dichloroethane	0.13
Benzo[a]pyrene	0.00025
Benzene	0.037
Chlorobenzene	0.007
Chloroform	0.029
Chlorophenol	0.00051
Dichloromethane	0.44
Endrin	0.00025
Ethylbenzene	0.058
2-Ethyltoluene	0.005
Hexachlorobenzene	0.0018
Isophorone	0.076
Napthalene	0.006
Polychlorinatedbiphenyls (PCBs)	0.00073
Pentachlorophenol	0.045
Phenol	0.38
1-Propenylbenzene	0.003
Tetrachloromethane	0.2
Toluene	0.41
Toxaphene	0.001
Trichloroethane	0.043
Vinylchloride	0.04
Xylenes	0.107
Dioxins/furans, toxic equivalent (TEQ)	0.32 ng

Sources: Rugge et al 1995; White et al 1995.

- Dissolved organic material (quantifies as COD and TOC), volatile fatty acids and fluvic and humic-like material.
- Inorganic macrocomponents including calcium, magnesium, sodium, potassium, ammonium, iron, manganese, chloride, sulphate and bicarbonate compounds.
- Heavy metals such as cadmium, chromium, copper, lead, nickel and zinc compounds.
- XOC which are compounds not degraded by organisms in the environment and include aromatic hydrocarbons, pesticides, plastisisers, chlorinated aliphatic compounds, etc. These compounds originate from household and industrial chemicals and are present in low concentrations.

The toxicity of leachate based on bioassay testing of leachate with various organisms has reported high toxicity levels in leachate, derived from municipal solid waste landfills (Kjeldsen et al 2002). Several studies of leachate toxicity report that ammonia, chloride, acidity or alkalinity and heavy metal concentrations, are the main toxic pollutant in landfill leachate. It has also been suggested that leachate from municipal solid waste landfill sites may be mutagenic and carcinogenic (Kjeldsen et al 2002).

4.12.1   Leachate management and treatment

It is a requirement of the EC Waste Landfill Directive (1999) that leachate is contained by a liner barrier system to protect the outside environment and that the leachate is collected and treated to remove pollutants to environmentally acceptable levels. A leachate management and treatment system would be required to collect the leachate emanating from the mass of waste and treat the leachate before discharge to sewer. The composition and management of leachate from waste landfill sites across Europe has shown that the most common form of active leachate management is abstraction and discharge to sewer, usually without pre-treatment, this is particularly the case for municipal solid waste leachate (Hjelmar et al 1995). On-site treatment of leachate is not common at landfill sites throughout Europe, but is an increasing practice, mostly at municipal solid waste landfills and at some smaller hazardous waste sites.

The generation rate of leachate is estimated, based on such factors as the rainfall, the amount of the rainfall infiltrating to the waste through the cover, the absorptive capacity of the waste, the input of co-disposed liquid waste, the weight of absorptive waste and any removal of the leakage via seepage or discharge (Figure 4.15, Waste Management Paper 26B, 1995). Estimates of leachate generation for a typical municipal solid waste landfill site using the estimation formula in Figure 4.15 are shown in Figure 4.16 (Waste Management Paper 26B, 1995). Figure 4.16 shows that, during the average production of leachate throughout the 30-year design life of the landfill, 276 l of leachate are generated per tonne of landfilled waste. Because of the uncertainties involved in the leachate generation process from real sites, the estimated leachate generation rate would include varied inputs to provide a worst-case scenario for sizing the leachate treatment system

$$Lo = [ER + LIW + IRA] - [LTP + aW + DL]$$

Lo = Free leachate retained at the site (leachate production minus leachate leaving the site)

ER = Effective rainfall (or actual on an active surface area); this may need to be modified to account for run-off, especially after capping

LIW = Liquid industrial waste (including any surplus water from sludges with a high moisture content)

IRA = Infiltration through restored and capped area

LTP = Discharge of leachate off-site

a   = Unit absorptive capacity of wastes

W   = Weight of absorptive waste

DL   = Designed seepage (if appropriate)

Figure 4.15 *Equation to estimate leachate generation from waste landfills. Source: Waste Management Paper 26B, 1995.*

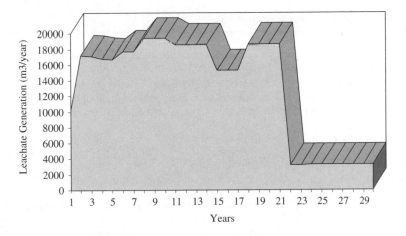

Figure 4.16 *Example of the estimation of leachate generation from waste landfills. Source: Waste Management Paper 26B, 1995.*

and discharging consents to sewer. For example, the leachate produced from a landfill in terms of volume is subject to large seasonal variations.

The leachate management system consists of a leachate drainage, collection and treatment system. The drainage of leachate is via gravity flow through drainage gradient paths which consist of a permeable granular system containing perforated pipes, to collection sumps at low points in the waste mass. The leachate collected in the sumps is then removed by either pumping, gravity drains, or side slope risers at the site perimeter.

The leachate generated from a landfill site will vary in volume and composition depending on the age of the site, the types of waste landfilled and for biodegradable wastes, the stages of biodegradation reached. A consequence of the changes in leachate composition with time is that the leachate control systems should adapt to the changes. Leachate treatment is required to remove the components of the leachate to standards whereby they can be released to sewer, water course, land or tidal water. Before release a discharge permit would be required from the local water company, or the regulatory authority. The permit may cover a range of potentially polluting components, for example, pH, concentration of organic material, ammonium and nitrate, suspended solids and metal content and would stipulate the emission levels which should be obtained before discharge.

Treatment processes for leachate are shown in Table 4.19 (Waste Management Paper 26B, 1995). There are a range of processes available to treat leachate in order to reduce the pollutants to environmentally acceptable levels. These may be broadly categorised as physico-chemical processes, attached and non-attached growth microbiological processes, anaerobic and aerobic treatment, leachate recirculation back into the landfill (controlled flushing bioreactor) or, for low-contaminant leachate, to land spreading.

Monitoring of the leachate is required by the EC Waste Landfill Directive (1999) in the form of sampling and analysis for a variety of pollutants, including monitoring outside the landfill site to assess any impact on the surrounding environment. The waste landfill permit issued and monitored by the regulatory authority will require leachate to be sampled at representative points throughout the landfill site and analysed separately for volume and composition. The detailed analysis of the leachate composition will be based on the type of waste landfilled and the category of landfill, whether hazardous, non-hazardous or

Table 4.19   *Treatment processes for leachate*

Process	Examples
Physico-chemical	
Air stripping of ammonia	Leachate pH adjusted to 11 followed by aeration and release of ammonia gas to a scrubbing unit or the atmosphere
Activated carbon adsorption	Highly porous activated carbon adsorbs organic components, used for final stages of treatment
Reverse osmosis	Ultrafiltration membranes concentrate pollutants into a concentrated solution for disposal, used for suspended material, ammoniacal nitrogen, heavy metals
Evaporation	Concentration of contaminants by evaporation or distillation for disposal
Oxidation	Addition of oxidising agents such as hydrogen peroxide or sodium hypochlorite solution. Used for sulphides, sulphite, formaldehyde, cyanide and phenolics
Wet-air oxidation	Used for high organic content leachates, based on combustion with air at temperatures up to 310 °C and 200 bar
Coagulation, flocculation and settling	Addition of reagents followed by mixing and settlement
Attached growth processes	
Trickling filters	Trickling or percolating filters allow the leachate to pass over a substrate containing aerobic micro-organisms which biodegrade the organic components of the leachate
Rotating biological contactors	Rows of rotating discs with attached aerobic micro-organisms alternatively exposed to air and leachate as they rotate
Non-attached growth processes	
Aereation in lagoons or tanks	Aerobic micro-organisms in suspension biodegrade the organic constituents of the leachate in aerated lagoons or tanks
Anaerobic treatment	
Anaerobic biodegradation	Utilises the anaerobic methanogenic type micro-organisms to biodegrade the constituents of the leachate, not effective for ammoniacal nitrogen
Anaerobic/aerobic treatment	
Reed Bed biodegradation	Reed bed plant systems stimulate the growth of aerobic micro-organisms at the root system and anaerobic micro-organisms in soil areas away from the roots. The range of micro-organisms biodegrade the leachate components, other contaminants may be immobilised or absorbed by the plants
Land treatment	
Spraying leachate onto land	Spray irrigation of leachate to grassland and woodland, used for low contaminant leachate. Treatment processes include, micro-organism biodegradation in the soil, plant uptake of contaminants, oxidation, absorption, transpiration, evaporation, precipitation, nitrification and denitrification
Leachate recirculation	
Recirculation of leachate	Recirculation of the leachate through the mass of waste, utilising the landfill as an uncontrolled reactor for further biodegradation

Source: Waste Management Paper 26B, 1995.

inert and the requirements would be determined by the regulatory authority (Waste Landfill Directive 1999). Analyses of leachate composition are required at quarterly stages during the operational phase of the landfill. In addition, sampling and analysis of the surrounding groundwater would be required to determine any impact from the possible contamination of leachate. After closure of the landfill, monitoring of leachate volume and composition is required at six-monthly intervals until the site is stabilised and the leachate is within environmentally permitted levels. The groundwater outside the site would also be required to be sampled and analysed in the post-closure period.

4.13   Landfill Capping

Final cover or capping of the landfill site is required after the final waste has been deposited. The purpose of the cap is to contain and protect the waste, to prevent rainwater and surface water from percolating into the site and influencing the generation of leachate, to control the release of landfill gas, and to prevent ingress of air which would disrupt the anaerobic biodegradation process. In addition, the final cover is landscaped and provides a soil for the establishment of the restored site plant materials. The design of the cover system of lining materials used to cap the site depends on the nature of the waste deposited, for example, if it is a hazardous, non-hazardous or inert landfill site (EC Waste Landfill Directive 1999). Figure 4.17 shows various components which may be used in a capping system (Waste Management Paper 26B, 1995). Overlying the main body of waste may be the gas collection layer, depending on the nature of the waste. The gas collection layer is a porous material such as geotextile, geonet or coarse sand, through which the gas can easily permeate to the gas collection and control system. A barrier layer is a low permeability layer such as a plastic polymer geomembrane, geosynthetic clay liner of bentonite/geotextile fabric, or compacted natural clay. The barrier layer serves a two-fold purpose: to prevent ingress of water and the egress of landfill gas. The barrier layer may have a protective geotextile layer above and below. The drainage layer/pipework zone may be required to minimise the amount of percolating water reaching the barrier layer. The water is drained off through porous pipes set in a porous layer of coarse sand or gravel, geotextile, geonet, etc. The protection layer protects the underlying liner system from plant root systems, burrowing animals and man-made intrusions. The protection layer consists of soils and may be an extension of the restoration layer. The restoration layer is the top soil, which may be landscaped, contoured for ease of surface water run-off and used for growing plants, depending on the end use of the restored site.

4.14   Landfill Site Completion and Restoration

At the end of the life of a landfill, the landfill operator must demonstrate that the site has physically, chemically and biologically stabilised and no longer poses a risk to the public or the local environment. When a site is deemed as complete, post-closure pollution controls and leachate and landfill gas control systems would no longer be required. Stabilisation is defined in terms of the quantity and composition of the leachate and landfill

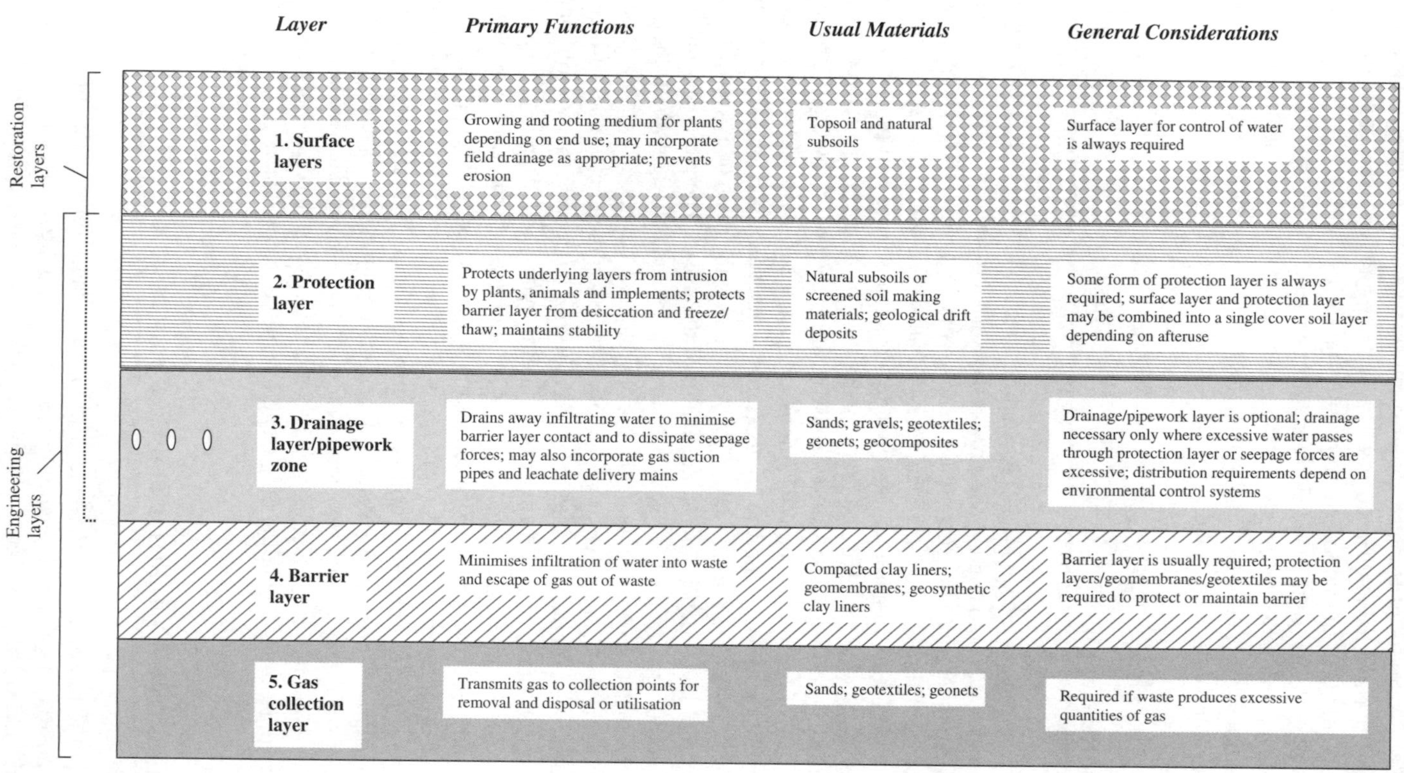
Layer
Primary Functions
Usual Materials
General Considerations
Restoration layers
Engineering layers
1. Surface layers
Growing and rooting medium for plants depending on end use; may incorporate field drainage as appropriate; prevents erosion
Topsoil and natural subsoils
Surface layer for control of water is always required
2. Protection layer
Protects underlying layers from intrusion by plants, animals and implements; protects barrier layer from desiccation and freeze/thaw; maintains stability
Natural subsoils or screened soil making materials; geological drift deposits
Some form of protection layer is always required; surface layer and protection layer may be combined into a single cover soil layer depending on afteruse
3. Drainage layer/pipework zone
Drains away infiltrating water to minimise barrier layer contact and to dissipate seepage forces; may also incorporate gas suction pipes and leachate delivery mains
Sands; gravels; geotextiles; geonets; geocomposites
Drainage/pipework layer is optional; drainage necessary only where excessive water passes through protection layer or seepage forces are excessive; distribution requirements depend on environmental control systems
4. Barrier layer
Minimises infiltration of water into waste and escape of gas out of waste
Compacted clay liners; geomembranes; geosynthetic clay liners
Barrier layer is usually required; protection layers/geomembranes/geotextiles may be required to protect or maintain barrier
5. Gas collection layer
Transmits gas to collection points for removal and disposal or utilisation
Sands; geotextiles; geonets
Required if waste produces excessive quantities of gas

Figure 4.17 *Components of a waste landfill capping barrier system. Source: Waste Management Paper 26B, 1995. Crown copyright material is reproduced with the permission of the Controller of HMSO and the Queen's Printer for Scotland.*

gas produced at the site. To demonstrate that the landfill is stabilised and the leachate and landfill gas volumes and concentrations pose no threat to the environment requires a monitoring and sampling programme for the post-closure period of the landfill. A further consideration is the settlement of the site and the possibility of physical instability of the waste or retaining structures. It has been suggested that to reach environmental target levels of leachate and landfill gas concentrations, time scales of decades and perhaps over 100 years for the treatment of leachate, may be required (Stegmann et al 2003).

Assessment of completion depends on the type of landfill site. For example, sites which have taken only inert wastes pose a low risk to human health and the environment since only low or zero levels of leachate and landfill gas are likely to be generated. For hazardous waste landfill sites and non-hazardous sites, i.e., those accepting biodegradable wastes such as municipal solid waste, then a full assessment of the leachate composition

Table 4.20 *Example of completion criteria for landfill leachate (mg/l)*

Parameter	Concentration
pH	6.5–8.5
Conductivity	4000
Chloride	2000
Sulphate	2500
Calcium	1000
Magnesium	500
Sodium	1500
Potassium	120
Aluminium	2
Nitrate	500
Nitrite	1
Ammonia	5
TOC	10
Iron	2
Manganese	0.5
Copper	1
Zinc	1
Phosphorus	10
Fluoride	10
Barium	1
Arsenic	0.5
Cadmium	0.05
Cyanides	0.5
Chromium	0.5
Mercury	0.01
Nickel	0.5
Lead	0.5
Phenols	0.005
Organo chlorine compounds	0.01
Pesticides	
individually	0.001
collectively	0.005
Polycyclic aromatic hydrocarbons	0.002

Source: Waste Management Paper 26A, 1995.

and gas volume and the potential future generation rates together with an assessment of the waste settlement would be required.

To assess whether stabilisation has occurred and completion has been reached, a sampling and analysis programme for leachate and landfill gas is implemented. The monitoring programme then identifies whether completion conditions have been reached, which equates to the point where the composition of leachate and the volume of gas have reached defined criteria of low levels. Assessment of the settlement of the waste is carried out by engineers. This is to assess the settlement of the waste under its own weight due to consolidation, as degradation takes place.

Table 4.20 shows an example of the completion criteria which should be reached for leachate where there is a likelihood of the leachate entering groundwater (Waste Management Paper 26A, 1995). In terms of landfill gas, completion is deemed to have been reached when the maximum gas concentrations of methane and carbon dioxide remain less than concentration levels set by the regulatory authority for a significant time period.

The stabilisation of the landfill may take decades to complete which also represents decades of responsibility and liability for leachate and landfill gas management and represents a major liability for the operator. Financial liabilities go hand-in-hand with technical liabilities and the provision of monitoring systems, collection and treatment systems for gas and leachate have to be provided long after the site has accepted the final load of waste. The EC Waste Landfill Directive (1999) therefore requires the site operator to demonstrate to the regulatory authority that the owner is financially secure to ensure that commitments to safeguard the environment are in place for the future.

The range of options available for the post-closure after-use of a landfill site includes agriculture, woodland, amenity/conservation and built developments, structures and hard standing areas.

4.15   Energy Recovery from Landfill Gas

The development of larger and larger landfill sites throughout many countries has provided for economies of scale and the economic viability of utilisation of landfill gas for energy recovery. The modern site is seen in this context as a 'bioreactor', used to stabilise waste and produce landfill gas for energy recovery. Therefore, whilst landfill sites exist, which are used for disposal without energy recovery, the modern purpose-built site would normally incorporate a landfill gas extraction system for the recovery of energy.

Estimates of the amount of landfill gas generated throughout the lifetime of the landfill site are highly variable with estimates of between 39 to 500 m^3/tonne (McBean et al 1995). For the estimation of landfill gas throughout the lifetime of a site for the assessment of energy recovery from landfill gas utilisation, values of between 150 and 250 m^3/tonne are typically used (Loening 2003). Annual rates of gas production have been estimated for a typical municipal solid waste landfill at between 6 and 8 m^3/tonne/year but much higher rates of over 25 m^3/tonne/year have been recorded (Characterisation of 100 UK landfill sites 1995). This allows the potential amount of energy which could be generated from the site, knowing that undiluted landfill gas can have a calorific value of between 15 and 21 MJ/m^3, compared to the calorific value of natural gas at about 37 MJ/m^3 (Waste Management Paper 27, 1994). The calorific value of the gas depends on the percentage composition of combustible gases such as methane, and non-combustible gases such as carbon dioxide. The presence of carbon dioxide results in reduced flame temperatures and burning rates, a narrower range of flame stability and thus lower combustion efficiency (Qin et al 2001). The carbon dioxide is also regarded as 'inert', in that it does not combust and therefore does not contribute to the energy content of the landfill gas.

Figure 4.18 shows a schematic diagram of a landfill gas energy recovery project (Brown and Maunder 1994). The energy recovery technology is based around the gas collection system and the pre-treatment and power generation technology. Gas collection is via either vertical gas wells or horizontal well collection systems, depending on the type of site, site filling techniques, depth of waste and leachate level. The gas is collected in a series of perforated gas pipelines connected to a central pipeline. The spacing of the wells for optimum gas collection depends on a number of factors including the rate of gas generation, but would typically be between 20 and 50 m apart. A condensate removal system is required, since the gas is at temperatures above ambient and is saturated with water vapour and organic vapours. As the gas cools the water vapour condenses to form water in the pipe, which reduces the efficiency of gas collection and transport. The condensate system used to remove the water vapour consists of baffled or expansion chambers which cool and condense the water. Condensate systems both below and above ground may be required to de-water the gas. A filter would also be included to remove fine particulate material from the gas flow. The gas is then compressed and possibly passed to a pre-treatment section if a greater degree of clean-up is required, for example, to remove corrosive trace gases and vapours from the gas stream. Such possible pre-treatments may include further filtration, gas chilling to condense certain constituents, absorption and adsorption systems to scrub the gases, and other gas clean-up systems such as membranes and molecular sieves to remove trace contaminants. A large proportion of the landfill gas consists of carbon dioxide which is non-combustible and therefore reduces the overall calorific value

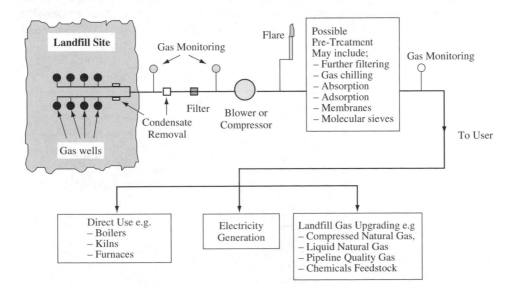

Figure 4.18 *Schematic diagram of a landfill gas energy recovery scheme. Source: Brown and Maunder 1994. Reprinted with kind permission from IWA publishing.*

of the gas. Therefore, for utilisation systems requiring a high specification gas or a high calorific value, then clean-up systems to remove carbon dioxide may be required. Such systems include water scrubbing, absorption on zeolites and membrane separation and are expensive to install and maintain (Brown and Maunder 1994; Stegmann 1996).

The utilisation of the landfill gas is via direct use as substitute fuel in boilers, kilns and furnaces, for electricity generation, or by upgrading to produce CNG, LNG substitute natural gas or for use as a chemical feedstock. Direct utilisation in boilers, kilns and furnaces close to the landfill site represents the easiest and cheapest option, since minimal modifications to the burner system of the combustion unit are required and transport costs are minimised. Power generation is produced from spark ignition engines, diesel engines and gas turbines, with availabilities of about 95% and load factors of over 80%. The engines are used with pure landfill gas or co-fuelled with natural gas. In some cases modifications to the engines are required before operation on landfill gas. The modifications are to take account of the presence of carbon dioxide, which lowers the calorific value and ignitability of the gas compared to natural gas. The main advantage of electricity generation is that the end-user does not have to be located close to the landfill site, since the electricity can be transported via the national grid. Upgraded landfill gas has also been used as a fuel for vehicles used on the landfill site itself. This requires upgrading of the gas for the high specifications required of a vehicle. If the gas is to be used as substitute natural gas for direct input into natural gas pipelines, the gas has to be thoroughly cleaned up to comply with natural gas industry specifications. For example, a minimum calorific value would be specified which would mean removal of carbon dioxide to a certain low level. In addition, fine particulate material, trace components, hydrogen sulphide, etc., would be required to be removed and the gas should attain a consistent composition. It is also technically possible to use the gas as a chemical feedstock with a wide range of products being potentially available from the methane. Table 4.21 shows the major landfill gas end-uses, their limitations and the required landfill gas treatments (Gendebien et al 1992; Stegmann 1996).

A further consideration for combustion of landfill gas is the presence of chlorinated organic compounds at trace level in the gas. The combustion of such chlorinated

Table 4.21   *Major landfill gas end-uses, pre-treatments and advantages/disadvantages*

Landfill gas application	Required pre-treatment	Advantages/disadvantages
1. Direct use, e.g., boilers, kilns, furnaces	Removal of condensate Removal of particulate Dehydration (raw)	Changes required to burner design Must be consumed at or close to the landfill site or transported short distances only
2. Power Generation, e.g., spark ignition engines, diesel engines, gas turbines	Removal of condensate Removal of particulate Dehydration (raw) Removal of halocarbons (if in high concentration)	Can be transported via pipeline, moderate distances Relatively high maintenance costs for spark ignition engines Gas turbines used mainly for large gas throughputs exceeding $2500\,m^3/h$

	Removal of condensate Removal of particulate Partial removal of CO_2 Thorough dehydration Removal of halocarbons (if in high concentration)	Can be transported, via pipeline, moderate distances and mixed with natural gas at low ratios Relatively high maintenance costs for spark ignition engines Gas turbines used mainly for large gas throughputs exceeding 2500 m^3/h
3. Vehicle fuel	Removal of CO_2 Particulate removal Removal of halocarbons (if in high concentration) Compression of the gas	Limited to on-site use
4. Chemical feedstock	Removal of condensate Removal of particulate Removal of CO_2 Removal of H_2S Removal of halocarbons (if in high concentration)	Expensive pre-treatment
5. Injection to the national gas grid	Removal of condensate Removal of particulate Removal of CO_2 Removal of H_2S Removal of halocarbons (if in high concentration)	Expensive pre-treatment

Sources: Gendebien et al 1992; Landfill Gas Development Guidelines 1996; Christensen et al 1996(a).

compounds may lead to the formation of dioxins and furans in the exhaust from the system. High chlorine contents in landfill gas may also lead to high levels of hydrogen chloride gas which condenses to hydrochloric acid, resulting in corrosion down-stream of the combustion section of the system (Allen et al 1997; Loening 2003). In the same way, high sulphur contents may cause similar problems due to sulphuric acid formation.

In many cases landfill gas is flared without energy recovery to destroy the methane and organic micro-pollutants as a means of gas hazard and odour control. In addition, the flare may be required to burn off any excess gas or to act as a standby for any plant shutdowns. The flare may be an exposed open flame, usually on a pedestal, or enclosed in a ceramic furnace. The open-type flare has to maintain a flame even under extremes of weather conditions. Enclosed flares have greater control of the combustion conditions and also enable longer residence times to completely burn out the organic compounds in the landfill gas. The stability of the flame is related to the gas composition, weather conditions, burner design, etc. However, the flame will be stable at methane concentrations of between 30 and 60%. Where flaring is used to dispose of the gas, minimum flame temperatures of between 850 and 1100 °C, and a minimum residence time of 0.3 s are recommended, to destroy any hazardous trace components (IEA Bioenergy 2000). The temperature of the flame is determined by the amount of air added to the landfill gas and the methane composition of the landfill gas. The basic requirements of a landfill gas flare are a flame arrestor, failsafe valve, ignition system and a gas compressor, to increase the pressure of gas to the burner. More advanced flares may also incorporate process control systems, emission monitoring and further safety features (IEA Bioenergy 2000).

The accurate assessment of landfill gas generation from a site, is a major factor in deciding whether the site will be developed for the recovery of energy via landfill gas. Assessment of the landfill gas generation curve over the lifetime of the site then becomes a basis for financial investment in a landfill gas utilisation project. The assessment would include both predictive computer modelling and physical site assessments. Difficulties arise in the assessment of the potential for landfill gas utilisation using physical assessment methods due to the heterogeneous nature of waste and often poor records of waste emplaced. A sample well or probe at one part of the site may give completely different results from one at another part of the site, even at sample points close to each other. Table 4.22 summarises the physical assessment techniques to estimate landfill gas generation (Landfill Gas Development Guidelines 1996). Other factors influencing the generation of landfill gas include the waste type deposited, and the site size and geometry.

Table 4.22   *Physical assessment techniques to estimate landfill gas generation*

Technique	Basis	Advantages	Disadvantages
Waste sampling from trial pits and boreholes	Qualitative assessment of nature of waste by inspection	Rapid, low cost. Can provide useful information about nature and location of wastes in the absence of records	Small samples taken from a large site may not be representative of the whole
Waste analysis	Chemical analysis of waste samples. Volatile solids or organic carbon content indicate landfill gas potential. Samples incubated in the laboratory can be used for gas generation	Rapid, low cost. Can provide useful information about nature and location of wastes in the absence of records	Small samples taken from a large site may not be representative of the whole
Measurement of passive landfill gas emissions	Measurement of flow of landfill gas from boreholes or into flux boxes installed within the landfill	Measures landfill gas generated under field conditions	Variable pressure fields in landfills can introduce errors in the scale-up computation. Atmospheric pressure influences flow
Pumping trials	Pumping of landfill gas from boreholes until steady state is achieved, with estimate of volume of waste affected by measuring pressure drop in the landfill	Measures landfill gas generated and extracted under field conditions. Provides early indications of the problems likely to be encountered later (eg, low permeability waste, air leakage)	More costly than other methods. Measurement of the radius of influence can lack precision[1]

[1] Landfill gas is drawn towards the test well under the influence of the pressure gradient within the landfill. The pressure difference decreases with increasing distance from the well until a point is reached where the applied suction at the test well has no influence on the pressure in the landfill. The horizontal distance to this point is termed the radius of influence.

Source: Landfill Gas Development Guidelines 1996. Crown copyright material is reproduced with the permission of the Controller of HMSO and the Queen's Printer for Scotland.

Landfill gas utilisation would only normally be considered for large sites with a minimum of between 200 000 and 500 000 tonnes of biodegradable municipal solid waste. In addition, higher gas recovery rates are obtained from large, deeper sites rather than shallower sites.

Estimation of landfill gas production involves not only the total amount of gas potentially available, but also the rate and duration of gas production (Cossu et al 1996). Modelling techniques used to estimate the gas generation rates from a landfill site are based on the assumption that a certain unit mass of biodegradable waste will produce a certain quantity of landfill gas. As the waste is further and further biodegraded, the rate of landfill gas production will decrease in proportion to the quantity remaining to be produced. Figure 4.19 shows the most commonly applied equation to describe the rate of landfill gas generation (Landfill Gas Development Guidelines 1996). Cossu et al (1996) have reviewed landfill gas production models and the factors influencing the models.

The typical composition of landfill gas was shown in Table 4.14, consisting mainly of methane and carbon dioxide with lower concentrations of hydrocarbons. Combustion of the methane in the landfill gas will produce mainly carbon dioxide and water vapour and other minor pollutants. Table 4.23 shows the concentration of pollutants produced from

$$Rate = kL_0e^{-kt}$$

Rate = rate of landfill gas production

k = rate constant, represents the decay value or half-life of the waste

L_0 = ultimate yield of landfill gas

t = time

Figure 4.19 *Equation to estimate the production of landfill gas from waste landfills. Source: Landfill Gas Development Guidelines 1996.*

Table 4.23 *Pollutant emissions from the combustion of landfill gas from three UK landfill sites in different types of power plant*

Emission	Dual fuel diesel engine[1] (mg/m^3)	Spark ignition engine[1] (mg/m^3)	Gas turbine[2] (mg/m^3)
Particulate matter	4.3	125	9
Carbon monoxide	800	~10000	14
Unburnt hydrocarbons	22	>200	15
Nitrogen oxides	795	~1170	61
Hydrogen chloride	12	15	38
Sulphur dioxide	51	22	6
Dioxins (ng/m^3)	0.4	0.6	0.6
Furans (ng/m^3)	0.4	2.7	1.2

Gas pre-treatments
[1] Drying, filtering, compression and cooling of gas prior to use.
[2] Wet scrubbing, compression, cooling, filtering and heating to 70 °C prior to use.
Sources: Young and Blackey 1993; Young and Blakey 1996.

the combustion of landfill gas in a dual-fuelled diesel engine, spark ignition engine and a gas turbine (Young and Blakey 1993). The emissions from the combustion of landfill gas will vary depending not only on the type of combustion system, but also the composition of the gas used.

4.16   Old Landfill Sites

Since landfilling of wastes has occurred throughout history, there are innumerable old landfill sites. For the majority of old landfill sites, very little is known of the input of the waste, monitoring of leachate and landfill gas. Table 4.24 shows some of the problems associated with old waste landfill sites (Jefferis 1995). Early landfill designs were based on the attenuate and disperse type of landfill design system. Attenuate and disperse landfills rely on the very slow 'uncontrolled' release of leachate into the surrounding geological and hydrogeological environment. The leachate is first diluted by the groundwater in the surrounding environment. Secondly, the action of the biological, physical and chemical processes on the leachate, as it migrates through the surrounding geological strata, reduces the polluting potential of the leachate (McBean et al 1995; Waste Management Paper 26, 1986; Bagchi 1994; Westlake 1995). This process is known as attenuation. Theoretically, the leachate levels become so low as to pose no significant hazard to the environment. The attenuate and disperse site was common throughout history where the environmental implications of landfill gas generation, and in particular leachate, were not realised. Following the rise in environmental concern throughout the 1970s, and 1980s, fuelled by a series of highly publicised pollution incidents associated with waste disposal, the attenuate and disperse type of landfill came under increased criticism as a landfill design and as an environmentally acceptable form of waste disposal. Even so the attenuate and disperse type of site is still found throughout the world, particularly in developing countries. There are a number of uncertainties associated with the attenuate and disperse landfill site including the level of leachate pollutants which can be accepted into the surrounding environment before unacceptably high migration occurs. In addition, there is great uncertainty as to the degree to which the leachate pollutants are reduced by interaction with the surrounding soils and geological strata. A further factor impacting on the

Table 4.24   *Problems associated with old waste landfill sites*

Parameter	Problem
Waste	Data on waste quantities and composition inputs are usually scarce
Siting	Old landfills tend to be sited for convenience rather than on grounds of geological and hydrogeological criteria
Surrounding environment	Old landfills are often sited close to conurbations, or new developments have been close to old landfills
Waste degradation	The waste degradation process is slow, resulting in long-term generation of leachate and landfill gas after the site is closed
Leachate and gas migration	Leachate and landfill gas may be migrating from the site

Source: Jefferis 1995.

use of attenuate and disperse landfill designs was the introduction by the EC of the Ground-water Directive in 1980 (EC Directive 80/68/EEC 1980). The Directive prohibited the direct or indirect discharge of a whole range of pollutants into groundwater (Waste Man agement Paper 4, 1996; Waste Management Licensing Regulations 1994). The range of pollutants were defined in 'List I' or 'List II' depending on their polluting potential. The specified pollutants include ammonia and nitrates, organohalogens and organophosphates, mercury, cadmium, zinc, copper and lead. Many of these compounds are found in landfill leachates and, consequently, further restricted the use of the attenuate and disperse type of landfill. In addition, it is also important to understand the processes involved in leachate interaction with the surrounding geological and hydrogeological environment, in order to assess the impact of any possible liner failure in containment-type landfills lined with a barrier system.

Attenuate and disperse-type landfills would produce leachate which migrates through the mass of waste and into the groundwater where the attenuation and dispersion process is transferred from within the waste mass to the surrounding geological and hydrogeological environment. The physical, chemical and biological attenuation processes acting on the leachate within the landfill mass and outside in the surrounding environment, depend on the nature and quantity of waste and the surrounding geological, geochemical and hydro-geological conditions.

Physical attenuation processes As the leachate migrates from the landfill site it is diluted by the surrounding groundwater. A plume of leachate spreading from the landfill site will disperse in the direction of groundwater movement and also laterally. Dispersion takes place both on a macroscale and microscale. For example, differential dispersion will occur in different types of rock, it will also be influenced by rock grain size, pore size distribution and concentration of clay minerals in the rock, etc. Leachate may be absorbed or adsorbed with the waste itself and the surrounding geological environment. The degree of absorption or adsorption increases with higher contents of clay minerals or increased concentrations of organic carbon in soils. Such processes are easily reversible and cause easy removal of the pollutants at later stages as conditions alter (Waste Management Paper 26, 1986; McBean et al 1995; Bagchi 1994; Westlake 1995).

Chemical attenuation processes Chemical attenuation processes rely on interaction between leachate and the surrounding geochemical environment to chemically alter or fix the leachate. Interaction of cations and anions in the leachate with those in the soil or rock, may occur via ion exchange. For example, heavy metals may ion exchange with cations found naturally in soils, clays or different consolidated rock types. Metals may also be removed by precipitation reactions, for example, many metal carbonates, hydroxides and sulphides are insoluble. Chemical reaction between metals in the leachate and such anions in solution, results in precipitation. Co-precipitation may also adsorb or occlude trace metals from the leachate within the primary precipitate. Acidic conditions tend to solubilise metals, whereas more alkaline conditions induce precipitation. The formation of large ion complexes, which include metal cations in the structure, will effectively remove the metals from the environment by fixation in a large complex molecule. Oxidation–reduction reactions may also occur between inorganic species in the leachate and the surrounding geological and soil environment. Some elements and compounds can exist in more than one oxidation state, and changes in the redox potential may influence their mobility into and out of solution. For example, iron is readily oxidised to the ferric, Fe^{3+}, state in alkaline and mildly acidic conditions and conditions may then cause precipitation of ferric hydroxide from solution (Waste Management Paper 26, 1986; McBean et al 1995; Bagchi 1994; Westlake 1995).

Biological attenuation processes   As the leachate passes into the surrounding geological and hydrogeological environment, the aerobic and anaerobic bidegradation of the organic materials in the leachate will continue. The circulation of groundwater serves to disperse the micro-organisms and nutrient organic material and remove degradation products (Waste Management Paper 26 1986; McBean et al 1995; Bagchi 1994; Westlake 1995).

Rugge et al (1995) have reported on the leachate plume emanating from an old landfill site in Denmark, which accepted municipal solid waste from the 1930s until its closure in the 1970s. They measured the 'plume' of leachate which penetrated the subsurface to a down-gradient distance of between 200 and 250 m. The extent of the plume was detected by increased concentrations of inorganic compounds, such as chloride. More than 15 organic compounds of potential harm to the environment, including benzene, toluene, xylenes, chlorinated hydrocarbons and polycyclic aromatic hydrocarbons were identified at the subsurface border of the waste and sub-rock. However, 60 m down-gradient of the landfill, the plume of hydrocarbons contained negligible concentrations of these hydrocarbons, due to the natural dispersion and attenuation processes operating in the subsurface environment.

The main problems associated with old landfill sites are the uncontrolled migration of leachate and gas into the surrounding environment. The control of leachate and landfill gas migration from old sites is via slurry trench cut-off walls (Jefferis 1995). A trench is excavated in the ground around the boundaries of the landfill and filled with a slurry of bentonite clay, ground granulated blast furnace slag, Portland cement and water. This sets to form a low-permeability barrier. The permeabilities of such barrier walls are of the order of 10^{-9} m/s hydraulic conductivity. In some cases a polymeric plastic membrane may also be added, to reach lower permeabilities.

The creation of a barrier beneath an old landfill is more complicated and requires drilling and high-pressure injection of the slurry. The slurry is injected in a series of overlapping 'V' cuts, to create an interlocking layer of low permeability barrier. The gas and leachate collection and control systems are introduced by excavating wells or boreholes for collection, extraction and treatment of the leachate and landfill gas. There are several other methods available for the lining of old landfill sites, but they all tend to be very expensive and would only be economically viable in special cases (Eichmeyer et al 1997). Finally, in many cases, old landfills are not capped to the high standards of current landfills, but the site may be much more easily accessible than the sides and beneath the landfill and so can be easily capped with suitable barrier systems and top soils (Jefferis 1995).

A new development associated with old landfill sites is landfill mining and reclamation where, after stabilisation, previously deposited wastes are excavated for processing to obtain recyclable materials, a combustible fraction and soil and also to create a new void space for further landfilling (The World Resource Foundation 1995). In addition, landfill mining and reclamation aids the remediation of poorly designed landfills or upgrade landfills that do not meet current environmental standards or operational practices. A number of landfill mining and reclamation schemes have been developed throughout the World, for example, in the USA. Figure 4.20 shows a possible landfill mining operation

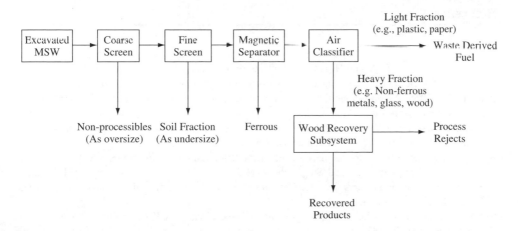

Figure 4.20 *Schematic diagram of a landfill mining and reclamation operation. Source: The World Resource Foundation, 1995. Reproduced by permission of R.C. Strange.*

involving extensive processing (The World Resource Foundation 1995). Excavation involves similar practices to open-cast mining, using excavators, clamshell grabs and lorries or conveyor belts to transport the waste to the processing facility, which may be either on-site or off-site. Processing involves the sizing of the material using a coarse screen to remove large items which cannot be processed. A further fine screen, such as a trommel screen, allows the separated fine material to be recovered as soil. The material passing through the screen is processed via magnetic separation to remove ferrous metals. Air classification of the non-ferrous metal fraction separates the light organic materials which may be used to produce a refuse-derived fuel, and the residual heavy material such as non-ferrous metals, glass and wood, may be further fractionated.

The factors involved in landfill mining and reclamation schemes involves the recovery efficiency and quality of recyclable materials from the site. Table 4.25 shows the typical recovery efficiencies of the available product and the purities of the product streams from landfill mining and reclamation (The World Resource Foundation 1995). The majority of the material recovered from landfill mining and reclamation is the soil fraction which can be typically 50–60% of the material mined. Table 4.25 shows that, whilst a high proportion of the available recyclable materials in the landfill can be recovered, the data on product purity indicates that impurities in the product may limit the market demand for these materials.

Table 4.25 *Typical recovery efficiencies and purities of the product streams from landfill mining and reclamation*

Product	Recovery efficiency (%)	Purity (%)
Soil	85–95	90–95
Ferrous metal	70–90	80–95
Plastic	50–70	70–90

Source: The World Resource Foundation 1995.

References

Allen A. 2001. Containment landfills: the myth of sustainability. Engineering Geology, 60, 3–19.

Allen M.R., Braithwaite A. and Hills C.C. 1997. Environ. Sci. Technol., 31, 1054–1061.

Bagchi A. 1994. Design Construction and Monitoring of Landfill. John Wiley & Sons Ltd, New York.

Beaven R.P. 1996. A hydrogeological approach to sustainable landfill design. In, Harwell Waste Management Symposium: Progress, Practice and Landfill Research. Harwell, Oxfordshire, May.

Bishop D.J. and Carter G. 1995. Waste disposal by landfill lining systems. In, Waste Disposal by Landfill R.W. Sarsby (Ed.). Balkema Press, Rotterdam.

Blakey N., Archer D. and Reynolds P. 1995. Bioreactor landfill: a microbiological review. In, Christensen T.H., Cossu R. and Stegmann R. (Eds.), Sardinia 95, Fifth International Landfill Symposium (Cagliari, Sardinia, October 1995), CISA-Environmental Sanitary Engineering Centre, Cagliari, Sardinia.

Brown K.A. and Maunder D.H. 1994. Exploitation of landfill gas: A UK perspective, Water Science and Technology, 30, 143–151.

Chaiampo F., Conti R. and Cometto D. 1996. Morphological characterisation of MSW landfills, Resources, Conservation and Recycling, 17, 37–45.

Characterisation of 100 UK landfill sites. 1995. Report CWM 015/90, Wastes Technical Division, Department of the Environment, HMSO, London.

Christensen T.H., Cossu R. and Stegmann R. 1996(a). Landfilling of Waste: Barriers. E & FN Spon, London.

Christensen T.H., Kjeldsen P. and Lindhardt B. 1996(b). Gas generating processes in landfills. In, Landfilling of Waste: Biogas, Christensen T.H., Cossu R. and Stegman R. (Eds.), E & FN Spon, London.

COM (97) 105 Final. 1997. Commission of the European Communities, Official Journal of the European Communities.

Commission Decision. 2000. 2000/532/EC Commission Decision of 3 May 2000, Official Journal of the European Communities, L226/1, 6.9.2000, Brussels, Belgium.

Cossu R., Andreottola G. and Muntoni A. 1996. Modelling Landfill Gas Production. In, Christensen T.H., Cossu R. and Stegmann R. (Eds.), E & FN Spon, London.

Council Directive 96/61/EC, 1996. Integrated Pollution Prevention and Control. Official Journal, L257, Brussels.

Council Directive 85/337/EEC, 1985. The Assessment of the Effects of certain Public and Private Projects on the Environment. Official Journal, L175, Brussels.

DEFRA. 2004. Sewage treatment in the UK: Implementation of the EC urban waste water treatment Directive, re-use and disposal of sewage sludge. Department of the Environment, Food and Rural Affairs, UK.

Diaz L.F. and Savage G.M. 2002. Developing Landfill. Waste Management World, International Solid Waste Association, Copenhagen, July–August.

EC Directive. 1980. 80/68/EEC, Council Directive of 17 December 1979 on the protection of groundwater against pollution from certain dangerous substances, Official Journal of the European Communities L020, January, Brussels, Belgium.

El-Fadel M. and Khoury R. 2000. Modelling settlement in MSW landfills: a critical review. Critical Reviews in Environmental Science and Technology, 30, 327–361.

Eichmeyer H., Boehm W. and Bredel-Schurmann S. 1997. In, Advanced Landfill Liner Systems, August H., Holzlohner U. and Meggyes T. (Eds.), Thomas Telford Publishers, London.

Energy from Waste: Best Practice Guide. 1996. Department of Trade and Industry, HMSO, London.

Environmental impacts from landfills accepting non-domestic wastes. 1995. Report CWM 036/91, Wastes Technical Division. Department of the Environment, London, published 1995.

Eunomia. 2003. Eunomia Research and Consulting, Costs for Municipal Waste Management in the EU, Final Report to Directorate of General Environment, European Commission, Brussels, Belgium.

European Commission. 2003. Waste Generated and Treated in Europe, European Commission, Office for Official Publications of the European Communities, Luxembourg.

European Environment Agency. 2002(a). Review of selected waste streams: sewage sludge, construction and demolition waste, waste oils, waste from coal fired power plants and biodegradable municipal solid waste. European Environment Agency, Copenhagen.

European Environment Agency. 2002(b). Biodegradable municipal waste management in Europe, Part 1 Strategies and Instruments. European Environment Agency, Copenhagen.

Fifield F.W. and Haines P.J. 1995. Environmental Analytical Chemistry. Blackie Academic and Professional, London.

Freeman H.M. 1998. Standard Handbook of Hazardous Waste Treatment and Disposal. McGraw-Hill, New York.

Gendebien A., Pauwels M., Constant M., Ledrut-Damanet M.J., Nyns E.J., Willumsen H.C., Butson J., Fabry R. and Ferrero G.L. 1992. Landfill gas: from environment to energy, Commission of the European Communities, Contract No. 88b-7030-11-3-17. Commission of the European Commission, Luxembourg.

Ham R.K. 1993. Overview and implications of US sanitary landfill practice, Air and Waste, 43, 187–190.

Haughey R.D. 2001. Report: landfill alternative daily cover: conserving air space and reducing landfill operating cost. Waste Management and Research, 19, 89–95.

Hickman L.H., 1996. MSW Management; State of practice in the USA. International Solid Waste Association, Yearbook, Paris.

Hjelmar O., Johannessen L.M., Knox K., Ehrig H.J., Flyvbjerg J., Winther P. and Christensen. 1995. Composition and management of leachate from landfills within the EU. In, Christensen T.H., Cossu R. and Stegmann R. (Eds.), Sardinia 95, Fifth International Landfill Symposium (Cagliari, Sardinia, October 1995), CISA-Environmental Sanitary Engineering Centre, Cagliari, Sardinia.

IEA Bioenergy. 2000. Biogas flares; state of the art and market review. IEA Bioenergy. AEA Technology, Abingdon, UK.

Jefferis S.A. 1995. In, Waste Disposal by Landfill, R.W. Sarsby (Ed.). Balkema press, Rotterdam.

Johannessen L.M. and Boyer G. 1999. Observations of solid waste landfills in developing countries: Africa, Asia and Latin America. The World Bank, Washington DC, US.

Joseph J.B. and Mather J.D. 1993. Landfill – Does current containment practice represent the best option? In, Christensen T.H., Cossu R. and Stegmann R. (Eds.), Sardinia 93, Fourth International Landfill Symposium (Cagliari, Sardinia, October 1993), p. 99–107. CISA Environmental Sanitary Engineering Centre, Cagliari, Sardinia.

Juniper. 2004. Mechanical Biological Treatment Systems. Juniper Consultancy Services Ltd, Uley, Gloucestershire, UK.

Kjeldsen P., Barlaz M.A., Rooker A.P., Baun A., Ledin A. and Christensen T.H. 2002. Critical Reviews in Environmental Science and Technology, 32, 297–336.

LaGrega M.D., Buckingham P.L. and Evans J.C. 1994. Hazardous Waste Management. McGraw-Hill, New York.

Landfill Gas Development Guidelines. 1996. Energy Technology Support Unit. Department of Trade and Industry, HMSO, London.

Lee F.G. and Jones-Lee A. 1993. Landfill and groundwater pollution issues: 'Dry tomb' vs Wet-cell landfills. Proceedings Sardinia 93, Fourth International Landfill Symposium, Cagliari, Italy, 11–15 October 1993, Christensen T.H., Cossu R. and Stegmann (Eds.), 1777–1796.

Loening A. 2003. Predictions and projections; looking at the power generation potential of landfill gas, Waste Management World. International Solid Waste Association, Copenhagen, November–December.

Maier T.B., Steinhauser E.S., Vasuki N.C. and Pohland F.G. 1995. Integrated leachate and landfill gas management. In, Christensen T.H., Cossu R. and Stegmann R. (Eds.), Sardinia 95, Fifth International Landfill Symposium (Cagliari, Sardinia, October 1995). CISA-Environmental Sanitary Engineering Centre, Cagliari, Sardinia.

McBean E.A., Rovers F.A. and Farquhar G.J. 1995. Solid Waste Landfill Engineering and Design. Prentice-Hall, New Jersey.

Mosher B.W., Czepiel P.M., Harriss R.C., Shorter J.H., Kolb C.E., McManus J.B., Allwine E. and Lamb B.K. 1999. Environ. Sci. Technol., 33, 2088–2094.

Moss H. 1997. Dynamotive Technologies UK Ltd, Bedford (Personal communication).

Pavelka C., Loehr R.C. and Haikola B. 1993. Hazardous waste landfill leachate characteristics, Waste Management, 13, 573–580.

Pescod M.B. (Ed.) 1991–93. Urban Solid Waste Management. World Health Organization, Copenhagen.

Petts J. and Eduljee G. 1994. Environmental Impact Assessment for Waste Treatment and Disposal Facilities. John Wiley & Sons Ltd, Chichester.

Porteus A. 1992. Dictionary of Environmental Science and Technology. John Wiley & Sons Ltd., Chichester.

Qin W., Egolfopoulos F.N. and Tsotsis T.T. 2001. Chemical Engineering Journal, 82, 157–172.

Redfearn A. and Roberts D. 2002. Issues in Environmental Science and Technology, 18, 103–139.

Reinhart D.R. 1996. Full scale experiences with leachate recirculating landfills: Case studies, Waste Management and Research, 14, 347–365.

Reinhart D.R. and Al-Yousfi A.B. 1996. The impact of leachate recirculation on municipal solid waste landfill operating characteristics, Waste Management and Research, 14, 337–346.

Reinhart D.R., McCreanor P.T. and Townsend T. 2002. Waste Management and Research, 20, 172–186.

Rettenberger G. and Stegmann R. 1996. Landfill gas components. In, Christensen T.H., Cossu R. and Stegmann R. (Eds.), Landfilling of Wastes: Biogas. E & FN Spon, London.

Rugge K., Bjerg P.L. and Christensen T.H. 1995. Distribution of organic compounds from municipal solid waste in the groundwater downgradient of a landfill (Grinsted Denmark), Environmental Science and Technology, 29, 1395–1400.

Sarsby R.W. (Ed.) 1995. Waste Disposal by Landfill. Balkema, Rotterdam.

Scheff P., Casten C., Ruesch P. and Friedl M. 2001. In, Environmental Health Risk, Fajzieva D. and Brebbie C.A. (Eds.), WIT Press, Southampton UK.

Sharma H.D. and Lewis S.P. 1994. Waste containment systems, waste stabilisation, and landfills: design and evaluation, John Wiley & Sons Inc., New York.

Smith A., Brown K., Ogilve S., Rushton K. and Bates J. 2001. Waste management options and climate change, Final Report to the European Commission-DG Environment. European Commission, Brussels.

Stegmann R. 1996. Landfill gas utilisation: An overview. In, Landfilling of Waste: Biogas. Christensen T.H., Cossu R. and Stegmann R. (Eds.), E & FN Spon, London.

Stegmann R., Ritzkowski M., Heyer K.-U. and Hupe K. 2003. Landfill aftercare and completion criteria. Waste Management World, International Solid Waste Association, Copenhagen, September–October.

Tchobanoglous G. and Burton F.L. 1991. Waste water Engineering, treatment, Disposal and Re-Use (Metcalf and Eddy Inc.). McGraw-Hill, New York.

Tchobanoglous G. and O'Leary P.R. 1994. Landfilling. In, Handbook of Solid Waste Management, Kreith F. (Ed.). McGraw-Hill, Inc., New York.

The World Resource Foundation. 1995. Technical Brief, Landfill Mining. The World Resource Foundation, Tonbridge, Kent.

The World Resource Foundation. 1996. Technical Brief, Landfill Techniques. The World Resource Foundation, Tonbridge, Kent.

Townsend T.G., Miller W.L., Lee H.-J. and Earle J.F.K. 1996. Acceleration of landfill stabilisation using leachate recycle, Journal of Environmental Engineering, April, 263–268.

Van den Brook B., Lambropoulos N.A. and Haggett K. 1995. Bioreactor landfill research in Australia. In, Christensen T.H., Cossu R. and Stegmann R. (Eds.), Sardinia 95, Fifth International Landfill Symposium (Cagliari, Sardinia, October 1995), CISA-Environmental Sanitary Engineering Centre, Cagliari, Sardinia.

Ward R.S., Williams G.M. and Hills C.C. 1996. Changes in major and trace components of landfill gas during subsurface migration, Waste Management and Research, 14, 243–261.

Waste Landfill Directive. 1999. Council Directive 1999/31/EC, L 182/1, 26.4.1999. Landfill of Waste Official Journal of the European Communities, Brussels, Belgium.

Waste Management in Japan. 1995. Department of Trade and Industry, Overseas Science and Technology Expert Mission Visit Report. Institute of Wastes Management, Northampton.

Waste Management Licensing Regulations. 1994. HMSO, London.

Waste Management Paper 4, 1996. Licensing of Waste Management Facilities. Department of the Environment, HMSO, London.

Waste Management Paper 26. 1986. Landfilling Wastes. HMSO, London.

Waste Management Paper 26A. 1995. Landfill Completion. Department of the Environment, HMSO, London.

Waste Management Paper 26B. 1995. Landfill Design, Construction and Operational Practice, Department of the Environment, HMSO, London.

Waste Management Paper 27. 1994. Landfill Gas. Department of the Environment, HMSO, London.

Waste Management Paper 26E. 1996. Landfill restoration and post closure management, Consultation Draft, August 1996. Environment Agency, Department of the Environment, London.

Waste Management Paper 26F, 1994. Landfill Co-disposal. Draft for External Consultation. Waste Technical Division, Department of the Environment, HMSO, London.

Waste Not Want Not. 2002. Strategy Unit, HMSO.

Westlake K. 1995. Landfill Pollution and Control. Albion publishing, Chichester.

White P.R., Franke M. and Hindle P. 1995. Integrated Solid Waste Management: A Lifecycle Inventory. Blackie Academic and Professional, London.

Williams G.M. and Aitkenhead N. 1991. Lessons from Loscoe: The uncontrolled migration of landfill gas, Quarterly Journal of Engineering Geology, 24, 191–207.

Young C.P. and Blackey N.C. 1993. Emissions from power generation plants fuelled by landfill gas. In, Third International Landfill Symposium, Sardinia, Italy, October.

Young C.P. and Blakey N.C. 1996. Emissions from landfill gas power generation plants. In, Christensen T.H., Cossu R. and Stegmann R. (Eds.), Landfilling of Wastes: Biogas. E & FN Spon, London.

Yuen S.T.S, Styles J.R. and McMahon T.A. 1995. An active landfill management by leachate recirculation: A review and an outline of a full-scale project. In, Christensen T.H., Cossu R. and Stegmann R. (Eds.), Sardinia 95, Fifth International Landfill Symposium (Cagliari, Sardinia, October 1995), CISA-Environmental Sanitary Engineering Centre, Cagliari, Sardinia.

5

Waste Incineration

Summary

This chapter is concerned with incineration, the second major option for waste treatment and disposal in many countries throughout the world. The various incineration systems are discussed. Concentration is made on mass burn incineration of municipal solid waste, following the process through waste delivery, the bunker and feeding system, the furnace, and heat recovery systems. Emphasis on emissions formation and control is made, with discussion of formation and control of particulate matter, heavy metals, toxic and corrosive gases and products of incomplete combustion, such as polycyclic aromatic hydrocarbons (PAHs), dioxins and furans. The wastewater and bottom and flyash arising from waste incineration are discussed. The dispersion of emissions from the chimney stack are described. Energy recovery via district heating and electricity generation are discussed. Other types of incineration including fluidised bed incinerators, starved air incinerators, rotary kiln incinerators, cement kilns, liquid and gaseous waste incinerators and the types of waste incinerated in the different incinerators, are discussed.

5.1   Introduction

For most countries throughout the world, landfill is the dominant route for waste treatment. As an alternative to landfill, wastes containing combustible material may be incinerated or combusted. Incineration is the oxidation of the combustible material in the waste to

Waste Treatment and Disposal, Second Edition   Paul T. Williams
© 2005 John Wiley & Sons, Ltd   ISBNs: 0-470-84912-6 (HB); 0-470-84913-4 (PB)

produce heat, water vapour, nitrogen, carbon dioxide and oxygen. Depending on the composition of the waste, other emissions may be formed including, carbon monoxide, hydrogen chloride, hydrogen fluoride, nitrogen oxides, sulphur dioxide, volatile organic carbon, dioxins and furans, polychlorinated biphenyls, heavy metals, etc. (European Commission 2004). The removal of such pollutant emissions from the flue gases of the incinerator involves extensive, complex and expensive gas clean-up systems. Whilst there are stringent EC emissions legislation in place (EC Waste Incineration Directive 2000) to control the emissions from waste incineration, there still remains firm public opposition to the incineration of waste. Incineration is a treatment route which can be applied to a wide variety of wastes. For example, Figure 5.1 shows the percentage of municipal solid waste incinerated in various European countries (European Commission 2003(a)). For most European countries, incineration of municipal solid waste remains a minority route for disposal. Municipal solid waste incineration has historically been seen in terms of a means of waste disposal. However, modern incinerators would now include a means of energy recovery as an economic necessity. Energy recovery is usually by the generation of electricity from high-temperature steam turbines or through district heating schemes. Figures 5.2–5.4 show the percentage of hazardous waste, industrial waste and sewage sludge incinerated in selected European countries (European Commission 2003(a)). Incineration of some commercial and industrial wastes which are hazardous and have low throughputs, use incineration as a means of disposal, and energy recovery is often a secondary objective. Sewage sludge incineration generates heat which is often used to dry the input sewage sludge to levels where the combustion is self-sustaining.

Incineration of waste has a number of advantages over landfill.

- Incineration can usually be carried out near the point of waste collection. In some cities, the number of landfill sites close to the point of waste generation are becoming scarcer, resulting in transport of waste over long distances.
- The waste is reduced into a biologically sterile ash product which for municipal solid waste is approximately 10% of its pre-burnt volume and 33% of its pre-burnt weight.

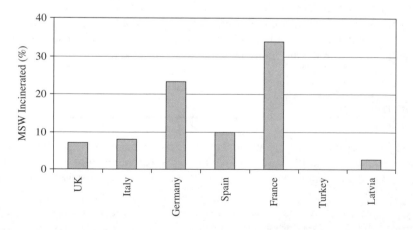

Figure 5.1 *Municipal solid waste incineration in selected countries. Source: European Commission 2003.*

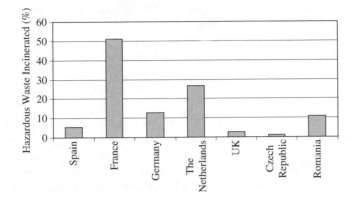

Figure 5.2   *Hazardous waste incineration in selected countries. Source: European Commission 2003.*

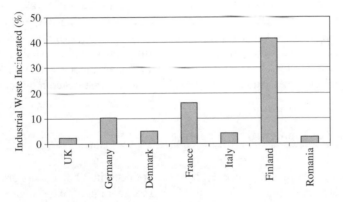

Figure 5.3   *Industrial waste incineration in selected countries. Source: European Commission 2003.*

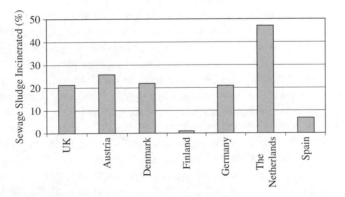

Figure 5.4   *Sewage sludge incineration in selected countries. Source: European Commission 2003.*

- Incineration Produces no methane, unlike landfill. Methane is a 'greenhouse gas' and is a significant contributor to global warming.
- Waste incineration can be used as a low-cost source of energy to produce steam for electric power generation, industrial process heating or hot water for district heating, thereby conserving valuable primary fuel resources.
- The bottom ash residues can be used for materials recovery or as secondary aggregates in construction.
- Incineration is the best practicable environmental option for many hazardous wastes such as highly flammable, volatile, toxic and infectious waste.

However, there are also disadvantages.

- Generally there are much higher costs and longer pay-back periods, due to the high capital investment.
- There is sometimes a lack of flexibility in the choice of waste disposal options once the incineration route is chosen; because of the high capital cost the incinerator must be tied to long-term waste disposal contracts.
- The incinerator is designed on the basis of a certain calorific value for the waste. Removal of materials such as paper and plastics for recycling may reduce the overall calorific value of the waste and consequently may affect incinerator performance.
- Whilst modern incinerators comply with existing emissions legislation there is some public concern that the emitted levels may still have an adverse effect on health.
- The incineration process still produces a solid waste residue which requires management.

5.2   EC Waste Incineration Directive

The EC Waste Incineration Directive (2000), introduced in 2000, covers emissions to the atmosphere. The regulation of wastewater from the cleaning of exhaust gases is set out in the Directive. In addition, the regulation of the wastewater derived from the waste incineration process is also covered by other EC Directives (Council Directive 91/271/EEC amended by 98/15/EC and 76/464/EEC). The disposal of ash to land is covered by the EC Waste Landfill Directive (EC Waste Landfill Directive 1999).

There is great public opposition to the incineration of waste centred on the emissions to air. The EC Waste Incineration Directive seeks to allay some of the fears related to waste incineration by setting stringent emission limits to prevent or reduce, as far as possible, pollution of the air, water and soil environment and consequent risk to human health caused by waste incineration. The Directive sets out the limit values for emission to air for a whole range of toxic gases including heavy metals such as mercury, cadmium, chromium and lead, dioxins and furans, carbon monoxide, dust, hydrogen chloride, hydrogen fluoride, sulphur dioxide, nitrogen oxides and gaseous organic compounds (expressed as TOC). Table 5.1 sets out the emission limits to air (EC Waste Incineration Directive 2000). The measurement of nitrogen oxides, carbon monoxide, dust, TOC, hydrogen chloride, hydrogen fluoride and sulphur dioxide are required on a continuous basis. Heavy metal concentrations and concentrations of dioxins and furans, because of the complexity of the analytical process, are required to be reported twice per year.

The emission-limit values for discharges of wastewater from the cleaning of exhaust gases, concentrates on the total suspended solids in the wastewater and the levels of heavy metals including mercury, cadmium, arsenic, lead and chromium. The permitted maximum concentration of dioxins and furans in the wastewater is also stipulated. The emission-limit values for wastewater from the cleaning of flue gases are shown in Table 5.2.

The solid residues or ash resulting from the waste incineration plant are also covered by the EU Waste Incineration Directive. The residues should be minimised in their amount and harmfulness and should be recycled where appropriate. Prior to determining the routes for disposal or recycling, appropriate tests should be carried out to establish the physical and chemical characteristics and polluting potential of the different incineration residues.

The incineration of waste is regarded by the EC Commission as an industrial process and as such will be subject to Integrated Pollution Prevention and Control (IPPC) and will require a permit which is issued and monitored by the environment agencies or competent authorities of the Member States. The permit sets out the categories and quantities of waste which can be incinerated, the plant capacity and the sampling and measurement procedures which are to be used (Europa 2003). Because incineration of waste comes under IPPC regulations, a Best Available Techniques (BAT) reference document is produced, which details the range of incinerator systems available, and the operational, gas cleaning, wastewater cleaning, sampling, monitoring and reporting requirements, etc. for the incineration of waste (European Commission 2004). The design and operation of the plant would have to comply with the recommendations of the IPPC document for waste Incineration before a permit would be issued by the regulatory authority.

With the aim of preventing the environmental impact of waste incineration in the European Union, the Directive sets strict requirements for the operational conditions and

Table 5.1 *Emission limits for waste incineration as set out in the EU Waste Incineration Directive, 2000 (Reference Conditions; 273 K, 101.3 kPa, 11% oxygen, dry gas)*

Substance	Emission limits (mg/m^3)
Total dust	10 (daily average)
TOC (gaseous and vapours)	10 (daily average)
HCl	10 (daily average)
HF	1 (daily average)
CO	50 (daily average)
SO_2	50 (daily average)
NO_x (expressed as NO_2) (New plant)	200 (daily average)
NO_x (expressed as NO_2) (Existing plant 6–16 t/h)	400 (daily average) (until 2010)
NO_x (expressed as NO_2) (Existing plant 16–25 t/h)	400 (daily average) (until 2008)
Cd and Tl	0.05 (new plant) 0.1 (existing plant to 2007)
Hg	0.05 (new plant) 0.1 (existing plant to 2007)
Sb, As, Pb, Cr, Co, Cu, Mn, Ni, V (Total)	0.5 (total – new plant) 1.0 (existing plant to 2007)
Dioxins and furans (TEQ) ng/m^3	0.1

Source: EC Waste Incineration Directive 2000.

Table 5.2   *Emission-limit values for the discharges of wastewater from the cleaning of exhaust gases*

Substance	Emission limits (mg/l)
Total suspended solids	30 (95% values do not exceed)
Total suspended solids	45 (100% values do not exceed)
Hg	0.03
Cd	0.05
Tl	0.05
As	0.15
Pb	0.2
Cr	0.5
Cu	0.5
Ni	0.5
Zn	1.5
Dioxins and furans (TEQ)	0.3

Source: EC Waste Incineration Directive 2000.

technical requirements of waste incinerator plants. For example, the delivery and reception of waste requires procedures which minimise environmental pollution, including odours and noise, in addition to any negative effects on the air, water and land environments. The mass and type of waste is required to be monitored and recorded and, in the case of hazardous waste, full source and compositional documentation is required.

The efficient combustion of the waste to ensure destruction of the combustible materials is controlled through the Directive. The Directive states that the gases derived from incineration are to be raised to a temperature of 850 °C for 2 s. If chlorinated hazardous waste is used with a chlorine content of over 1%, then the temperature has to be raised to 1100 °C. To ensure that such conditions are met throughout the incineration process including plant start-up and shut-down, auxiliary burners are required, which must automatically be switched on if the temperatures fall below those required. The efficient incineration of the waste represented by the burn-out of combustible material is controlled by the operating condition requirements of the Waste Incineration Directive. The Directive states that waste incinerators should achieve a level of incineration such that the slag and bottom ashes shall have a TOC content of less than 3%. The TOC represents the degree of complete burnout of the waste organic materials. The procedures to be undertaken if an unavoidable stoppage, disturbance or failure occurs, are also stipulated.

The main aims of the air emission-limit values for waste incineration and the requirements to use best available technology, are to minimise the emissions to air as far as possible. Even so, flue gases containing very low levels of pollutants will be emitted from the stack. The Directive further states that the stack design and height shall be such that ground-level concentrations of pollutants shall not give rise to significant environmental impact or harm to human health.

The Directive concerns the incineration of all types of waste including municipal solid waste, hazardous waste, sewage sludge, tyres, clinical waste, waste oils and solvents, etc. The Directive also covers the co-incineration of waste with fossil fuels. A co-incineration plant refers to any plant whose main purpose is the generation of energy or production of material products and which uses wastes as a regular or additional fuel or in which waste is thermally treated for the purpose of disposal.

The emission-limit values to air for co-incineration are set down in the Directive. The Directive states that the co-incineration of waste in plants, not primarily intended to incinerate waste, should not be allowed to cause higher emissions of polluting substances in that part of the exhaust gas volume, resulting from such co-incineration, than those plants permitted for dedicated incineration. For the co-incineration of waste, the air emission-limit value is determined by a mixing rule formula. The formula is based on the exhaust gas volume resulting from the incineration of waste alone (V_{waste}) multiplied by the pollutant gas emission limit for waste incineration (C_{waste}), plus the exhaust gas volume based on normal (fossil) fuel operation ($V_{process}$) multiplied by the emission-limit value for the process excluding wastes ($C_{process}$). That is,

$$\frac{(V_{waste}) \times (C_{waste}) + (V_{process}) \times (C_{process})}{(V_{waste}) + (V_{process})}$$

However, the Directive also states that, in the case of co-incineration of untreated mixed municipal waste, the emission-limit values for waste incineration will apply. Also, if in a co-incineration plant more than 40% of the resulting heat release comes from hazardous waste, the emission-limit values for waste incineration will apply. Waste incineration in cement kilns, large combustion plants and other combustion systems co-incinerating wastes, have special emission-limit values which are set down in the Directive, specific to those industries.

Public participation in the waste incineration process is encouraged through access to information prior to the granting of the permit. The functioning parameters of the incinerator and the measured emissions produced from the incinerator must also be available to the public.

5.3 Incineration Systems

The modern incinerator is an efficient combustion system with sophisticated gas clean-up which produces energy and reduces the waste to an inert residue with minimum pollution. Incineration plants may be classified on a variety of criteria, for example, their capacity, the nature of the waste to be combusted, the type of system, etc. However, a broad classification may be made between mass burn incineration and other types.

Mass burn incineration   Large-scale incineration of municipal solid waste in a single-stage chamber unit in which complete combustion or oxidation occurs. Typical throughputs of waste are between 10 and 50 tonnes per hour.

Other types of incineration   Other types of incineration involves smaller scale throughputs of between 1 and 2 tonnes per hour of wastes such as clinical waste, sewage sludge and hazardous waste. Typical examples of such systems include fluidised bed, cyclonic, starved air or pyrolytic, rotary kiln, rocking kiln, cement kiln, and liquid and gaseous incinerators.

5.3.1   Mass Burn Incineration

Mass burn incineration is used for the treatment and disposal of municipal solid waste throughout the world. Within Europe, the amount of municipal solid waste incineration undertaken varies between countries (Table 5.1). In addition, the number of incinerator plants and their capacities also varies (Table 5.3, European Commission 2004). For example, the average waste incinerator plant size in the Netherlands is more than 480 000 tonnes per year throughput, whereas in Italy and Norway, the average size is less than 100 000 tonnes per year throughput (European Commission 2004). The economic viability of incineration as a waste treatment and disposal route for municipal solid waste, depends on the recovery of energy from the process to offset the high costs involved in incineration. The properties and composition of municipal solid waste were discussed in Chapter 2. The composition and characteristics of the waste will influence the combustion properties and emissions produced from the combustion system. A typical calorific value for municipal solid waste is approximately $9000 \, kJ \, kg^{-1}$. Ash and moisture contents tend to be high and thus, in terms of a fuel, the waste would compare poorly with coal, for example. Table 5.4 shows typical properties of municipal solid waste. Of particular importance are the 'fuel' properties of the waste, the proximate analysis (ash, moisture, volatile contents) and the ultimate (elemental) analysis which can be used to assess how the waste will burn in the incinerator and the emissions which are likely to result. Moisture content is obviously important since ignition will not occur if the material is wet and moisture also diminishes the gross calorific value of a fuel. Volatile matter contains the combustible fraction of the waste and consists of gases such as hydrogen, carbon monoxide, methane, ethane, etc., a more complex organic hydrocarbon fraction, and an aqueous phase derived by decomposition of water-bound compounds. The ash content is important since a high ash percentage will lower the calorific value of the waste and must be removed and disposed of after combustion. Waste ash is highly heterogeneous and contains inert

Table 5.3   *Number of municipal solid waste incinerators and average plant capacity for selected European countries*

Country	Number of incinerators	Average plant capacity (1000 tonnes/year)
Austria	3	178
Belgium	17	141
Denmark	32	114
France	210	132
Germany	59	257
Italy	32	91
Norway	11	60
Portugal	3	390
Spain	9	166
Sweden	30	136
The Netherlands	11	488
UK	17	246

Source: European Commission 2004.

Table 5.4 Typical properties of municipal solid waste

Composition	Wt%	Elemental analysis	Wt%
Paper/board	33.0	Carbon	21.5
Plastics	7.0	Hydrogen	3.0
Glass	10.0	Oxygen	16.9
Metals	8.0	Nitrogen	0.5
Food/garden	20.0	Sulphur	0.2
Textiles	4.0	Chlorine	0.4
Other	18.0		
		Metals	**mg/kg**
Proximate analysis	**Wt%**	Copper	200–700
Combustibles	42.1	Chromium	40–200
Moisture	31.0	Mercury	1–50
Ash	26.9	Cobalt	3–10
		Arsenic	2–5
Calorific value	**kJ/kg**	Cadmium	1–150
CV	9000	Lead	100–2000
		Zinc	400–1400
Trace organics	**µg/kg**	Vanadium	4–11
PCB	200–400	Nickel	30–50
PCDD/PCDF	0.050 0.150	Manganese	250
		Thallium	<0.1

Sources: Waste Management Paper 28, 1992; Buekens and Patrick 1985; European Commission 2004.

non-combusted material such as glass and metal cans. The municipal solid waste also contains significant concentrations of heavy metals such as cadmium, lead, zinc and chromium and will influence the emissions of such metals. Similarly, the sulphur and chlorine content will produce emissions of sulphur dioxide and hydrogen chloride.

In most cases, waste incinerator operators have limited control of the precise composition of the incoming waste. Consequently, mass burn incinerators are designed to be sufficiently flexible to cope with the wide range of waste compositions that they may receive (European Commission 2004). The composition of waste may be generally represented in a ternary diagram, shown in Figure 5.5 and shows the range of analyses acceptable to the combustion system (Buekens and Patrick 1985; Hall and Knowles 1985). The shaded area represents the typical composition of municipal solid waste which can sustain combustion without the requirement for auxiliary fuel. The area encloses the minimum acceptable calorific value and the maximum permissible moisture content. In addition, the influences of pre-treatment of the waste, prior to arrival at the waste incinerator, may influence the composition and properties such as the metal content and calorific value (European Commission 2004). For example, removal of glass and metals for recycling would increase the calorific value of the waste and reduce the emission of metals to either the flue gases or bottom ash. Recovery of paper, card and plastic would decrease the calorific value of the incoming waste. Recycling of organic food and garden waste, for example to composting, would reduce the moisture content of the municipal solid waste and thereby increase the net calorific value.

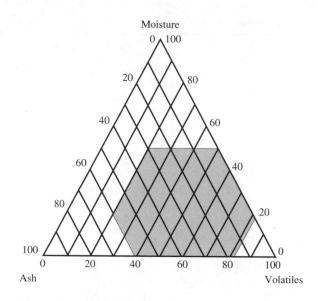

Figure 5.5 *Suitability of municipal solid waste composition for incineration. Sources: Buekens and Patrick 1985; Hall and Knowles 1985.*

A typical modern municipal waste incineration plant with energy recovery is shown in Figure 5.6 (Energy from Waste: Best Practice Guide 1996). The incinerator may be divided into five main areas:

1. waste delivery, bunker and feeding system;
2. furnace;
3. heat recovery;
4. emissions control;
5. energy recovery via district heating and electricity generation.

5.3.1.1   Waste Delivery, Bunker and Feeding System

The waste is usually delivered by collection vehicles, although in some European incinerators, barges or trains may be used. The collection vehicles are weighed on arrival and departure to provide accurate weights of the waste throughput for determining the fees to be charged for disposal and for incinerator operational control. The incinerator may handle a variety of wastes from households, commercial sites and industry and these would be monitored not only to differentiate the fees charged, but also since they may have very different combustion properties which would influence incinerator performance. Odour may result from the waste due to biodegradation and handling, and therefore plants are normally kept under a slight negative pressure, because the combustion air is taken from the waste storage area, which prevents escape of odour. The EC Waste Incineration Directive (2000) sets out the requirements for the handling of waste at the incinerator to minimise the environmental impact of waste incineration and hazard to human health.

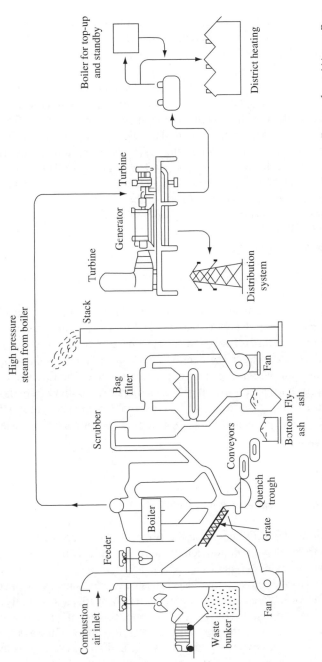

Figure 5.6 *Schematic diagram of a typical mass burn municipal solid waste incinerator. Source: Energy from Waste: Best Practice Guide 1996. Crown copyright material is reproduced with the permission of the Controller of HMSO and the Queen's Printer for Scotland.*

The bunker is large enough to allow for storing the waste to ensure a balance between the uneven delivery of the waste and the continuous operation of the plant. Therefore, the bunker would be designed to hold about 2–3 days equivalent of weight of waste which would be typically 1000–3000 tonnes of waste. Longer periods of storage are undesirable due to the rotting of the waste and consequent bad odours. The waste is delivered to the bunker which may be divided into different sections in separate unloading bays to allow for the mixing of the waste of different calorific values and combustion properties by the crane operator. The crane is of a travelling type and the crane operator will not only mix the wastes, but will also extract any bulky or dangerous items from the refuse for separate treatment. The operator then loads the waste to the feeding system. The crane grab can hold up to $6\,m^3$ of waste.

The feeding system is a steel hopper where the waste is allowed to flow into the incinerator under its own weight and is fed into the grate system by a hydraulic ram or other conveying system without bridging or blocking. The hoppers are kept partly filled with waste to minimise air leakage into the furnace and to ensure there is no interruption of feed to the grate. Monitors are used to measure the level of waste in the hopper. To prevent the fire in the furnace from burning back up into the feeding hopper, hydraulic shutters are used to seal the hopper at the furnace entrance. Also, the feed chute may be water-cooled or refractory-lined to prevent fire.

5.3.1.2   Furnace

Figure 5.7 shows a schematic diagram of a typical furnace system for a mass burn municipal solid waste incinerator (Clayton et al 1991; European Commission 2004). Each incinerator may have several furnaces fed by the operator from the waste bunker. For example, a typical 50 tonne/hour incinerator might have five separate 10 tonne/hour furnaces. The use of multiple furnaces allows for down time of the furnace for repair and regular maintenance. During the start-up of the incinerator, auxiliary burners are used to raise the temperature of the gases to initiate waste combustion. The waste is fed into the furnace usually by an independently controlled ram. In the furnace the waste undergoes three stages of incineration:

1. drying and devolatilisation;
2. combustion of volatiles and soot;
3. combustion of the solid carbonaceous residue.

In practice the various stages merge, since the components of the waste stream differ in moisture content, thermal degradation temperature, volatile composition and ignition temperature and carbon (fixed) content.

As the waste enters the hot furnace, the waste is heated up via contact with hot combustion gases, pre-heated air, or radiated heat from the incinerator walls, and initially moisture is driven off in the temperature range 50–100 °C. The water content of waste is very important since heat is required to evaporate the moisture, and therefore more of the available calorific value of the waste is lost in heating up the wet waste and so less energy is available. In addition, the rate of heating up of the waste, and therefore the rate of thermal decomposition, will also be affected by the water content of the waste. Water contents of municipal solid waste can vary between 25 and 50%. After moisture release, the waste then undergoes

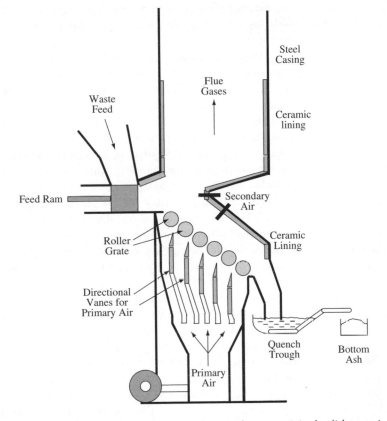

Figure 5.7   *Schematic diagram of the furnace of a mass burn municipal solid waste incinerator.*
Sources: Clayton et al 1991; European Commission 2004.

thermal decomposition and pyrolysis of the organic material such as paper, plastics, food
waste, textiles, etc., in the waste which generates the volatile matter, the combustible
gases and vapours. The volatile components of organic material in municipal solid waste
comprise typically between 70 and 90% and are produced in the form of hydrogen,
carbon monoxide, methane, ethane and other higher molecular weight hydrocarbons.
Devolatilisation takes place over a wide range of temperatures from about 200–750 °C
with the main release of volatiles between 425 and 550 °C. Thermal decomposition of the
waste and volatile release will also be dependent on the different components present in
the waste. For example, polystyrene decomposes over the temperature range 450–500 °C
and yields almost 99% volatiles, whereas wood decomposes over the temperature range
280–500 °C and produces about 70% volatiles. In addition to composition, the physical
state of the waste will influence the rate of thermal decomposition. For example, cellulosic material in thin form such as paper will decompose in a few seconds, whereas in the
form of a large piece of wood may take several minutes to decompose totally.

The combustion of volatiles to produce the flames of the fire takes place immediately
above the surface of the waste on the grate and in the combustion chamber above the grate.
Complete combustion of the gases and vapours requires sufficiently high temperature,

adequate residence time and excess turbulent air to ensure good mixing. The EC Waste Incineration Directive (2000) stipulates that the gases derived from incineration of the waste are to be raised to a temperature of 850 °C for 2 s to ensure complete burn-out of the volatile hydrocarbons. The volatile gases and vapours released, immediately ignite in the furnace since the furnace gas temperature will be typically between 750 and 1000 °C, but can occur at temperatures up to 1600 °C (European Commission 2004). The ignition temperature of the volatiles derived from the waste are well below these temperatures. Combustion chamber temperatures above about 1200 °C are avoided, as above this temperature ash fusion is likely to occur leading to a build-up of slag on refractory material. Typical mean residence times of the gases and vapours in the combustion chamber are 2–4 s, to comply with the EC Waste Incineration Directive which compares with typical burn-out times for volatile hydrocarbons of the order of milliseconds. Secondary air is blown in through nozzles above the grate to ensure excess air for combustion and to provide turbulence. Excess secondary air is required to avoid areas of zero oxygen levels, which serve to pyrolyse rather than combust the hydrocarbons as this can produce potentially hazardous high molecular weight hydrocarbons and soot. Therefore the distribution and turbulence characteristics of the secondary air are important factors in minimising the formation of pollutants in the combustion chamber.

After the drying and devolatilisation stages, the residue consists of a carbonaceous char and the inert material. The carbonaceous char which is defined as the fixed carbon rather than the volatile carbon contained in the volatile gases such as methane, ethane and other hydrocarbons, combusts on the grate and may take between 30 and 60 min for complete burn-out. The ash and metals residue is discharged continuously at the end of the last grate section into a water trough and quenched or air-cooled. The handling equipment is subject to heavy wear, due to the moist and abrasive nature of the material. The ash residue should be completely burnt out and biologically sterile. The EC Waste Incineration Directive (2000) sets a TOC content for the ash, of less than 3%, to ensure complete burn-out of the waste. The ash, known as bottom ash, is removed continuously or periodically via a conveyor and is disposed of in landfill sites or used for recycling as secondary aggregate for construction projects such as road building and concrete production. The bottom ash comprises about 30% of the total mass of waste input. The lighter flyash is transported through the system as particulate material which will also adsorb metals and organic material as it cools through the incinerator heat recovery and gas clean-up system. The flyash is collected in the cyclones, electrostatic precipitators and bag filters of the gas clean-up system. Because the flyash contains heavy metals, polycyclic aromatic hydrocarbons and dioxins and furans it has no recyclable value, is regarded as hazardous waste and is therefore landfilled. Flyash comprises only a few percent of the waste mass input (European Commission 2004).

At the heart of the incinerator is the grate, and a number of different types of furnace grate exist for municipal waste incineration, for example, the roller system (Figure 5.7) and the rocker, stoker, forward reciprocating systems and reverse reciprocating systems (Figure 5.8, Clayton et al 1991; European Commission 2004).

The grates are automatic and serve to move the waste from the inlet hopper end to the discharge end, whilst providing agitation or tumbling of the waste to stoke the furnace fire and loosen the combusting materials (European Commission 2004). The grate has a variable speed drive to adjust the residence time of the waste in the combustion zone to allow for changes in composition. For example, the roller grate system (Figure 5.7) has a roller diameter of 1.5 m and a circumferential speed of between 5 and 15 m/h. The grates in the roller system are inclined at about 30°, which assists in the movement of the burning waste through the furnace. Rocker grates have alternative rows of mechanical rockers which are pivoted or rocked to produce an upward and forward motion, advancing and agitating the waste (European Commission 2004). Horizontal stoker-type or travelling grates are generally arranged in sections of drying, ignition and burn-out, and also assist the distribution and control of primary air. Reciprocating grates consist of three or more sections with a step of about 0.5–1 m between sections. Each section consists of a series of fixed bars and moveable bars in a staircase-like arrangement (Figure 5.8). The movement of the mobile bars serves to agitate and move the waste down the grate. Most grates are cooled, most often with air. Control of the air supply to the furnace and combustion chamber is essential for efficient combustion. Primary air is blown evenly through the waste bed via the underside of the grate through slits in the grate which assists in combustion and cooling of the grate. Overgrate or secondary air is introduced through nozzles above the fuel bed, and in some plants tertiary air is added to cool the flue gases before gas cleaning treatment. Some fine material, referred to as riddlings, fall through the grate and may be recycled back to the incinerator or removed for disposal.

The size and shape of the combustion chamber itself are both important in determining optimum combustion efficiency and a number of different designs exist (Figure 5.9) (European Commission 2004). The size determines the mean residence time of the volatiles

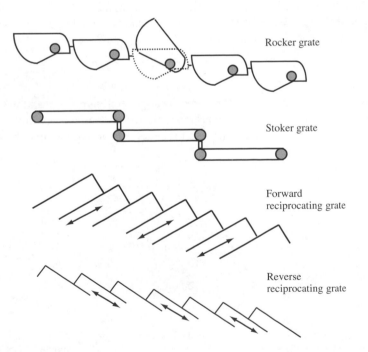

Figure 5.8 *Types of municipal solid waste incinerator grate. Sources: Clayton et al 1991; European Commission 2004.*

affecting their burn-out. The shape affects the heating pattern of the incoming waste from hot flue gases and furnace-wall heat radiation, which influences drying time, ignition time and burn-out time. Shape also influences the gross flow patterns within the chamber including recirculation and bulk mixing, which in turn influence combustion. Counter flow, medium flow and uni-directional flows of gases can be produced depending on design (Figure 5.9). The initial drying and devolatilisation of municipal solid waste can be very malodorous and the flow pattern of the gases and vapours should be through the hottest part of the furnace to completely combust and therefore destroy the odorous organic compounds.

The furnace and combustion chamber are lined throughout with refractory materials within a steel outer casing. Between the outer casing and the refractory of the combustion chamber may also be contained the water tubes of the boiler, which generate the steam for energy recovery. The main boiler tubes are located in the main boiler chamber above the combustion chamber, through which the hot flue gases flow. The refractory material of the furnace and combustion chamber essentially contains the combustion process in an area which will not fail due to thermal stress or degradation from high-temperature corrosion and abrasion. Refractories also re-radiate heat to accelerate drying, ignition and combustion of the incoming waste. The type of refractories used vary with the different parts of the furnace and combustion chamber. This is because at each stage the temperatures and fluctuations in temperature, oxidation and reduction conditions, abrasion from hard objects, erosion from dust laden flue gases, and corrosion from gases and slags, will require different properties from the refractories. For example, the alumino-silicates with high alumina refractories, or silicon carbide bricks are used in the hotter, grate-level part of the furnace, and the upper walls of the combustion chamber would be lower specification alumino-silicate firebricks (Turner 1996).

5.3.1.3   Heat Recovery

The combustion of waste is an exothermic or heat-generating process and the majority of the heat generated is transferred to the flue gases (European Commission 2004). The potential for heat recovery from the incineration process is due to the fact that the combustion

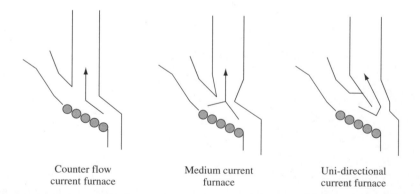

| Counter flow | Medium current | Uni-directional |
| current furnace | furnace | current furnace |

Figure 5.9   *Types of incinerator furnace design. Source: European Commission 2004.*

gases must be cooled before they can be discharged through the flue gas cleaning system. The temperature of the gases leaving the combustion zone, at typically 750–1000 °C is too high for direct discharge since gas temperatures below 250–300 °C are required for the gas cleaning equipment such as electrostatic precipitators, scrubbers and bag filters. Cooling is by the integral boiler and boiler chamber system in the modern municipal waste incinerator, although older incinerators where heat recovery was not practised, used water injection and air cooling. The heat of the flue gases is transferred to the water in the boiler tubes to produce steam. The boiler consists of banks of steel tubes through which water flows. The integral type or water-wall boilers are constructed around, and integrated with, the combustion chamber. This stage is usually an empty shaft since the flue gases are corrosive and high in particulate matter with typical temperatures of between 650 and 700 °C and fusion of hot flyash at the boiler tube surface may occur (European Commission 2004). The main bank of boiler-tube bundles is located in separate boiler chambers, the first pass of hot flue gases being across superheater boiler tubes which allow for higher temperature heat transfer, followed by evaporator-tube bundles which operate at lower temperatures. After the boiler there may be an economiser, which is a heat-exchange system, to heat water in a tube bank in order to produce further hot water from the flue gases before they enter the flue gas cleaning system. The boiler water/steam flow arrangement through the boiler tubes is from the economiser, to the evaporator, to the superheater, gradually producing hotter and hotter steam (European Commission 2004). The generated steam may be used for electrical power generation, district heating or may also be used within the plant to provide power and space heating.

In an incineration plant, waste is burned at a more or less constant rate, generally near to the design capacity, and therefore the output of energy cannot be varied to meet fluctuating demand. For space or district heating utilisation of this energy may be a problem, whereas electricity generated may be sold to the mains grid. Therefore if continuous output of heat is required, a back-up furnace is required with consequent additional investment and maintenance costs. Similarly, if heat demand is reduced, for example, in the summer months, alternative uses for the heat or a system of flue gas cooling is required so that waste incineration may be continued. The boiler is designed to ensure good heat transfer with the optimum circulation of the water without the occurrence of excessive fouling, allow for the cleaning of the boiler surfaces and to be mechanically stable under operating conditions. Figure 5.10 shows a typical boiler configuration with four passes of the hot gases over the boiler tube bundles (Darley 2003). Progressively, the temperature of the flue gases is reduced from about 1000–1200 °C just above the grate, until eventually the gases leave the fourth boiler/economiser stage at about 250 °C.

A major factor in the efficient operation of the boiler is the fouling of the tubes with deposits from the flue gases which contain flyash, soot, volatilised metal compounds, etc. The deposits stick to the boiler tubes and therefore reduce the transfer of heat from the hot flue gases to the water in the steel tubes, and hence the generation of steam and recovery of energy (Darley 2003). The rate at which tube fouling deposits build-up depends on the dust loading of the flue gases, the stickiness of the flyash, which in turn depends on temperature, flue gas velocity and tube bank geometry. The boiler tubes should be arranged parallel to the gas flow to minimise fouling and corrosion. The adherence of flyash to boiler tubes is mainly determined by the presence of molten salts such as calcium, magnesium and sodium, sulphates, oxides, bisulphates, chlorides, pyrosulphates, etc., in the flyash,

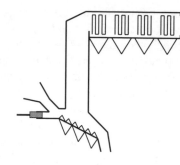

(a) Single pass, horizontal
boiler system

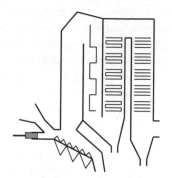

(b) Vertical pass boiler
system

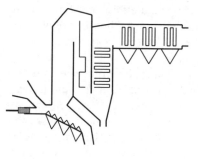

(c) Hybrid boiler system

Figure 5.10 *Typical boiler configurations for energy recovery. Sources: Darley 2003; European Commission 2004.*

and the presence of SO_3 and HCl (Buekens and Schoeters 1984). Scale deposits can be partially removed by means of soot blowers (using superheated steam), shot cleaning (dropping cast iron shot on the tubes to knock off the deposits), or by rapping the tubes (rapping the tube banks to knock off the deposits). Soot blowers are the most common and are usually operated once per operational shift. When the outlet temperature of the flue gases reaches a pre-determined maximum value, the operation has to be halted for a thorough mechanical or wet cleaning of the boiler, approximately every 4000 h of operation (Darley 2003).

Corrosion is another primary consideration in the design and operation of incinerator boilers (Krause 1991; Darley 2003). The formation of HCl by the combustion of chlorine-containing wastes such as paper and board and plastics such as PVC, may cause serious corrosion of tubes due to low-temperature acid corrosion. Critical control of temperature is required to prevent high-temperature and low-temperature corrosion of the boiler. High-temperature corrosion involves superheater boiler tubes in the boiler chamber at temperatures above 450 °C, and involves a series of chemical interactions between tube metal, tube scale deposits, slag deposits and flue gases. The rate of corrosion is influenced by temperature, the presence of low melting phases such as alkali bisulphates and pyrosulphates, acid gases such as HCl and SO_3, the nature of the tube metal and the periodic occurrence of reducing conditions. Low-temperature corrosion is due to condensation of acid gases such as HCl and H_2SO_4 formed as the temperature falls below the dew point. The dew point for H_2SO_4 is between about 40 and 155 °C and for HCl it is between 27 and 60 °C depending on gas concentration and water content in the flue gas (Krause 1991). Therefore, gas temperatures of more than 200 °C are required to minimise downstream dew-point corrosion.

Erosion of boiler tubes may also be a consideration. Flue gases from waste incineration are high in particulate or dust concentration and the dust particles can be very hard, causing erosion of the tubes by abrasion.

5.3.1.4   Emissions Control

The emissions to air and water and the solid residues arising from the incineration of waste are highly regulated by the EC Waste Incineration Directive (2000). In particular, the gas clean-up system required to meet the requirements of the Directive now constitutes a major proportion of the cost, technological sophistication and space requirement of an incinerator.

Of the pollutant emissions arising from the incineration of waste, those emitted to the atmosphere have received most attention from environmentalists and legislators. Table 5.1 shows the emission limits to air as set out in the EC Waste Incineration Directive (2000). There are a wide variety of emissions limits, but it is clear that the emissions of most concern are total particulate or dust, acidic gases such as hydrogen chloride, hydrogen fluoride and sulphur dioxide, and heavy metals such as mercury, cadmium and lead. In addition, the combustion efficiency is controlled by limits on the emission of carbon monoxide and organic carbon. There are also limits on the emission of dioxins of 0.1 ng TEQ/m^3 (TEQ = toxic equivalent). Table 5.1 shows emissions in terms of reference conditions. All legislative emission limits and plant emissions data across the world are related to a set of reference conditions, such as 7% O_2, 9% O_2, 9% CO_2, 11% O_2, 11% CO_2, etc., so that emissions can be compared from different plants which may be actually operated at very different conditions. In fact this makes the emission limit data for most countries very similar. Municipal solid waste incinerators generally produce flue gas volumes typically between 4500 and 6000 m^3 per tonne at 11% O_2, reference conditions (European Commission 2004).

Table 5.5 shows typical concentration ranges for emissions *before* any gas clean-up treatment for a range of European municipal solid waste incineration plant (European Commission 2004). The emissions are very much higher than are legally permitted under the EC Waste Incineration Directive (2000, Table 5.1) which emphasises the need for efficient and sophisticated gas clean-up to reduce the emissions to below legislated values. The layout of a hypothetical gas clean-up system for a municipal waste incinerator is shown in Figure 5.11. The particulate material is first removed by an electrostatic precipitator, and pre-collector, then the acid gases are removed by a lime scrubber which, may be of the dry-lime or wet-lime type. After the lime scrubber, and addition of an additive, such as activated carbon and lime, to adsorb mercury and dioxins and furans, there is a fabric filter to remove the fine particulate and activated carbon with the adsorbed

Table 5.5 *Typical concentration ranges of emissions from municipal solid waste mass burn incineration after the boiler and before gas clean-up*

Emission	Units	Range
Total dust	mg/m^3	1000–5000
TOC	mg/m^3	1–10
Hydrogen chloride	mg/m^3	500–2000
Hydrogen fluoride	mg/m^3	5–20
Carbon monoxide	mg/m^3	5–50
Sulphur oxides	mg/m^3	200–1000
Nitrogen oxides	mg/m^3	250–500
Cadmium + thallium	mg/m^3	<3
Mercury	mg/m^3	0.05–0.50
Other heavy metals	mg/m^3	<50
Pb, Sb, As, Cr, Co, Cu, Mn, Ni, V, Sn		
Dioxins and furans (PCDD/PCDF)	ngTEQ/m^3	0.5–10

Source: European Commission 2004.

pollutants. Finally, the oxides of nitrogen are removed by addition of ammonia to form inert nitrogen.

The main emissions control equipment used in the incineration of waste is described in Boxes 5.1–5.7. Box 5.1, cyclones; Box 5.2, electrostatic precipitators; Box 5.3, fabric filters; Box 5.4, wet scrubbers; Box 5.5, dry scrubbers; Box 5.6, semi-dry scrubbers; and Box 5.7, de-NO$_x$ systems.

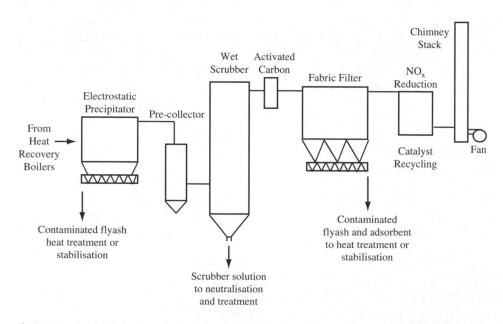

Figure 5.11 *Hypothetical advanced gas clean-up system for a municipal solid waste incinerator. Source: Wade 2003.*

Box 5.1
Cyclones

Cyclones are most effective in removing particles of larger than 15 µm, but they are much less efficient when it comes to the finer particles. Incinerator particulates have a significant proportion of particles in the less than 15 µm size. In addition, it is the finer grain size that has a concentration of heavy metals and organic micropollutants. Therefore, cyclones are only used as a preliminary collector, prior to an electrostatic precipitator or fabric filter, to improve the collection efficiency of these more effective trapping systems. The principle of the cyclone is that the dust-laden gas stream enters the cyclone tangentially or axially, forms a vortex and rotates in a helical path down the tube. The particles are flung to the inner wall of the cyclone by centrifugal force and then drop down to the bottom of the cyclone where they are collected. The cleaned gas forms a second vortex which flows up the middle of the cyclone and out through the central inner cylinder(a). Cyclones may be operated in banks of smaller sized cyclones as collection efficiency for smaller particles is improved with smaller inlet-orifice cyclones.

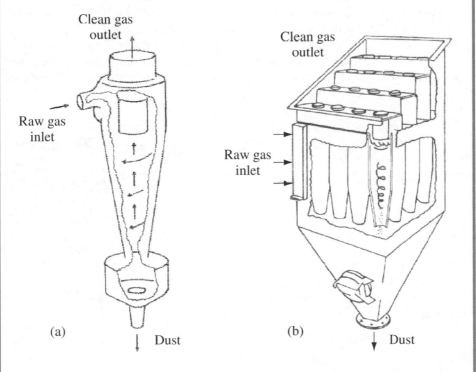

Sources: Porteous 1992; Clayton et al 1991; European Commission 2004.
Figure (a) reproduced by permission from Professor A. Porteous.
Figure (b), Crown copyright material is reproduced with the permission
of the Controller of HMSO and the Queen's Printer for Scotland.

Box 5.2
Electrostatic Precipitators

Electrostatic precipitators are a common form of particulate removal system for municipal waste incinerators. The dust-laden gas stream enters the electrostatic precipitator where the particles are first charged by negative ions produced by corona discharge in an intense electrostatic field. Typical voltages used are about 50 kV. The particles entering the field become negatively charged by the free electrons. The charged particles are electrically attracted to the collector electrode plates, which have an opposite charge, and the collected particles accumulate. The plates are regular cleaned by 'rapping' with a rotating hammer system which dislodges the layers of accumulated particles. The dust from the plates falls to the bottom of the precipitator where it collects in hoppers. Typically, the electrostatic precipitator consists of an array of wires or thin metal rods, which form the charging electrodes, with the collector plates running between them and separated by a distance of about 25 cm. The size of a typical electrostatic precipitator would be about 7 m³. Electrostatic precipitators can remove 97–99.5% of the particulates in a gas stream and are extremely efficient down to sub-micron size. In addition, the mechanism of electrostatic precipitation applies to liquid and tar droplets, and therefore they can also remove other pollutants in addition to particulates. Electrostatic precipitators must be operated under design conditions since their efficiency can be markedly affected by flue gas temperature, humidity and the layers of particulate accumulated on the collector electrodes.

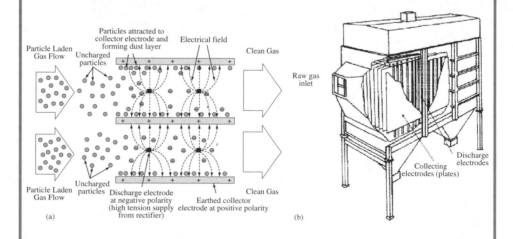

Sources: Porteous A. 1992; Clayton et al 1991; European Commission 2004.
Figure (a) reproduced by permission from Professor A. Porteous.
Figure (b), Crown copyright material is reproduced with the permission
of the Controller of HMSO and the Queen's Printer for Scotland.

Box 5.3
Fabric Filters

Fabric filters consist of a series of elongated, permeable fabric bags through which the particulate-laden gas flows. The fine fabric filters out the particles from the gas stream. The fabric can be designed with a range of pore sizes, although the filtering process will also include impaction, diffusion and electrostatic attraction, in addition to simple filtration. The build-up of particulate matter will also aid removal. The bags, which may number up to 100, are housed in a casing. Because of the fabric construction the operation temperature of fabric filters is usually low, and acid gases can cause damage. They are effective for removal of particulates down to submicron size and can remove particles from the gas stream to levels of less than $10\,\text{mg/m}^3$ which easily meets current EC legislation. Increasingly, fabric filters are used as the final gas clean-up after the electrostatic precipitator and acid gas scrubber with a pre-addition of additives, such as lime and activated carbon, to remove heavy metals such as mercury, and organic micropollutants such as dioxins and furans. Alternatively, the bags themselves can be coated with additives where chemical adsorption or absorption takes place, or a catalyst impregnation where destruction of organic compounds would then take place. The bags are cleaned regularly to remove the accumulated particulate matter by an air pulse, which rapidly expands the bag and releases the dust into a hopper at the base of the baghouse.

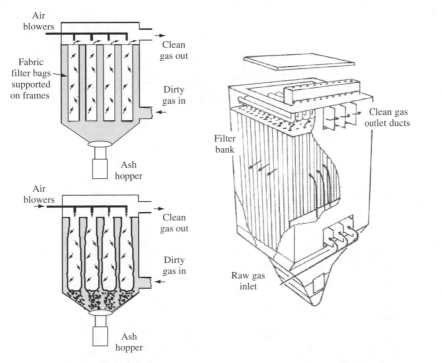

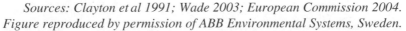

Sources: Clayton et al 1991; Wade 2003; European Commission 2004.
Figure reproduced by permission of ABB Environmental Systems, Sweden.

Box 5.4
Wet Scrubbers

Wet scrubbers are used in incinerator clean-up systems to remove soluble acid gases such as hydrogen chloride, hydrogen fluoride and sulphur dioxide. The gases are passed countercurrent through thin films or sprays of liquid in a tower, thus providing a very large surface area for reaction of the gas. The most common liquid used to trap the acid gases is an alkaline solution such as lime (calcium hydroxide) or sodium hydroxide. Wet scrubbers can also remove other pollutants such as particulates and heavy metals. For example, a typical scrubber design would have a quench unit prior to the scrubber tower to cool the gases to about 60 °C. The gases are passed to the scrubber tower with a water spray in the first stage, which would absorb the hydrogen chloride and hydrogen fluoride gas to produce hydrochloric and hydrofluoric acid. The acid solution passes down through the tower, removing heavy metals which are soluble in acid solution. In the second stage, which may be in the upper portion of the tower or a separate tower, the alkaline solution is used to remove any remaining hydrogen chloride and sulphur dioxide. After a de-mister stage to remove liquid carryover, the gases leave the scrubber. Acid gas removal efficiencies are very high at over 95% for hydrogen chloride and efficiencies for removing heavy metals are also high, of the order of 99% for lead and 92% for cadmium.

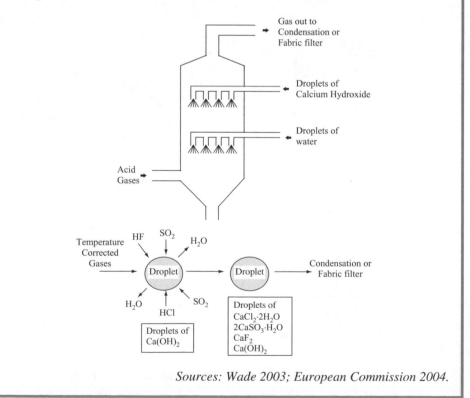

Sources: Wade 2003; European Commission 2004.

Box 5.5
Dry Scrubbers

The main disadvantage of wet scrubbers is the need for extensive treatment of the derived highly polluted liquid, which increases the costs of gas clean-up. Therefore, the dry and semi-dry scrubbers were developed for the treatment of acid gases. The dry system utilises dry fine-grained powder, for example, dry calcium hydroxide, which is sprayed onto the flowing gases. The incoming gases to a dry scrubber tower are cooled to about 160 °C. The reaction of the acid gases takes place in the dry state, to produce, for example, calcium chloride and calcium sulphate with hydrogen chloride and sulphur dioxide, respectively, which pass to the bottom of the tower for treatment or recycling. The system is used in conjunction with a down-stream fabric filter, which traps the particulate material. In addition, the dry alkaline powder trapped on the fabric filter presents an additional surface for reaction of the acid gases as they pass through the filter. The dry system is used for acid gases, but also can be designed to include heavy metals such as cadmium and mercury and organic micropollutants such as dioxins, by the addition of activated carbon to the calcium hydroxide.

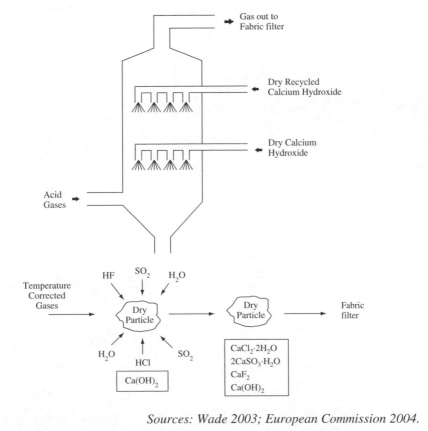

Sources: Wade 2003; European Commission 2004.

Box 5.6
Semi-dry Scrubbers

The semi-dry scrubber utilises a spray of droplets of calcium hydroxide which react with the acid gases, but as the droplets pass through the scrubber tower the water in the droplets evaporates to form fine-grained solid particles. Again the semi-dry scrubber is used in conjunction with a down-stream fabric filter. Dry and semi-dry scrubbers used in conjunction with fabric filters are very effective in the removal of acid gases and when used together with an activated carbon additive, volatile heavy metals and dioxins and furans.

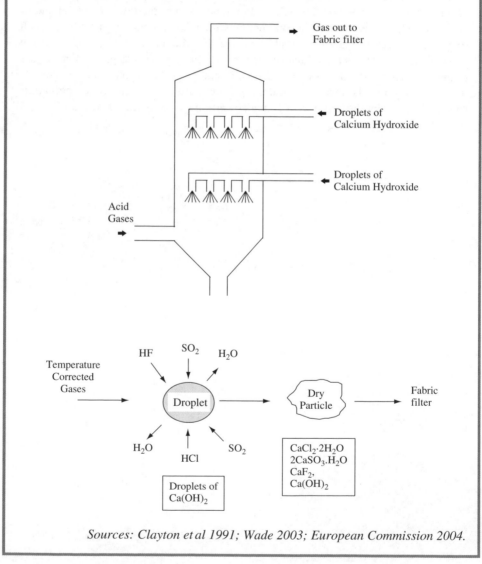

Sources: Clayton et al 1991; Wade 2003; European Commission 2004.

Box 5.7
De-NO$_x$ Systems

Nitrogen oxides (NO$_x$) are formed in combustion processes, and because municipal waste incinerators are 'large combustion plant' they are subject to NO$_x$ regulation. NO$_x$ emissions in flue gases are most commonly as NO (nitrogen oxide, typically about 90%) and NO$_2$ (nitrogen dioxide, typically about 10%). NO$_x$ forms from nitrogen in the air used for combustion and in the fuel itself, with higher levels of NO$_x$ formed at higher combustion temperatures and higher fuel nitrogen. NO$_x$ contribute to acid rain and act as a photochemical oxidant in the atmosphere. Reduction of NO$_x$ can be achieved by restricting its formation by control of combustion conditions. Lower temperatures and lower oxygen levels reduce NO$_x$ formation. Non-catalytic reduction of NO$_x$ can be achieved by introduction of ammonia in a narrow temperature range of 850–950 °C. The ammonia reduces the NO$_x$ to nitrogen and water. At higher temperatures the ammonia itself breaks down to produce NO$_x$ and at lower temperatures the reaction is too slow to be effective. Flue gas NO$_x$ are more often controlled by a selective catalytic reduction (SCR) process in the presence of added ammonia, which reproduces the ammonia reduction reaction but at a lower temperature and wider temperature range. Typical catalysts are platinum, palladium, vanadium oxide and titanium oxide. The ammonia, which is added upstream of the catalyst bed, reacts with the NO$_x$ in the presence of the catalyst at 250–400°C to produce nitrogen and water. Since the catalyst can be de-activated by heavy metals, the de-NO$_x$ system is located after the fabric filter. Selective catalytic reduction can reduce NO$_x$ levels by over 90%.

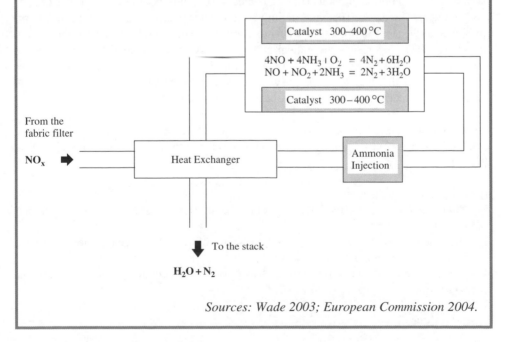

Sources: Wade 2003; European Commission 2004.

Table 5.6   *Emissions to the atmosphere from older municipal solid waste incineration plant*

Emission (units mg/m^3)	UK Older plant[1] (range)	Sweden Older plant[2] (range)
Dust	16–2800	1–90
Carbon monoxide	6–640	—
Hydrogen chloride	345–950	450–900
Sulphur dioxide	180–670	90–360
Hydrogen fluoride	—	4.5–9
Nitrogen oxides	—	180–360
Lead	0.1–50	0.45–2.7
Cadmium	<0.1–3.5	0.045–0.9
Mercury	0.21–0.39	0.27–0.36
Dioxin (TCDD ng/m^3)	0.73–1215	—
Furan (TCDF ng/m^3)	6.84–1425	—

Reference Conditions:
[1] STP dry gas, 273, K, 101.3 kPa.
[2] 9% CO_2, 273 K, 101.3 kPa, dry gas.
Source: Clayton et al 1991.

Table 5.6 shows a comparison of emissions to the atmosphere from older plant (Clayton et al 1991). Older plant built in the 1970s and 1980s represents older technology, where the gas clean-up consisted mainly of particulate removal systems such as electrostatic precipitators. There was little public or legislative concern over acidic gas emissions, heavy metals such cadmium and mercury or organic micropollutants such as dioxins and furans and polycyclic aromatic hydrocarbons. Consequently, municipal waste incinerators deservedly developed a poor image as pollution sources. However, the modern incinerator is an efficient combustion plant with a gas clean-up system complying with the highest environmental standards. The older polluting incinerator plants have largely been decommissioned as the introduction of stringent emissions legislation has been enacted throughout the world. Table 5.7 shows the annual average emissions of clean gas from European municipal solid waste incinerators after the emissions control system has cleaned the gases and shows that the emissions are well below those stipulated in the 2000 EC Waste Incineration Directive (Table 5.1) (European Commission 2004).

Formation and control of emissions   The emissions to the environment of most concern in relation to mass burn, municipal waste incinerators are those covered by the legislation and are listed below.

1. Dust (particulate matter).
2. Heavy metals such as mercury, cadmium, lead, arsenic, zinc, chromium, copper, nickel, etc.
3. Acidic and corrosive gases such as hydrogen chloride, hydrogen fluoride, sulphur dioxide and nitrogen oxides.
4. Products of incomplete combustion such as polycyclic aromatic hydrocarbons, dioxins and furans.
5. Wastewater.
6. Ash residue.

Table 5.7 *Range of clean gas emissions from some European municipal solid waste mass burn incinerators. Data are annual averages at reference conditions: 273 K, 101 kPa, 11% oxygen, dry gas*

Emission	Units	Range
Total dust	mg/m^3	0.1–4
TOC	mg/m^3	0.1–5
Hydrogen chloride	mg/m^3	0.1–6
Hydrogen fluoride	mg/m^3	0.01–0.1
Carbon monoxide	mg/m^3	2–30
Sulphur oxides	mg/m^3	0.2–20
Nitrogen oxides	mg/m^3	30–180
Cadmium + thallium	mg/m^3	0.0002–0.03
Mercury	mg/m^3	0.0002–0.05
Other heavy metals	mg/m^3	0.0002–0.05
Pb, Sb, As, Cr, Co, Cu, Mn, Ni, V, Sn		
Dioxins and furans (PCDD/PCDF)	ngTEQ/m^3	0.0002–0.08
Polycyclic aromatic hydrocarbons (PAHs)	mg/m^3	<0.01
Polychlorinated biphenyls (PCBs)	mg/m^3	<0.005

Source: European Commission 2004.

Each will be discussed in turn in relation to their formation, environmental impact and control.

Dust or particulate matter   The combustion of waste is a very dusty process. The agitation of the waste as it tumbles down the grate, the blowing of primary air through the bed, the high ash content of the waste, and the heterogeneous nature of the waste, all serve to produce a high particulate loading in the flue gases (Buekens and Patrick 1985). The design of the incinerator also influences the particulate loading in the flue gases. Such design factors include the size of the incinerator, the grate type, and the combustion chamber design. Particulate emissions from incinerators are the most visual to the public and require efficient and high levels of removal so that complaints do not arise. The particulate is largely composed of ash. However, in addition, pollutants of a more toxic nature such as heavy metals and dioxins and furans are associated with particulate matter, either as individual solid particles or adsorbed on the surface of the particles.

The particulates may also contain carbon and adsorbed acidic gases such as hydrochloric, sulphuric or even hydrofluoric acid to produce corrosive acid 'smuts'. Soot is formed when carbon-containing wastes are combusted in conditions of high temperature and low oxygen content. Polycyclic aromatic hydrocarbons (PAHs) have been cited as chemical intermediaries in soot formation, an alternative proposed mechanism is via acetylene radicals which build to form large soot molecules (Williams 1990a). The control of soot formation is via adequate residence time for the combustion process to completely burn out any soot being formed, with good mixing of the primary and secondary combustion air.

The emission of untreated flue gases would give rise to a dark plume and the deposition of dust downwind of the incinerator stack. The size range of incinerator particulates found in the flue gases is from <1 μm to 75 μm, the larger particles tending to settle out prior to the flue (Niessen 1978). It is the ultrafine particles that are of particular concern in assessment of health effects since they contain ash and adsorbed acid gases, heavy metals and organic micropollutants which, because of their size, can pass deep into the respiratory system of humans. There is currently some concern that the important factor in determining the deleterious effects of fine particles is not particularly their composition but their ultrafine nature and the fact that they can penetrate deep into the lungs. A separate size category of particulate matter of environmental concern, PM_{10} (particulate matter) has been designated for particulate matter of less than $10\,\mu m$ in size (Donaldson et al 2001). Exposure to PM_{10} particles is associated with both acute and chronic health effects (IEH 1997). These include increased rates of bronchitis, reduced lung function, respiratory symptoms and cancer (EPAQS 2001). A large fraction of municipal waste incinerator particulates are of such a small size (Niessen 1978). The human health concern relates to the interaction of the respiratory system with the high surface area, ultrafine particles and their associated adsorbed heavy metals and organic pollutants (IEH 1997). In addition, the small size of waste incinerator particulate emissions, promotes both short and long-range dispersion from the chimney stack into the environment (Denison and Silbergeld 1988). However, Rabl and Spadaro (2002) have assessed the risk to human health from the particulates emitted from municipal solid waste incinerators and have concluded that the health risks are insignificant. The environmental and human health impact of particulates from municipal solid waste incinerators were calculated to be insignificant compared with other sources of particulates in the atmosphere.

The selection of gas cleaning equipment used for the control of particulates in municipal solid waste incinerator flue gases is determined by a range of parameters. These include: the particle load in the gas stream; the average particle size; the particle-size distribution; the flow-rate of gas; the flue gas temperature; the required outlet gas concentration and the other components of the overall clean-up system used (European Commission 2004). Particulate emissions from mass burn incinerators are controlled by a range of possible equipment which are effective in removing particulate material. These include cyclones, electrostatic precipitators and fabric filters (Boxes 5.1, 5.2 and 5.3). The most common initial particulate removal apparatus is an electrostatic precipitator, since fabric filters will be found down stream in most incinerators because of the need to trap heavy metals and organic micropollutants which are in the solid phase or adsorbed on fine ash particles. Electrostatic precipitators can reach low exit gas particulate concentrations, typically in the range 15–$25\,mg/m^3$ and even lower exit gas values of less than $5\,mg/m^3$ are possible, where additional electrostatic fields or larger volume equipment is used (European Commission 2004). Cyclones are not common in municipal solid waste incinerator systems because of their lower capture efficiency compared with electrostatic precipitators and fabric filters for the particular size range of incinerator flue gas dust emissions. However, they may be used as an initial stage to reduce the particulate load for other devices. Cyclones typically achieve exit gas concentrations of only between 200 and $300\,mg/m^3$ (European Commission 2004). Cyclones also have the advantage that they may be used at higher temperatures of above $500\,^\circ C$. Wet scrubbers are also effective in reducing the particulate load of flue gases with about 50% capture efficiency (European Commission 2004). Fabric filters can achieve very low exit gas particulate concentrations, of typically less than $5\,mg/m$, and are increasingly used in most municipal solid waste incinerators. There are a range of materials available from which the fabric filters are made, the choice depending on the operational temperature required and their resistance to acid or alkali gas attack. Materials include polypropylene which has a maximum operational temperature of 95 °C, wool at 100 °C, polyester at 135 °C, Nylon at 205 °C and fibreglass at 260 °C.

Heavy Metals   Metals and metal compounds are present in the components of raw waste. For example, municipal refuse may contain lead from lead-based paints, mercury and cadmium from batteries, aluminium foil, lead plumbing, zinc sheets, volatile metal compounds, etc. Table 5.4 shows the range of trace components found in European municipal solid waste (European Commission 2004). High levels occur and the concentrations are very variable. The behaviour of metals during the incineration process are illustrated schematically in Figure 5.12 (Barton et al 1990). Within the environment of the furnace of the incinerator, the release of heavy metals from the waste and incorporation into the flue gases is a function of many factors, including volatility, combustion conditions and ash entrainment. Metals and metal compounds may evaporate in the furnace to condense eventually in the colder parts of the flues and generate an aerosol of submicron particles, or they may become adsorbed onto flyash particles through a range of processes. The extent of evaporation of these metals and metal compounds in the furnace depends on complex and interrelated factors such as operating temperature, oxidative or reductive conditions and the presence of scavengers, mainly halogens such as chlorine (Buekens and Patrick 1985). The volatility of these metals and salts is low, for example: Cd, 765 °C; Hg, 357 °C; As, 130 °C; $PbCl_2$, 950 °C; and $HgCl_2$, 302 °C. However, for some compounds the volatility temperatures are not known. As these metals enter the incinerator in the waste they are subject to combustion temperatures of anything between 800 and 1400 °C, well above the boiling points of metals such as mercury and cadmium and metal compounds such as lead chloride. The metals therefore enter the gas phase. The metals may also react with hydrogen chloride or oxygen to form compounds which are more volatile than the metal. Zhang et al (2001) have shown that there is a direct relationship between temperature and the

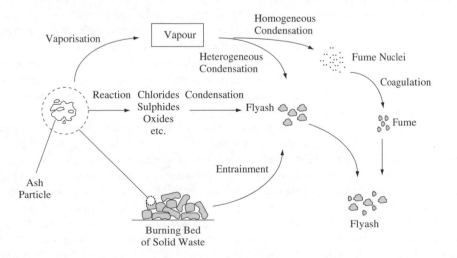

Figure 5.12   *Metals behaviour during waste incineration. Source: Barton et al 1990. Reproduced by permission of Taylor & Francis Ltd.*

amount of volatilisation of a whole range of heavy metals as the temperature of municipal solid waste is raised from 500 to 1000 °C.

The partitioning of the heavy metals in the incinerator system is a function of their physico-chemical properties. For example, cadmium and mercury being the more volatile of the heavy metals, with high vapour pressures and low boiling points, are most likely to be found in the flue gas (Brunner and Monch 1986; Morf et al 2000). Other metals with low vapour pressure and high boiling points, such as iron and copper, are almost completely trapped in the bottom ash. In addition to the physical properties of volatilisation, simultaneous chemical processes such as decomposition, chlorination, oxidation and reduction may take place (Mulholland and Sarofim 1991(a,b)). For example, it has been shown that chlorine influences the volatility of heavy metals via the formation of chlorides. Nickel, because of its low vapour pressure and high boiling point, will not vaporise under the conditions of incinerator furnaces, but will do so in the presence of chlorine (Seeker 1990; Davis et al 1998). Also, cadmium is easily volatilised during incineration and is oxidised in the presence of hydrogen chloride to form mainly cadmium chloride (Vogg et al 1986). A further route for the heavy metals to enter the flue gas stream is via entrainment of fine ash particles containing the metal either as the metal itself or as metal compounds. Entrainment is a function of the size, shape and density of the ash particles as well as the incinerator operating conditions. Changes in the oxidising and reducing conditions within the incinerator can also influence the volatilisation of heavy metals. Low-volatility metal compounds may also react under reducing conditions to form metal compounds which are more readily volatilised (Barton et al 1990).

As the furnace off-gases cool on passing through the flue gas system, the heavy metals are subject to a series of condensation reactions involving homogeneous nucleation to form a fine fume of metal particles and heterogeneous deposition onto flyash (Seeker 1990). Homogeneous nucleation occurs, for example, when the partial pressure of an inorganic vapour species exceeds a certain critical value (Ho et al 1992). The incineration gases may become supersaturated as a result of rapid cooling of the gas or rapid formation of a new metal species of lower volatility. Heterogeneous deposition involves flyash particles in the flue gases providing sites for condensation of the cooling metal vapour. It has been shown that the relative rates of homogeneous nucleation and heterogeneous deposition also depend on the time/temperature gradient experienced by the metal-containing flue gas (Davis et al 1998). Following homogeneous nucleation and heterogeneous deposition, particles will subsequently grow by coagulation.

The distribution of the metals in the various outputs from municipal waste incinerators has been investigated by a number of workers (Brunner and Monch 1986; Morf et al 2000). Figure 5.13 shows the percentage distribution of heavy metals as a mass balance into and out of an incinerator, equipped with an electrostatic precipitator, as the only gas clean-up measure, in terms of that fraction either emitted to the flue gas, or captured in the electrostatic precipitator (flyash) or the bottom ash from the furnace (Brunner and Monch 1986). It is suggested that the partitioning is a function of the physico-chemical properties of the elements and their derived compounds, such that volatile mercury and cadmium compounds with high vapour pressures and low boiling points are most likely to be found in the flue gas. Metals with a medium vapour pressure and boiling points such as lead and zinc are retained better in the slag and are less concentrated in the electrostatic

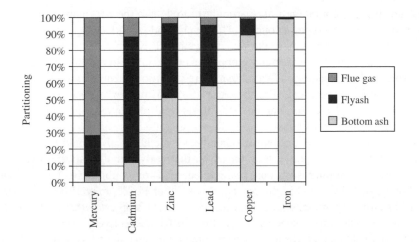

Figure 5.13   *Partitioning of heavy metals during municipal solid waste incineration. Source: Brunner and Monch 1986.*

precipitator dust. Other metals with low vapour pressure and high boiling points, such as iron and copper, are almost completely trapped in the slag.

The speciation of these metals, whether metal or metal compounds, in the incinerator off-gas is strongly influenced by the presence of compounds of chlorine, sulphur, carbon, nitrogen, fluorine and others during combustion and gas cooling. The form of the metal in the flue gases is important in determining the extent to which it will be captured by the gas-cleaning system. For example, mercury chloride is more easily captured than mercury metal is by wet scrubber systems. The off-gases containing metals and chlorine species, particularly hydrogen chloride, leads to the formation of metal chlorides, many of which have lower boiling points than the parent metal (Zhang et al 2001; Morf et al 2000). For example, cadmium is easily volatilised during incineration and is oxidised in the presence of hydrogen chloride to give cadmium chloride as the main product (Morf et al 2000; Vogg et al 1986). Mercury has also been shown to be present largely in the halogenated form, predominantly mercury(II)chloride and to a lesser extent mercury (I)chloride. Whilst initially mercury is vaporised as the metal in the furnace, it quickly becomes oxidised to the halogenated form and only a small percentage is present as metal vapour (Vogg et al 1986).

Bergstrom (1986) also reported on theoretical calculations of equilibrium and mass transfer data and concluded that mercury initially exists as mercury metal (Hg°), but during the cooling of the combustion gases an increasing fraction reacts to form $HgCl_2$. Experiments on a pilot plant and an incinerator also suggested that only a small proportion of mercury is present in the flue gases as mercury metal, most of it is oxidised and is present as gaseous salts, with $HgCl_2$ being the most likely since its high vapour pressure permits existence at temperatures down to 140 °C. Horne and Williams (1997) reported thermochemical equilibrium studies of mercury with oxygen, HCl and SO_2. At high temperatures, above 700 °C, Hg° was the stable species, below 700 °C, the presence of $HgCl_2$ and HgO becomes significant and below 500 °C, $HgCl_2$ is the dominant species under simulated flue gas conditions.

Organo-mercury compounds are particularly toxic and their formation in waste incineration flue gases would be of concern. However, Lindquist (1986) suggests that organo-mercury compounds decompose during combustion. It is unclear whether recombination and/or formation of organo-mercury compounds takes place in the flue. Sampling of flue gases from coal combustion indicated that there were no measurable amounts of organo-mercury compounds present (Prestbo et al 1995). However, Chow et al (1994) have reported that methyl mercury can be 10% of the total mercury species emitted from the combustion of bituminous coal.

The form of the metal species for other heavy metals is also important. Uberoi and Shadman (1990) have reported that, based on thermodynamic equilibrium calculations, when chlorine containing wastes are incinerated, lead compounds exist predominantly as PbO, $PbCl_2$, and $PbCl_4$ and in the presence of sulphur $PbSO_4$ may also be formed. Wey et al (1999), using thermochemical equilibrium analysis, suggested that Cr_2O_3, $PbCl_2$ and $CdCl_2$ were the dominant species of Cr, Pb and Cd respectively. The thermodynamic analysis assumed that equilibrium had been attained in the reaction system. Seeker (1990) has suggested that the principal species for arsenic is As_2O_3.

The gas clean-up systems required to control heavy metals are dependent on the metal's volatility. Measures used to control total particulate emissions, such as electrostatic precipitators and fabric filters, will collect the associated heavy metals which are in the flyash, are adsorbed to the surface or are discrete heavy metal particles (European Commission 2004). The heavy metals are associated with the particulate, because of the volatilisation of metals during the combustion of the waste, subsequent condensation at lower temperatures and adsorption onto the fine particulates in the flue gas. The heavy metals tend to concentrate in the finer grained size fraction of the particulate (Walsh et al 2001; Bouscaren 1988; Greenberg et al 1978; Zhang et al 2001). For example, Walsh et al (2001) found heavy metal enrichment factors, including cadmium, zinc and lead, of more than ten-fold for fine particles emitted from waste incinerators compared with those larger particles retained on the air pollution control system as flyash. Zhang et al (2001), have also reported that the emitted heavy metals are concentrated in the fine particulate sub-micron size fraction from municipal solid waste incinerators.

Fabric filters are particularly effective for the removal of heavy metals, since they are operated at temperatures typically below 250 °C where the metals have condensed to form particulate material which are effectively trapped by the fabric filter. Exit gas concentrations of particulate from fabric filters are typically less than $5\,mg/m^3$ and therefore fabric filters efficiently and effectively remove heavy metals adsorbed to the particulate material (European Commission 2004). Mercury is present in the flue gases mainly as mercury and mercury chloride. Because mercury has a boiling point of 357 °C and mercury chloride has a boiling point of 302 °C, all the mercury will be in the gas phase as it exits the furnace and boiler system (Figure 5.13). Even at the operating temperatures of the fabric filter (and also electrostatic precipitator) mercury and mercury chloride, with their high vapour pressure and low boiling points, will pass through the particulate trapping system. Therefore, since the emission-limit value for total mercury set by the EC Waste Incineration Directive (2000) is $0.05\,mg/m^3$, most municipal solid waste incinerators require the addition of special gas-cleaning measures specifically for the capture of mercury, in order to meet the required emission limit (European Commission 2004).

Scrubber systems, used to remove acid gases, develop an acidic solution with a low pH which are effective at trapping mercury chloride at the level of about 95%, but for mercury metal the level is only 0–10% (European Commission 2004). Addition of sulphur additives to the scrubber solution enables an improvement in the trapping of mercury metal but other methods are required to enable the emission limits for mercury to be attained. Systems used to trap total mercury include addition of reagent up-stream of the scrubber system or fabric filter. Additives which have proven effective are sodium sulphide, TMT 15 (trimercapto-s-triazine) and activated carbon (Wade 2003). The additives add to the cost of gas clean-up with sodium sulphide being the most cost-effective, followed by activated carbon, which is approximately three times the cost, and TMT 15 at seven times the cost, of sodium sulphide.

Activated carbon is the most commonly used additive to trap mercury, coupled with a downstream fabric filter, and it also has the advantage of removing dioxins and furans from the gas stream. Activated carbon is a high surface area material produced as a fine carbon powder, typically with surface areas in the range of 1500–2000 m^2/g. The high surface area of the carbon surface adsorbs the mercury vapour by physical adsorption on the surface active sites or, where the activated carbon is impregnated with sulphur, by chemical adsorption as mercury sulphide. The activated carbon additive is added at concentration levels of between 0.1 and 0.5 g/m^3 of waste gas, and high removal efficiencies have been reported (European Commission 2004; Wade 2003). For example, capture efficiencies for total mercury of more than 95% have been achieved, resulting in emission levels of less than 0.03 mg/m^3. The activated carbon traps the mercury vapour in the flue gas stream and also on the surface of the fabric filter, where it deposits and acts as a further adsorbing filter deposit. There is a risk of fire with the use of fine-sized activated carbon powder and therefore the carbon is generally mixed with other reagents, such as calcium hydroxide, at a ratio of 10% activated carbon to 90% calcium hydroxide. The calcium hydroxide also acts to remove acid gases.

An alternative to gas clean-up of heavy metals is to eliminate them from the raw waste material, the recycling of batteries for the removal of cadmium and mercury has been shown to be effective in reducing the emissions of these metals (European Commission 2004; World Health Organization 1990).

The human health effects of heavy metals reported in the literature, relate to occupational health exposure studies and to accidental exposure or to animal studies. Heavy metals exert a range of toxic health effects including carcinogenic, neurological, hepatic and renal effects (IEH 1997). For example, cadmium represents a health risk via accumulation in living tissue and has been associated with an increased risk of lung cancer, emphysema and kidney damage and in extreme circumstances, damage to bones and joints (IEH 1997; Probert et al 1987). Mercury and mercury compounds give rise to toxic effects associated with the central nervous system, the major areas affected being associated with the sensory, visual and auditory functions as well as those concerned with co-ordination (World Health Organization 1990). Lead exposure has been associated with disfunction in the haematological system and central nervous system. Decreases in intelligence and behaviour have been reported in children subject to exposure of increased levels of lead (World Health Organization 1990). The primary route for human exposure to heavy metals released by incineration is the food chain. Of the heavy metals, cadmium, mercury and lead are deemed of most importance in relation to municipal waste incinerators since, whilst other metals do occur, their toxicities or emission levels are much lower. The health effects of heavy metals arising from incineration is increased because they are readily available to the body as they are concentrated on the finer size fraction, tend to be adsorbed to the surface of particles, and their fine size means they are more easily ingested (Bouscaren 1988; Greenberg et al 1978; Buekens and Schoeters 1984). Whilst the effects of heavy metals on human health are significant, the gas clean-up measures required to meet legislative limits of heavy metal emissions from incinerators do reduce emissions of a range of heavy metals to very low levels (EC Waste Incineration Directive 2000). Investigations into the influence on human health of the emissions of heavy metals from waste incinerators suggest that they do not pose a significant problem to the environment. Reports on the effects of heavy metals and human health studies, in relation to municipal solid waste incinerators, have concluded that no effects on health have been linked to the release of heavy metals from incineration plants (Royal Commission on Environmental Pollution 1993). A later review of epidemiological studies on the impact of a range of pollutants, including heavy metals, emitted from waste incinerators, on workers in the waste inciner-ation industry and surrounding population, also concluded that no consistent pattern of ill health had emerged from such studies (IEH 1997). This was attributed to the very low level of concentration and the effects of compounding and modifying factors, such as smoking and socio-economic effects. Hu and Shy (2001) reviewed several epidemiological studies investigating the health effects of waste incinerator emissions on incinerator workers and community residents. They reported that, whilst some studies showed increased body levels of some organic chemicals and heavy metals, there were no effects on respiratory symptoms or pulmonary function and the findings for cancer and reproduc-tion were inconsistent. Rabl and Spadaro (2002) undertook a risk assessment of the emissions from waste incinerators, including heavy metals, and concluded that the health impacts of municipal solid waste incinerators were insignificant compared with the background air quality, provided that the emissions complied with the emission limits of the EC Waste Incineration Directive (2000).

Acidic and corrosive gases   Municipal waste contains a range of compounds which contain chlorine, fluorine, sulphur, nitrogen and other elements which may result in the generation of acidic, toxic or corrosive gases. Nitrogen oxides also result from the nitrogen in the combustion air formed by reaction with oxygen at the high temperatures of the combustion zone and from nitrogen in the waste. Typical waste contains about 4000 mg/kg chlorine, 100–350 mg/kg fluorine, 2000 mg/kg sulphur and 5000 mg/kg nitrogen (Table 5.4, European Commission 2004). The waste chloride and fluoride are in the form of waste plastics, for example, PVC and PTFE (polytetrafluoroethylene), chlorides are also found in paper and board, rubber, leather and as sodium chloride, for example, from street sweepings after road salting in icy conditions (Buekens and Schoeters 1984). The sulphur content of municipal solid waste is low compared with coal sulphur contents.

Normally, because the combustion of the hydrocarbon volatile fraction in an incinerator is almost complete, the flue gases consist mainly of nitrogen, oxygen, water vapour and carbon dioxide. However, the combustible waste compounds which contain the chlorine, fluorine, sulphur or nitrogen during combustion, generate gaseous contaminants such as hydrogen chloride, hydrogen fluoride, sulphur oxides and nitrogen oxides:

$$\text{C, H, Cl, F, S, N} + O_2 \Rightarrow CO_2 + H_2O + HCl + HF + SO_2 + NO_x$$

At the high temperatures of the combustion zone, carbon dioxide and water vapour may partially dissociate, but the resulting carbon monoxide, hydrogen and oxygen recombine when the temperature decreases. Many of the other products, including hydrogen chloride, hydrogen fluoride and sulphur dioxide, are stable. Normally chlorine is not detectable in the furnace emissions since it is reduced by numerous gases or solid reducing agents to hydrogen chloride (Buekens and Schoeters 1984).

The origin of HCl in incinerator flue gas has been the subject of much research, due to the corrosive nature of HCl at low temperature, i.e., dew point corrosion and high-temperature corrosion and its implication in dioxin formation. One of the major sources of HCl is regarded as being PVC plastic at more than 50% (European Commission 2004). However, other sources such as metal chlorides like NaCl or $CaCl_2$ from paper and board, rubber, leather and vegetable matter are regarded as significant sources of HCl (Uchida et al 1988). PVC emits HCl by a gradual process of thermal decomposition, which takes place between 180 °C and 600 °C. Chlorides are also implicated in the formation of dioxins and furans. Hydrogen fluoride is even more reactive and corrosive than HCl and arises from combustion of fluorinated hydrocarbons, such as plastics like PTFE.

Municipal waste incinerators are regarded as only a minor source of sulphur dioxide (SO_2) emission when compared to power plants and industrial boilers firing heavy fuel oil or coal (Clayton et al 1991). About 1% of the SO_2 may be further oxidised to sulphur trioxide, SO_3, which reacts with water vapour to form highly corrosive sulphuric acid, H_2SO_4, in the flue gas.

Nitrogen oxide (which includes all the nitrogen oxides, but particularly nitric oxide, NO, and nitrogen dioxide, NO_2) from waste incineration arises mainly from the nitrogen in the waste ('fuel NO_x') and by direct combination of the nitrogen and oxygen present in the combustion air, which occurs more rapidly at high temperatures ('thermal NO_x'). In practice, thermal NO is formed almost exclusively at high temperatures in the flame, particularly under oxidising conditions. In reducing conditions, little NO is formed (Buekens and Patrick 1985). Fuel NO_x is formed from the nitrogen compounds in the waste, but may also form nitrogen gas instead of NO_x.

The potential problem of acid gases in the back-end of the incinerator plant is due to their low dew point, which results in corrosive damage to metals. The emission of the pollutant gases to the atmosphere produces the well-documented acid rain with its associated environmental damage, whilst NO, after atmospheric oxidation to NO_2, is active in the generation of photochemical smog. HCl, HF, SO_2 and NO_x all produce acids in the atmosphere, which contribute to acid rain, forming hydrochloric, hydrofluoric, sulphuric and nitric acids, respectively. Increased acidification of the atmosphere has resulted in damage to buildings, acidification of lakes, respiratory problems, die-back of forests, etc. The health effects associated with acid gases and NO_x include respiratory and bronchial problems and contribute to the general environmental deterioration of the urban atmosphere (IEH 1997).

The EC Waste Incineration Directive sets emission limits for hydrogen chloride, hydrogen fluoride, sulphur dioxide and nitrogen oxides (EC Waste Incineration Directive 2000). Wet, dry and semi-dry processes are used to remove the acid gases produced by waste combustion. Wet scrubbing systems use slurries and solutions at relatively low temperatures and produce a liquid or wet solid/sludge reaction product. Generally, wet scrubbers operate in two stages, a first stage which uses water to trap the highly soluble acid gases, hydrogen chloride and hydrogen fluoride, and a second stage which uses added alkali absorbents to trap the less soluble sulphur dioxide. The absorbents used include calcium hydroxide and sodium hydroxide. That is, alkali solutions which neutralise and scrub out the acid gases. Gas-liquid absorption of acid gases is very efficient, but the liquid or sludge product is highly polluted and difficult and expensive to treat. In addition, the slurry or solution used, requires to be made up to certain specifications which again adds to the cost (Wade 2003). A further disadvantage is that the flue gases produced are high in moisture content and may result in acid condensation and corrosion.

Consequently, developments have centred on new methods of control which generate a solid residue which is easier to handle. Dry systems use a dry powder such as calcium oxide (lime) or sodium bicarbonate and possibly upstream humidification to improve gas/sorbent reaction. The dosage rate of the dry powder is approximately three times the stoichiometric amount, that is, three times more than the amount required to exactly trap the acidic component of the acid gases. The reaction products are solid particles and the process increases the dust load of the flue gas stream and would have to be trapped out of the flue gases, usually by a fabric filter (European Commission 2004). The build-up of the scrubber powder on the fabric filter produces a layer of adsorbent, which also aids the acid gas removal process.

Gas-solid adsorption tends to be less efficient than gas-liquid absorption. Therefore, a development is the semi-dry system which is also called the semi-wet system or spray absorption. Semi-dry processes use an alkaline sorbent slurry or solution which is atomised into fine droplets into the flue gas, the droplets react and dry in the hot flue gases to produce a dry powder (European Commission 2004). Absorption of the acid gases takes place efficiently at the gas–liquid interface, but then the heat of the flue gases causes evaporation of the water to produce a dry powder which is easier to handle. The flue gases have an increased particulate content and have to be cleaned by a down-stream fabric filter. For the semi-dry systems, as was the case for the dry system, adsorption is improved by the use of a down-stream fabric filter which increases contact time between the gases and the alkaline filter cake formed on the filter by the adsorbent (Boxes 5.4–5.6).

Nitrogen oxides (NO_x) may be reduced by primary measures such as the temperature of combustion, recirculation of the flue gases, staged combustion, or by secondary down-stream clean-up measures such as ammonia addition (European Commission 2004; Buekens and Patrick 1985). Since the formation of NO_x is directly related to increasing temperature, the use of primary and secondary air to ensure good gas mixing minimises high-temperature excursions, which thereby results in lower formation of NO_x. The use of flue gas recirculation, where the flue gases are fed back into the combustion chamber as secondary air, results in the reduction of NO_x formation, since the recirculated flue gases have a lower nitrogen content and therefore produce lower thermal NO_x. A further method to reduce the nitrogen content of the air supply which is required for combustion, is to use either pure oxygen or enriched oxygen-air. Again, the reduced nitrogen in the incoming gas has less nitrogen available to be converted to thermal NO_x. Other combustion operational measures to reduce NO_x include staged combustion and re-burn techniques which are common in the fossil fuel combustion power generation industry. Staged combustion involves the supply of a reduced amount of combustion air, and therefore oxygen, to the combustion zone which reduces the amount of NO_x formed because of the lower levels of nitrogen. The combustion of the waste is less complete and volatile unburnt hydrocarbons are formed which are then burnt-out in a second stage where excess air is added. Reburning is a three-step process involving injection of a reburning fuel such as natural gas to a combustion zone above the burning bed of waste. NO_x formed in the primary combustion zone is reduced to molecular nitrogen by reaction with hydrocarbon fragments formed in the re-burn zone from the combustion of natural gas (McCahey et al 1999).

Nitrogen oxide, the main oxide of nitrogen found in flue gases, cannot be reduced by scrubbing because of its low solubility in scrubber systems (Brna 1990). Therefore, secondary techniques to control NO_x emission rely on addition of a reactant, such as ammonia (NH_3) or urea (a derivative of ammonia – H_2NCONH_2) to convert the NO_x to nitrogen and water by reaction. The ammonia or urea are added in aqueous solution. The secondary selective reduction takes place at either high temperature, known as the selective non-catalytic reduction (SNCR) process, or at lower temperature using a catalyst, known as the selective catalytic reduction (SCR) process. SNCR involves addition of the ammonia or urea into the incinerator furnace at temperatures typically between 850 and 1000 °C, with the optimum temperature range between 850 and 950 °C, where the reaction with NO_x is maximised. Higher levels of NO_x removal require higher inputs of ammonia or urea, but can result in 'ammonia slip' where some of the added ammonia or urea is not utilised and passes into the flue gases unreacted. SCR operates with the addition of ammonia or urea as an additive to form nitrogen and water from NO_x, but operates at temperatures, typically between 250 and 400 °C, with the use of a catalyst. The catalyst is usually in mesh form and generally consists of TiO_2 impregnated with V_2O_5 and WO_3 catalyst and lasts for approximately three to five years. High NO_x reduction rates of over 90% can be produced with SCR (European Commission 2004). For waste incineration emissions control systems, the SCR process is usually applied after the scrubber and particulate removal system and the flue gases may need to be re-heated to the temperature of the SCR unit (250–400 °C), resulting in consumption of energy.

Odours from incineration of waste may not pose a health hazard as such, since the concentration levels of compounds where the nose can detect odour are extremely low. However, odours cause nuisance to the local population and are often the most common cause of complaint. Odours associated with incineration of waste are usually complex mixtures of organic compounds and result from incomplete combustion. Threshold limit values, above which the odour can be detected, can be very low, for example, many organic compounds which have been associated with waste incinerators have a threshold limit value between 0.001 and 10 ppm (Brunner 1985). The odours released into the atmosphere will be influenced by the efficiency of the combustion process and the gas clean-up system and peaks of emission may last only a few seconds. Emission to the atmosphere is influenced by dilution in the surrounding atmosphere and the dispersion of the stack plume, which in turn will be influenced by meteorological conditions. Brunner (1985) has reviewed the odours from incineration of waste, the threshold limit values of common odours and their control.

Products of incomplete combustion: polycyclic aromatic hydrocarbons (PAHs), dioxins and furans   The volatile matter arising from the thermal degradation of waste is normally completely combusted by providing adequate residence time, post combustion temperature and turbulent mixing. However, it is a consequence of the incineration process that there will be some areas in the incinerator which allow incomplete combustion of the gases and vapours. These incompletely combusted vapours may contain volatile organic compounds such as polycyclic aromatic hydrocarbons (PAHs), dioxins and furans, tar and soot particles. Incomplete combustion may occur when the incinerator is improperly operated, for example, operation at excessively low temperatures, below 800 °C or overloading of the plant. The occurrence of incomplete combustion can be detected by monitoring the flue gas composition. For efficient waste combustion, the concentration level of carbon monoxide consistently remains below 0.1 volume %.

There are potentially a wide range of waste-derived thermally degraded organic compounds which may be found in incinerator flue gases. Of these organic micropollutants there is most public concern associated with the emissions of polycyclic aromatic hydrocarbons, dioxins and furans from the incineration of waste.

Polycyclic aromatic hydrocarbons   PAHs are compounds based on aromatic benzene rings which are fused to form two or more polycyclic rings. Figure 5.14 shows examples of PAHs. PAHs are known to occur naturally in the environment, for example, in sediments, fossil fuels and by natural combustion in forest fires (Lee et al 1981). The major sources of PAHs, however, are anthropogenic; examples include oil and coal-fired power generation plant, coke production, residential furnaces, diesel and gasoline engines and in waste combustion (Lee et al 1981; Williams 1990(b)). Concern over the emission of PAHs to the environment is centred on the associated health hazard, because PAHs comprise the largest group of carcinogens among the

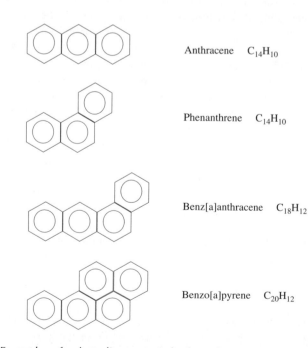

Anthracene $C_{14}H_{10}$

Phenanthrene $C_{14}H_{10}$

Benz[a]anthracene $C_{18}H_{12}$

Benzo[a]pyrene $C_{20}H_{12}$

Figure 5.14   *Examples of polycyclic aromatic hydrocarbons.*

environmental chemical groups (Lee et al 1981). PAHs adsorbed onto air-borne particles are believed to be a major contributory reason why death rates from lung cancer are higher in urban than in rural areas (Love and Seskin 1970). Cancers of the lung, stomach, kidneys, scrotum and liver have been associated with exposure to PAHs (Lee et al 1981; Gelboin and Tso 1978). The International Agency for Research on Cancer and the World Health Organization indicate that some polycyclic aromatic hydrocarbons are carcinogenic (EC Waste Incineration Directive 2000). PAHs are also important as they have been suggested as precursors to the formation of soot in combustion systems (Williams 1990(a); Richter and Howard 2000). In addition, PAHs have been implicated in the formation of dioxins and furans (PCDD and PCDF) on waste incinerator flyash (Wilhelm et al 2001; Weber et al 2001). The aromatic chemical structure of the PAHs, together with the flue gas composition, which includes the precursors for dioxin molecular chemistry, including oxygen and chlorine (in the form of hydrogen chloride), react with the flyash as a catalysing medium to form dioxins and furans (de-novo synthesis).

The EC Waste Incineration Directive (2000) does not specifically set emission limits for PAHs. However, the 2000 Waste Incineration Directive states that Member States of the European Union may set their own emission limits for PAHs (Waste Incineration Directive, 2000).

Table 5.8 shows the PAHs emission from municipal solid waste incinerators firing municipal solid waste (MSW) and refuse-derived fuel (RDF) (Parker and Roberts 1985;

Table 5.8 *Emissions of PAHs from municipal waste incineration ($\mu g/m^3$)*

Reference	MSW [1]	RDF [2]	MSW^1 [3]	MSW^2 [4]	MSW^3 [4]	MSW [5]
Fluorene	—	0.2	12.0	—	—	—
Methylfluorenes	—	0.09	—	—	—	—
Phenanthrene	—	—	43.0	—	—	—
Fluoranthene	0.58	0.2	11.0	—	—	—
Pyrene	1.58	0.09	6.8	—	—	—
Benzo[a]anthracene	⎧	0.1	1.1	4.6–0.6	<0.1	—
	⎩ 0.72					
Chrysene		0.5	3.0	—	—	—
Benzofluoranthenes	0.32	0.04	4.0	1.1–0.2	<0.2	—
Benzo[a]pyrene	⎧	0.04	—	1.2–<0.1	<0.1	<0.001
	⎩ 0.02					
Benzo[e]pyrene		0.04	0.7	—	—	—
Perylene	0.18	0.04	—	—	—	—
Benzo[ghi]perylene	0.42	0.04	—	0.8–<0.2	<0.2	—
Coronene	0.04	—	—	—	—	—
Total PAHs	—	—	—	7.7–0.6	0.2	<0.01

^{1} Incinerator emissions during plant start-up.
^{2} Incinerator emissions during plant start-up 270–730 °C.
^{3} Combustion 800–860 °C.

Sources: [1] Davies et al 1976; [2] Parker and Roberts 1985; [3] Colmsjo et al 1986; [4] Yasuda and Takahashi 1998; [5] EC Waste Incineration Directive 2002.

Davies et al 1976; Colmsjo et al 1986; Yasuda and Takahashi 1998; European Commission 2004). The RDF incinerator used RDF pellets in a fluidised bed, and reported that a large percentage of the PAHs were shown to be adsorbed on the surface of the flyash. Davies et al (1976), also reported that the PAHs emissions from waste incineration were associated with the flyash. Analysis of wastewater from the incinerator showed only low concentrations of PAHs reflecting their low solubility in water. Also shown in Table 5.8 are PAHs emissions during the cold start-up of a municipal solid waste incinerator in Sweden (Colmsjo et al 1986). The emissions are clearly much higher, reflecting the less than optimum combustion efficiency during start-up. Under normal operation the concentration of each individual PAH never exceeded 10ng/m^3. Also detected were a number of halogenated PAHs and chlorobenzenes and chlorophenols which are known to act as precursors for the formation of dioxins and furans. Yasuda and Takahashi (1998) also showed that, during the start-up of a waste incinerator, the initial stages at low temperature, where incomplete combustion may occur, can result in significantly high concentrations of PAHs. Their work showed that as the incinerator was operated at the lower start-up temperatures of between 270 and 330 °C, the total PAHs emission was $7.7 \, \mu\text{g/m}^3$ which decreased as the temperature was increased to levels of total PAHs of $0.9 \, \mu\text{g/m}^3$ at temperatures between 660 and 730 °C. Under operating combustion temperatures of 800–860 °C, the total PAHs decreased to $0.2 \, \mu\text{g/m}^3$. This was reflected in the concentrations of an individual PAH such as benzo[a]anthracene and benzo[a]pyrene. Benzo[a]pyrene is known as a marker carcinogen as it is regarded as highly carcinogenic (Lee et al 1981) and often used as a measure of the carcinogenicity of an environmental sample. Table 5.8 shows the emission of benzo[a]pyrene from European municipal solid waste incinerators at <0.001 and total PAHs at <0.01 (European Commission 2004). These very low emission data for PAHs from modern incinerators, indicate much higher combustion efficiencies and much greater emissions-gas clean-up efficiencies which produce significantly lower emissions of PAHs than the older incinerators emission levels reported in the 1970s and 1980s. In addition, to overcome the problem of increased emissions during the start-up period of the incinerator, the 2000 Waste Incineration Directive requires that auxiliary burners must be used to maintain temperatures within the incinerator at above 850 °C.

The PAHs reported from municipal waste incineration include some species known to be biologically active in human and bacterial cell tests, for example, benzo[a]pyrene, benzo[e]pyrene, phenanthrene, methylphenanthenes, fluoranthene and the methylfluorenes (Lee et al 1981; Longwell 1983; Barfnecht et al 1980). The levels of PAHs reported for incinerator emissions are similar to those reported for coal and oil-fired power stations and lower than those emitted from older diesel engines, industrial coal combustion, domestic heating and certain industrial processes, reflecting the combustion process in all its forms as a source of PAHs (Williams 1994; Dyke et al 2003). The physical and chemical properties of PAHs suggests that control mechanisms introduced for the control of dioxins and furans would easily control PAHs emissions from incinerators, as shown by Table 5.8 for the modern European municipal solid waste incinerator.

Dioxins and Furans   Polychlorinated dibenzo-p-dioxins (PCDD) or 'dioxins' and the closely related polychlorinated dibenzofurans (PCDF) or 'furans' constitute a group of chemicals that have been demonstrated to occur ubiquitously in the environment. They have been detected in soils and sediments, rivers and lakes, chemical formulations and

Table 5.9   *Concentrations of dioxins in soil, sediment and air for various European countries*

Country	Rural soil (10^{-9} g/m³)	Urban sediment (10^{-9} g/m³)	Urban air (10^{-15} g/m³)
Austria	—	—	—
Belgium	2.1–2.3	—	86–129
Germany	1–5	12–73	—
Italy	—	<1–23	48–277
Luxembourg	1.4	2.4–16	54–77
Spain	<1–8.4	<1–57	
Sweden	<1	—	<1–29
The Netherlands	2.2–16	—	4–99
UK	<1–20	2–123	0–810

Source: European Commission DG Environment 1999.

wastes, herbicides, hazardous waste site samples, landfill sludges and leachates (European Commission DG Environment 1999, Tiernan 1983, Table 5.9). PCDD and PCDF have a number of recognised sources, among which are their formation as by-products of chemical processes, such as the manufacture of wood preservatives and herbicides, the smelting of copper and scrap metal, the recovery of plastic-coated wire, and natural combustion such as forest fires (Steisel et al 1987, Pollution Paper 27 1989). More contentiously, they are found in combustion products, the ash, stack effluents, water and other process fluids from the combustion of municipal waste, coal, wood and industrial waste (Tiernan 1983). Historical chemical studies of sediment cores have shown that PCDD/PCDF were not present to any great extent before approximately 1935, suggesting that they are a recent phenomenon (Brzuzy and Hites 1996). They have been particular associated with the combustion of chlorinated wastes, including hazardous waste and municipal solid waste. This association of PCDD and PCDF with emissions from the incineration of waste gives rise to considerable concern from the public and environmental groups.

The generalised molecular structures of PCDD and PCDF are shown in Figure 5.15, tricyclic aromatic compounds containing two (dioxin) or one (furan) oxygen atoms. Each of these structures represents a whole series of discrete compounds having between one and eight chlorine atoms attached to the ring, for example, Figure 5.15 shows the tetra congeners with four chlorine atoms in the 2,3,7,8 positions, i.e., 2,3,7,8-tetrachlorodibenzo-p-dioxin (2,3,7,8 TCDD) and 2,3,7,8-tetrachlorodibenzofuran (2,3,7,8 TCDF). Since each chlorine atom can occupy any of the eight available ring positions, it can be calculated that there are 75 PCDD congeners and 135 PCDF congeners (Table 5.10). All the PCDD and PCDF are solids with high melting and boiling points, and with limited solubility in water. Many of these congeners are not readily available in pure form and hence the toxicology of many of the 210 congeners of PCDD and PCDF have not been assessed in great detail (European Commission DG Environment 1999). PCDD and PCDF are resistant to degradation in the environment, are persistent pollutants, and may be transported over long distances and give rise to trans-national exchanges of pollutants (European Commission DG Environment 1999).

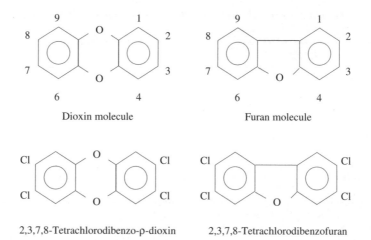

Figure 5.15   *Dioxin and furan molecules and the 2,3,7,8 congeners.*

Table 5.10   *Numbers of dioxin and furan congeners as a function of the number of chlorine substituents*

Number of chlorine atoms	Number of dioxin congeners	Number of furan congeners
1	2	4
2	10	16
3	14	28
4	22	38
5	14	28
6	10	16
7	2	4
8	1	1
Total	75	135

The potential threat of PCDD and PCDF to humans should be assessed, bearing in mind that the 75 PCDD congeners and 135 PCDF congeners have differing toxicities and are often present in multiple mixtures of the congeners. The toxicities of the PCDF congeners generally parallel those of PCDD (Tiernan 1983). The assessment of the toxicity of PCDD and PCDF mixtures has led to the development of the TEQ scheme. This uses the available toxicological and biological data to generate a set of weighting factors, each of which expresses the toxicity of a particular PCDD or PCDF in terms of an equivalent amount of the most toxic and most analysed PCDD which is 2,3,7,8 TCDD. For example, for the NATO/CCMS scheme, 2378 TCDD would have a TEQ of 1.0 and OCDD, for example, would have a TEQ of 0.001 (European Commission DG Environment 1999; Pollution Paper 27 1989). Table 5.11 shows two of the most common TEQ schemes (European Commission DG Environment 1999; Pollution Paper 27 1989). The TEQ system is based on the toxicity of the various congeners. Toxicity

Table 5.11 *Toxic Equivalents (TEQ) based on the Eadon and NATO/CCMS schemes*

PCDD/PCDF congener	Eadon (Eadon-TEQ)	NATO/CCMS (I-TEQ)
PCDD		
2,3,7,8-Tetrachlorodibenzodioxin	1	1
1,2,3,7,8-Pentachlorodibenzodioxin	1	0.5
1,2,3,4,7,8-Hexachlorodibenzodioxin	0.033	0.1
1,2,3,6,7,8-Hexachlorodibenzodioxin	0.033	0.1
1,2,3,7,8,9-Hexachlorodibenzodioxin	0.033	0.1
1,2,3,4,6,7,8-Heptachlorodibenzodioxin	0	0.01
Octachlorodibenzodioxin	—	0.001
PCDF		
2,3,7,8-Tetrachlorodibenzofuran	0.33	0.1
1,2,3,7,8-Pentachlorodibenzofuran	0.33	0.05
2,3,4,7,8-Pentachlorodibenzofuran	0.33	0.5
1,2,3,4,7,8-Hexachlorodibenzofuran	0.011	0.1
1,2,3,7,8,9-Hexachlorodibenzofuran	0.011	0.1
1,2,3,6,7,8-Hexachlorodibenzofuran	0.011	0.1
2,3,4,6,7,8-Hexachlorodibenzofuran	0.011	0.1
1,2,3,4,6,7,8-Heptachlorodibenzofuran	0	0.01
1,2,3,4,7,8,9-Heptachlorodibenzofuran	0	0.001
Octachlorodibenzofuran	0	0.001

Sources: European Commission DG Environment 1999; Pollution Paper 27, 1989.

depends on the number and position of the chlorine substituents and is highest for the tetra-, penta- and hexa-chloro compounds (Buser and Rappe 1984). The most toxic dioxin isomers belong to the 2378 group, this group includes the 2378 tetra, 12378 penta- and 123478, 123678 and 123789 hexa-chlorinated dibenzo-p-dioxins and chlorinated dibenzofurans. Hence the emphasis in the TEQ schemes in Table 5.11 for these congeners to be selected for the toxic equivalent determination. Congener-specific analyses of emissions from municipal solid waste incinerators are not common. Data from the UK suggest that they are dominated by the hepta- and octa-PCDD and PCDF, with significantly lower concentrations of the tetra-PCDD and PCDF (Alcock et al 2001). Ogura et al, reporting on Japanese municipal solid waste incinerators, showed a distribution of congeners which varied between incinerators and with different PCDD/PCDF ratios (Ogura et al 2001).

Although there is most concern about emissions of dioxins and furans, there is increasing awareness that emissions monitoring and regulation should also cover 'dioxin-like compounds' which have the same toxicological health hazard as dioxins and furans. The main dioxin-like group of chemicals is the polychlorinated biphenyls or PCBs, which were produced for decades before they were banned in 1985 and were used in applications such as electrical capacitors and transformers, hydraulic fluids, flame retardants, pesticide extenders, paints, etc. Their presence in the environment is mainly from their historical use and accumulation in the environment, but small quantities are produced during the combustion process. The health effects of PCBs have been discussed (IEH 1997). Of the 209 different PCB congeners, 36 are of environmental concern. PCBs are stable in the environment and can bio-accumulate in humans. PCBs have a significant impact on human health, and occupational studies of exposure to PCBs has shown a range of adverse effects including respiratory, cardiovascular, gastrointestinal, haematological, endocrine, neurological, reproductive and carcinogenic effects (IEH 1997). A system of TEQ factors has been suggested as being also applicable for PCBs (Dyke et al 2003). Measurements of PCBs in the emissions from the incineration of municipal solid waste are not common but have been shown to range from 0.001 ng/m^3 to more than 1.0 ng/m^3 (Dyke et al 2003). The data represent a range of incinerator systems and methods of sampling, analysis and reporting. Overall it was concluded that the emissions of PCBs were in general lower than those of PCDD and PCDF for equivalent incinerator systems.

The Fifth Environment Action Programme: Towards Sustainability (1993–2000) set out an objective that critical loads and levels of certain pollutants, such as dioxins and furans, should not be exceeded. Nitrogen oxides, sulphur dioxide and heavy metals were also included. The Programme also set out an objective of a 90% reduction of dioxin emissions of identified sources from the 1985 level by 2005 (EC Waste Incineration Directive 2000). To comply with the Environment Action Programme, required detailed analyses and identification of the major contributors to the overall burden of dioxins and furans in the environment and the compilation of a dioxin and furan inventory of emissions. A major monitoring and reporting exercise resulted in the publication of the EU 'dioxin inventory' (European Commission DG Environment 1999). The potential emission sources of dioxins and furans were reviewed, using data between the reference periods of 1993 and 1995 (for the EU countries). The reported emissions data were re-evaluated to account for the incomplete reporting of emissions data from certain industrial sectors and countries. The re-evaluated major sources of dioxin emissions to air, land and water are summarised in Table 5.12 (Quab et al 2000; European Commission DG Environment 1999). The major source of PCDD and PCDF emissions to air in Europe is waste inciner- ation, at a total estimated emission of 2464 gTEQ/year, including incineration of MSW at

Table 5.12   *European dioxin emissions inventory for emissions to air (1993–95 reference period)*

Industry	PCDD/PCDF Emission to Air (gTEQ/year)
MSW incineration	1437
Uncontrolled MSW incineration	174
Sinter plants	1010
Residential wood combustion	945
Residential coal combustion	40
Preservation of wood	381
Clinical waste incineration	816
Industrial waste incineration	37
Accidental fires	380
Non-ferrous metals production	136
Road transport	111
Cremation	17
Electric furnace steel plant	83
Others	161

Sources: Quab et al 2000; European Commission DG Environment 1999.

1437 gTEQ/year, uncontrolled waste incineration at 174 gTEQ/year, clinical waste incineration at 816 gTEQ/year and industrial waste incineration at 37 gTEQ/year.

The recognition of the contribution of PCDD and PCDF from the incineration of waste to the overall burden on dioxins and furans in the environment led to calls for the control of emissions of dioxins and furans from waste incinerators. The Protocol on Persistent Organic Pollutants, signed by the European Community within the framework of the United Nations Economic Commission for Europe, on long-range transboundary air pollution, sets legally binding limit values for the emission of dioxins and furans of $0.1 \, ng/m^3$ for large-scale municipal solid waste incinerators. Consequently, the EC Waste Incineration Directive (2000) contained an emission-limit value for PCDD and PCDF of $0.1 \, ng/m^3$ (TEQ).

It is however, important to note that the review period of the EU dioxin inventory was 1993–95, before the implementation of the EC Waste Incineration Directive (2000). Indeed, the earlier EC Waste Incineration Directive which was introduced in 1989 for new and existing municipal solid waste incinerators, applied to new incinerators as from 1989, but for existing incinerators from 1996 (Council Directive 89/429/EEC 1989; Council Directive 89/369/EEC 1989). Consequently, many of the older highly polluting waste incinerators, built in the 1970s, which were operating across Europe, and which had minimal gas-cleaning systems, were still in operation during the dioxin inventory review period of 1993–95. It is well known that such older incinerators produced very much higher emissions than those built to comply with the 2000 EC Waste Incineration Directive. Table 5.6 shows examples of the older type of municipal solid waste incinerators which ranged in PCDD (measured as TCDD) emissions from 0.73 to a very high level of $1215 \, ngTCDD/m^3$. The emission level for PCDF (measured as TCDF) was equally high, covering an emission range up to $1425 \, ngTCDF/m^3$. The 1989 European Directives did not have a specific dioxin control emission limit, but sought to minimise dioxin emissions by setting the combustion conditions of minimum gas temperature, residence time and minimum oxygen level, to ensure efficient burn-out of organic compounds. The combustion conditions set were $850 \, °C$ for 2 s with 6% minimum oxygen. However, to comply with the EC Waste Incineration Directive (2000) for PCDD and PCDF emissions sophisticated gas clean-up systems are required, and the modern municipal solid waste incinerator has emissions of between 0.0002 and $0.08 \, ngTEQ/m^3$ for PCDD/PCDF. Consequently, the PCDD and PCDF emissions from the incineration of waste, including municipal solid waste, hazardous waste, clinical waste, sewage sludge, etc., will result in a dramatic fall in the overall contribution of PCDD and PCDF from the incineration sector to the total European environmental burden of PCDD and PCDF. Indeed, Quab (2000) has estimated that the total emission to air from municipal solid waste incineration could be reduced to only 20 gTEQ/year, if all the municipal solid waste incinerators across Europe complied with the 2000 EC Waste Incineration Directive requirements of a $0.1 \, ngTEQ/m^3$ PCDD/PCDF emission level.

The EU has a Community Strategy, introduced in 2001, for the reduction of dioxins, furans and polychlorinated biphenyls in the environment (CEC 2001; CEC 2004). The Strategy covers a ten-year period and aims to reduce the presence of dioxins, furans and PCBs in the environment, in food and in animal feed. The role of the emission limits for PCDD and PCDF, set in the EC Waste Incineration Directive, are recognised as a major contribution to attaining the objectives of the Community Strategy.

The most important route for human exposure to PCDD and PCDF is via the food chain through food consumption, which accounts for between 95 and 98% of the total exposure (European Commission DG Environment 1999). The primary mechanism by which dioxins enter the food chain is via atmospheric deposition (Dioxin Risk 1995). It has been suggested by the World Health Organization that a tolerable daily intake of dioxins and furans and dioxin-like compounds is 1–4 picograms (10^{-12} gram) TEQ per kilogram of body weight (CEC 2001). However, the European Commission estimates that, for many parts of the European population, the dietary exposure to such compounds exceeds this recommended tolerable daily intake (CEC 2001). This fact is one of the reasons behind the implementation of the Community Strategy on dioxins, furans and PCBs (CEC 2001) which aims to reduce their concentrations in the environment significantly.

PCDD and PCDF have a reputation for being highly toxic. However, whilst there is strong evidence that a wide range of carcinogenic and non-carcinogenic effects are associated with PCDDs and PCDFs in animals, there is less compelling evidence of effects in humans. The concern over dioxins and furans arises from a number of animal studies which show that, for some species, they are highly toxic at very low levels of exposure (Tosine 1983; Oakland 1988; IEH 1997; Kogevinas 2001).

The effects of PCDD, PCDF and PCDF exposure in animals have shown a range of effects, including suppression of the immune system, reduced sperm count, abnormal testicular structure and feminisation in the male reproductive system, decreased fertility and ovarian dysfunction in the female reproductive system, birth defects, foetal death and a range of cancers (Allsopp et al 2001; IEH 1997; Kogevinas 2001). The toxicity of PCDD and PCDF depends on the number and position of the chlorine substituents. PCDD and PCDF which contain four, five and six chlorine atoms have higher toxicities (Buser and Rappe 1984). Also PCDD and PCDF with chlorine atoms in the 2,3,7 and 8 positions demonstrate the highest toxicities. The tetra-chlorinated dibenzo-p-dioxin 2,3,7,8-TCDD, is the most toxic dioxin and has been shown to cause lethal effects in certain laboratory animals at very low levels, for example, the LD_{50} (lethal dose where 50% of the species tested dies) for guinea pigs is 0.6 µg/kg body weight (European Commission DG Environment 1999; Baker 1981). However, it is clear that the health impact of PCDD and PCDF are very much species-dependent, for example, for animals of similar body weight to guinea pigs, such as mice and hamsters. The LD_{50} for mice is 284 µg/kg body weight and for hamsters it is between 1000 and 5000 µg/kg body weight (European Commission DG Environment 1999; Baker 1981).

The notoriety of dioxins and furans in relation to emissions from waste incinerators is associated with their high-profile health hazard in animals and the potential risk to human health from exposure to PCDD and PCDF. However, there is less information with respect to the toxic effects of PCDD and PCDF on humans, compared with the extensive data on animal tests, and most existing data has been derived from occupational exposure or industrial accident victims. PCDDs and PCDFs have been involved in a number of incidents in recent years which give them their notoriety. For example, the Seveso accident in Italy, 1976, the herbicide spraying program of Agent Orange in Vietnam in the late 1960s, the Yusho rice oil poisoning incident in Japan 1968 and the Times Beach Missouri land poisoning of 1982 (British Medical Association 1991; CEC 2001). The effects attributed to PCDD and PCDF for such exposures in humans, include a persistent skin acne condition known as chloracne and systemic effects such as digestive disorders and muscle and joint pains, also neurological disorders such as headaches and loss of hearing and psychiatric effects such as depression and sleep disturbance. Also, potentially connected to PCDD and PCDF exposure, are long-term health risks such as chromosome damage, heart attacks and cancer (Saracci et al 1991; Dioxin Risk 1995).

However, such studies on human exposure to PCDD and PCDF tend to be inconclusive. For example, an international steering group reporting on the Seveso incident in Italy where an uncontrolled release of 2,3,7,8 TCDD occurred from a plant manufacturing 245-trichlorophenol, concluded that no clear-cut adverse health effects attributable to 2,3,7,8 TCDD, besides chloracne, could be observed (Ahlborg and Victor 1987). However, another report on the health effects on the local population of Seveso suggested that there were elevated incidences of cancer (Bertazzi et al 1993). In addition, it has been suggested that human epidemiological studies are difficult to interpret since there have been problems in controlled methodologies, lack of information on compounding effects such as smoking, and inadequate information on intake and exposure mode and level of PCDD and PCDF (Dioxin Risk 1995). Also, occupational and accidental exposures have often been to mixtures of PCDDs and/or PCDFs and also in conjunction with other related and possibly hazardous compounds. The link of adverse human health effects and exposure to PCDD and PCDF from the incineration of municipal solid waste has concluded that incinerators operating within legislative limits posed no health risk to individuals who were exposed to plant emissions of dioxins or to the population in the surrounding environment (Environment Agency 1996). Reviews of the health effects of PCDD and PCDF have suggested that the link to the clear adverse health effects of PCDD and PCDF on animals and their effect on humans through such accidental and occupational exposures, are difficult (IEH 1997; European Commission DG Environment 1999).

However, the European Commission suggest that the toxic properties of PCDD and PCDF have been underestimated, rather than exaggerated (CEC 2001). Epidemiological, toxicological and mechanistic data suggest that PCDD, PCDF and some PCBs are linked to neurodevelopmental, reproductive and endocrine effects in humans. These data suggest a broader impact on human health than was previously suspected, even at low dosages, and in particular on vulnerable groups such as breast-fed infants and the foetus which are directly exposed to the accumulated body burdens of PCDD, PCDF and PCBs in the mother (CEC 2001). As a consequence, the European Commission introduced the ten-year Community Strategy on PCDD, PCDF and PCBs (CEC 2001)

A number of theories have been proposed for the formation of PCDD and PCDF during combustion and their formation route may be a combination of processes, depending on prevailing conditions (Lustenhouwer et al 1980; Tuppurainen et al 1998; European Commission DG Environment 1999; Eduljee 1999(a); European Commission 2004).

1. PCDD and PCDF occur as trace constituents in the waste and because of their thermal stabilities they survive the combustion process. Waste material has been shown to contain PCDD and PCDF at trace levels (Oakland 1988; Abad et al 2000; Quab et al 2000). Concentration levels of between 1.4 and 255 µgTEQ/tonne for PCDD and PCDF have been reported for various analyses of European municipal solid waste (Giugliano et al 2002). Mass balances have shown that higher concentrations have been found in the emissions than are found in the input (Commoner et al 1987). However, conditions undoubtedly exist for the thermally stable PCDD and PCDF to

survive the combustion process, particularly if poor gas mixing results in lower combustion temperatures in certain zones of the incinerator.

2. PCDD and PCDF are produced during the incineration process from precursors such as polychlorinated biphenyls (PCB), chlorinated benzenes, pentachlorophenols, etc. The in-situ synthesis of PCDD and PCDF occurs therefore via re-arrangement, free-radical condensation, dechlorination and other molecular reactions (Hagenmaier et al 1988). The precursor theory to the formation of PCDD and PCDF has arisen from laboratory studies and incinerator waste input–emission output studies, which show that these compounds can be formed from chlorophenols, chlorobenzenes, PCBs and brominated diphenylethers (Hutzinger et al 1982; Buser 1979). These precursors may be present in the waste or formed by the combustion of chlorinated plastics and other chlorinated organic materials in the waste (Oakland 1988; Probert et al 1987; Marklund et al 1987).

3. PCDD and PCDF are produced as a result of elementary reactions of the appropriate elements, i.e., carbon, hydrogen, oxygen and chlorine atoms. This reaction is called a de-novo synthesis of PCDD and PCDF. PCDD and PCDF have been shown to form on flyash containing residual carbon collected within a combustion system at temperatures in the region of 250–400 °C in the presence of flue gases containing HCl, O_2 and H_2O (Hagenmaier et al 1988; Eduljee 1999(a); European Commission DG Environment 1999). It is thought that the reaction is catalysed by various metals, metal oxides, silicates, etc., particularly copper chloride present in the flyash. This theory is borne out by the observation that low levels of PCDD have been observed in the furnace exit of incinerators, but levels 100 times greater were found in the electrostatic precipitator ash of the same plant (Hagenmaier et al 1988; Commoner et al 1987). Enhanced formation in electrostatic precipitators has also been reported by Cains and Eduljee (1997) and for fabric filters by Giugliano et al (2001).

However, it is thought that the dominant formation route for PCDD and PCDF in municipal solid waste incinerators is 'de-novo' synthesis, the reaction occurring in the down-stream boiler and pollution abatement system (Eduljee 1999(a); Giugliano et al 2001). It has been shown that the de-novo synthesis of PCDD and PCDF takes place in reaction times of the order of minutes (Huang and Buekens 2001). The other routes to formation are suppressed by the efficient operation of the waste incinerator since any waste containing PCDD or PCDF should be destroyed in the combustion conditions of the furnace. Also any chlorinated organic precursors in the waste or formed in the furnace should also be destroyed at the high temperatures, long residence times, high oxygen levels and turbulent gas mixing conditions of the furnace.

The control of PCDD and PCDF emissions may be approached by either restricting their formation, by efficient combustion control and/or by clean-up of the flue gases after they have formed.

The removal of the chlorine and hydrogen chloride-producing plastic components from the waste prior to incineration has been suggested as a mechanism of PCDD and PCDF control. However, experiments on hydrogen chloride formation in incinerators have shown that, even when all the plastic is removed from the waste, significant concentrations of hydrogen chloride are still produced in the flue gases from other sources of chlorine such as paper and board (Buekens and Schoeters 1984; Visalli 1987). In addition, no correlation between PVC plastic in the waste stream with PCDD and PCDF emissions from incinerators, has been reported (Nchida and Kamo 1983; Visalli 1987). Also, wood burning has been shown to produce chlorophenols and chlorobenzenes which can combine to form PCDDs and PCDFs without the presence of chlorine, hydrogen chloride or chlorinated plastic.

Combustion control to limit PCDD and PCDF emissions has centred on the destruction of PCDD and PCDF at high temperatures. Consequently, the recommended conditions are temperatures above 1000 °C residence times of >1 s, and turbulence to ensure good mixing with excess air. Decomposition increases exponentially with temperature, for example, at 1200 °C a residence time of only 1 ms is required for destruction. Incinerators have been shown to be particularly effective in destroying dioxins (Commoner et al 1987). The emission of carbon monoxide from incinerators is used as a measure of efficient combustion control, such that minimum carbon monoxide correlates with efficient combustion and therefore also with minimum PCDD and PCDF emission. However, de-novo synthesis has been shown to be the dominant formation route to PCDD and PCDF in many municipal waste incinerators (Hagenmaier et al 1988; Commoner et al 1987; Vogg et al 1987). Therefore PCDD and PCDF may be destroyed in the high temperature of the furnace with efficient combustion control, but the overall emission of PCDD and PCDF from the incinerator may not be affected by this destruction since formation of these compounds takes place in the cooler parts of the incinerator system, down-stream of the furnace. Furnace conditions are, however, optimised for efficient combustion since this also influences the production of the products of incomplete combustion, carbon content, chloride content and heavy metal content as well as surface activity of the ash particles which are essential for the de-novo synthesis and therefore should be minimised (European Commission DG Environment 1999). Efficient combustion of the municipal solid waste can be achieved by the design of the grate system of the furnace, the design and positioning of the primary and secondary air inlets and the feeding of the waste onto the grate (Darley 2003). In addition, since the temperature of formation by the de-novo route is maximised in the 250–400 °C range, incinerator flue gases are rapidly cooled from the furnace outlet to below this 'temperature window' before entering the clean-up system to minimise their formation (European Commission 2004). This may be achieved by, for example, addition of a spray quench tower. De-novo formation of PCDD and PCDF may still occur to some extent since, within the economisers, or electrostatic precipitator, or fabric filter of the clean-up system, ideal conditions for the synthesis will occur.

Post-combustion control of PCDD and PCDF has centred on the efficient collection of particulate since they are shown to be mostly found on flyash, either adsorbed or formed in-situ, they also exist at lower levels in the gas phase. Gas-phase removal of PCDD and PCDF by wet scrubbing systems is not so effective, since dioxins such as TCDD have very low solubilities in water. However wet/dry scrubbers with lime slurry as the active scrubbing agent have been shown to reduce PCDD and PCDF emissions (Nielsen et al 1986). Also, calcium oxide and calcium hydroxide addition, used to remove hydrogen chloride, will also influence PCDD and PCDF emissions by removing the chlorine as a building block of the PCDD and PCDF molecule in de-novo synthesis. The most effective system for removal of PCDD and PCDF has been shown to be the addition of small quantities of activated carbon as a fine powder, injected independently, or added to the calcium oxide adsorbent prior to a fabric filter (Wade 2003; Atkins 1996; European

Commission 2004). The activated carbon is used with wet, semi-dry and dry scrubbers for PCDD and PCDF removal. After adsorption on activated carbon the activated carbon, with flyash particulate, is then captured by the fabric filter and with the build-up of the activated carbon on the fabric also provides a surface for further adsorption on the filter cake on the fabric filter surface (Chang and Lin 2001). PCDD and PCDF emissions from incinerators using such gas clean-up systems have reported levels well below legislated limits (European Commission 2004).

Novel methods of dioxin removal incorporating catalytic reaction (Maaskant 2001; Liljelind et al 2001) and catalytically coated fabric filters, have also been commercialised (Pranghofer and Fritsky 2001; Bonte et al 2002). Catalytic systems, for example, may be used in association with the selective catalytic NO_x reduction techniques but with a catalyst that also removes PCDD and PCDF (Maaskant 2001). The flue gases are forced to pass through a catalyst of high metal content, high surface area and porosity and high activity, which traps the PCDD and PCDF and at the same time destroys them. The catalyst is a titanium/vanadium oxidative-type catalyst. The system operates at low temperature, and has been shown to achieve PCDD and PCDF destruction efficiencies of over 99.9% at 230 °C and achieves levels of less than 0.1 TEQng/m^3. Catalytically coated fabric filters have been used to trap and at the same time destroy PCDD and PCDF (Pranghofer and Fritsky 2001; Bonte et al 2002). The fabric filter traps the fine particulate, but any PCDD and PCDF react to form carbon dioxide, water vapour and hydrogen chloride. The system allows for the trapping and/or destruction of PCDD and PCDF in both the solid particulate phase and the gaseous pahse.

A further concern in the control of the emissions of PCDD and PCDF from municipal solid waste incinerators is the influence of a 'memory effect', which leads to increased emissions derived via desorption of previously formed PCDD and PCDF. The PCDD and PCDF may form and desorb from residual carbon and flyash from the duct walls which contain PCDD and PCDF precursors and metal catalysts, and thereby increase their concentration in the flue gases (Chang and Lin 2001). Also, cleaning the residual carbonaceous flyash deposited on the flue duct walls of waste incinerators on a regular basis, has been shown to reduce the emissions of PCDD/F (Wevers and De Fre 1998). It has been shown that changes in PCDD/F emissions data, from municipal solid waste incinerators, tends to react slowly to changes in municipal waste composition, operating conditions or boiler fouling (Adams et al 2001). Some of this inertia has been attributed to PCDD and PCDF formation in carbonaceous flyash surface deposits in the various parts of the incinerator system during start-up, bag house bypass, unstable combustion conditions, etc. These are then slowly released later, increasing the flue gas concentration of PCDD/F by as much as one or two orders of magnitude (Zimmerman et al 2000). In addition to the carbonaceous flyash as a major source of desorbed PCDD/F, the plastic material of the wet scrubber has also been identified as a potential source of desorbed PCDD and PCDF (Hunsinger et al 1998; Kreisz et al 1996).

Sampling and analysis of emissions to air   The compliance of an incinerator with the legislated emission limits requires sampling and analysis of the emissions (Clarke 1998). The EC Waste Incineration Directive (2000) stipulates which emissions are required to be analysed on a continuous or periodic basis. Continuous measurement of nitrogen oxides, carbon monoxide, total dust (particulate), TOC, hydrogen chloride, hydrogen fluoride and sulphur dioxide, is required. Combustion parameters such as oxygen, temperature, and water vapour monitoring, is also stipulated to be carried out on a continuous basis. At least two measurements per year of heavy metals and PCDD and PCDF are required, but more frequent monitoring is required in the first twelve months of operation. There are variations in the requirements, subject to the authorisation of the regulatory authority.

The point of analysis is within the stack by means of in-situ measurement or else a sample of the stack gas is withdrawn for analysis, for example, for heavy metals and PCDD and PCDF. Where a sample is withdrawn from the stack, the sample is usually conditioned, i.e., water vapour, particulates, etc., are removed and the gas is cooled. Some analyses require a heated sampling system. Acidic gases such as HCl should be maintained at temperatures above the dew point to prevent acid corrosion in the instrument. Water in the sample may be removed with a drying tube, alternatively, chilling the sample may be appropriate.

Particulate analysis is required on a continuous basis and the main method of analysis is usually by infra-red absorption, measured by a reduction of light intensity in relation to concentration of dust. The particles absorb the light and the reduction in intensity, compared to clean flue gas, is proportional to the concentration of particles in the gas stream. The continuous measurement system is normally calibrated by a periodic method, such as a gravimetric system.

The analysis of hydrogen chloride is required on a continuous basis. Instruments which fulfil this criteria are infra-red and electrochemical analysers. Hydrogen fluoride analysis may be made on a periodic basis, provided it can be demonstrated that control measures for hydrogen chloride will not be exceeded. Sulphur dioxide and carbon monoxide may be analysed on a continuous basis using infra-red analysis as described for hydrogen chloride.

The preferred method of analysis for TOC is the flame ionisation detector. The sample is burnt in a hydrogen/air flame and the organic carbon molecules produce ions. An electric potential difference across the ionised molecules causes a current to flow between the electrodes and is proportional to the mass flow of carbon atoms.

Nitrogen oxides, for most purposes, can be regarded as NO and NO_2. Infra-red and electrochemical systems can be used, but perhaps the preferred method is chemiluminescence. Nitric oxide (NO) is reacted with ozone produced in the analyser in a reaction chamber, and is oxidised to nitrogen dioxide (NO_2) in a chemically excited state. The excited molecule loses energy and reverts to the ground state by emission of energy as light. The emitted light energy is detected and measured by a photomulitiplier tube. The chemiluminescence reaction is specific to NO and NO plus NO_2 is detected and reported as NO_x by passing the sample gas through a catalytic converter which reduces the NO_2 to NO prior to the reaction chamber.

Heavy metals are monitored by periodic gravimetric analysers followed by laboratory analysis. Isokinetic sampling traps the coarse and fine particulates on a glass fibre filter paper held in an oven. Mercury may largely be in the vapour phase and a vapour-trapping system should also be used in conjunction with the filter system. The filter paper is weighed before and after sampling at constant humidity and temperature to determine the mass of particulate, per metre cubed, of flue gas. The weighed filter is digested in hot concentrated acids in either an open or closed vessel in order to take the metals into solution for analysis. The resulting solution is made up to a known volume and analysed, usually by atomic absorption spectrometry or inductively couped plasma spectrometry.

Monitoring of PCDD and PCDF is a time consuming, expensive and meticulous process requiring great analytical skill. The sampling of PCDDs and PCDFs is required at a minimum time period of 6 h and maximum of 8 h using standardised sampling protocols which trap the particulate and vapour phase PCDD and PCDF (Waste Incineration Directive 2000; Clarke 1998). At each stage of the sampling and analysis procedure, radio labelled standard PCDD and PCDF are added to the sample to measure the recovery of the sampling, sample clean-up and analysis efficiency. The analysis of PCDD and PCDF is difficult, since they occur at very low concentrations, in a sample matrix which contains other chlorinated hydrocarbons, and there are many congeners of PCDD and PCDF. Analysis is performed in three stages. Extraction of the PCDD and PCDF from the sample which may be ash, particulate adsorbed on glass fibre filter paper, or condensed water, each requiring a different extraction procedure. The extract is analysed in a series of steps to eliminate compounds which interfere with the analysis and to remove the PCDD and PCDF as a separate fraction from the extracted sample which may contain many chemical groups in high concentration, in addition to the PCDD and PCDF. Final analysis of the fraction containing PCDD and PCDF is most frequently carried out by coupled capillary gas chromatography/mass spectrometry which separates the individual PCDD and PCDF congeners and detects them using mass spectrometry.

Wastewater   Water pollution from incinerators is not generally regarded as an important problem, because of the limited amount of wastewater generated. A typical European municipal solid waste incinerator would generate approximately 0.15–$0.3 \, m^3$/tonne of wastewater depending on the type of gas cleaning system used (European Commission 2004). The main sources of wastewater from incinerators are from flue gas treatment as flue gas scrubber water, and alkaline scrubbing of the gases to remove acid gases and the quenching of incinerator ash. Other minor sources include, for example, scrubber water pre-treatment and the purification of boiler feedwater where a boiler plant is installed. Such water is contained in a closed system and would not come into contact with the pollutant flue gases.

Where the flue gases are scrubbed or cooled with water, the absorbed acid gases will make the water very acidic and will also consequently contain significant quantities of heavy metals which are soluble in the acidic solution. Where the flue gases are scrubbed with an alkaline solution, such as sodium hydroxide or calcium hydroxide to remove acid gases, the scrubber water will be very alkaline. Recirculation of wastewater in the wet scrubbing system can result in a substantial reduction in the amount of wastewater (European Commission 2004).

The EC Waste Incineration Directive (2000) regulates the emission of heavy metals, suspended particulate, dioxins and furans in wastewater which is derived from the cleaning of flue gases from waste incineration. These are shown in Table 5.13. Also shown in Table 5.13 are typical compositions of wastewater from the first and second stages of a two-stage wet scrubber system for the clean-up of flue gases and after treatment in a wastewater treatment plant for a 4×12 tonnes/h municipal solid waste incinerator plant (IAWG 1997). The results show very low levels of wastewater emissions.

Table 5.13   *Emission-limit values for the discharges of wastewater from the cleaning of exhaust gases*

Polluting substance	Emission limit (mg l^{-1})	Wastewater 1st stage	Wastewater 2nd stage	Treated wastewater both stages
Total solids	30[1]	—	—	—
Total solids	45[2]	—	—	—
Hg	0.03	0.051	0.02	0.03–0.27
Cd	0.05	<0.004	<0.004	—
Tl	0.05	—	—	—
As	0.15	—	—	—
Pb	0.2	2.6	0.46	<0.10–0.21
Cr	0.5	3.2	0.74	0.10–0.47
Cu	0.5	3.0	0.79	0.002–0.015
Ni	0.5	34	2.0	<0.02–0.68
Zn	1.5	—	—	<0.01–0.11
Dioxins (TEQ)	0.3	—	—	—

[1] 95% values do not exceed.
[2] 100% values do not exceed.

Sources: EU Waste Incineration Directive 2000; IAWG 1997.

The bottom ash from the incinerator grate is removed in a unit which serves to cool the ash and also maintains a partial vacuum in the incinerator chamber. Bottom ash wastewater is alkaline and contains only low levels of dissolved heavy metals, below permitted sewerage discharge levels (Reimann 1987). Normally this wastewater need not be discharged from the incineration site (European Commission 2004).

The main pollutants in incinerator wastewaters are heavy metals. Clean-up of the heavy metals is usually through neutralisation via precipitation with calcium hydroxide with an additive of TMT15 (trimercaptotriazine) (Reimann 1987). The calcium hydroxide causes the major part of the pollutants to be precipitated as hydroxide sludges. However, more than 60% of the mercury and other heavy metals remain in the wastewater and the addition of TMT15 is required, which precipitates the mercury and heavy metals down to levels well below the permissible limits (Reimann 1987). Other methods of heavy metal removal include flocculation using flocculation agents such as polyelectrolytes and ferric chloride (European Commission 2004). The wastewater pH is adjusted and the flocculation additives added to produce a precipitate. The precipitate resulting from the wastewater treatment is typically mixed with flyash and other air pollution control residues and landfilled (IAWG 1997). In some cases, the wastewater is evaporated leaving a solid residue, for example, in a spray drier, where the particulate formed would be then trapped on a fabric filter.

Whilst there is most concern over the presence of heavy metals in wastewater, the presence of organic pollutants such as PAHs and dioxins and furans should also be suspected. However, reported levels of PAHs and PCDD and PCDF are very low as dissolved compounds in wastewater (Reimann 1987; Ozvacic et al 1985), but have been detected in significant concentrations in suspended particles, which therefore are required to be filtered out (Ozvacic et al 1985; Bumb et al 1980). Where removal of organic pollutants such as PCDD, PCDF, PCBs and PAHs are required, the wastewater may be passed through a bed of activated carbon, which adsorbs the pollutants. Sand filtration may also be used, which filters out the suspended solids which contain adsorbed pollutants (European Commission 2004).

Ash Residue   If the incinerator is operating correctly, the residue or ash should be completely burnt out and biologically sterile. The EC Waste Incineration Directive (2000) specifies that the carbon content of the bottom ash should be less than 3 wt% as a measure of the burn-out efficiency of the incinerator. Bottom ash from the furnace grate represents the bulk of total ash and is composed mainly of mineral oxides. Flyash comprises only a few percent of the waste mass input (European Commission 2004). Three types of solid residue or ash may be distinguished as the solid residues of a municipal solid waste incinerator, these are bottom ash collected at the bottom of the grate, boiler ash collected in the heat recovery boiler system of the incinerator, and flyash collected from the air pollution control system.

Bottom ash from the furnace grate represents the bulk of total ash and is a heterogeneous mixture of slag, ferrous and non-ferrous metals, ceramics, glass, other non-combustible material and uncombusted organic material. The bottom ash consists mainly of silicates, oxides and carbonates. Various compounds and mineral species have been identified in bottom ash, for example, $Ca_2Al_2SiO_7$, $Ca_2MgSi_2O_7$, SiO_2, Fe_3O_4, Fe_2O_3, $CaCO_3$, $MgCO_3$, $Ca(OH)_2$, $CaSO_4$, $NaCl$ and KCl and elemental Fe, Al and Cu (Pfranf-Stotz and Reichelt 1997). Bottom ash contains metals which are less volatile than those released during the high temperatures of the furnace. For example, iron and nickel concentrations tend to be higher in the bottom ash than the flyash. More volatile metals such as cadmium, mercury, zinc and lead tend to be higher in concentration in the flyash compared with the bottom ash. The bottom ash PCDD and PCDF concentrations are much lower than the flyash. When the TEQ of the PCDD and PCDF congeners is calculated, the bottom ash TEQ is very low compared with the flyash. PCDD and PCDF tend to be much lower in bottom ashes since they are quickly quenched before significant PCDD and PCDF production can take place (Ozvacic et al 1985; Shaub 1989).

Bottom ash recycling, for aggregate use in the construction industry and road building is common in Europe (Sakai et al 1996; European Commission 2004). For example, in Germany, approximately 80% of bottom ash is utilised, in the Netherlands more than 90%, Denmark 90%, France more than 70% and in the UK more than 50% (European Commission 2004; York 2003). Other countries, such as Austria, Switzerland, Portugal, Italy and Norway recycle less than 10% of the bottom ash, the large majority going to waste landfill. The main use in Denmark is for development of a granular sub-base for car parking, bicycle paths and paved and un-paved roads, etc. In Germany, utilisation has been through, for example, sub-base paving applications and, in the Netherlands, bottom ash has been used in construction for granular base, or in-fill road base, embankments and noise and wind barriers. In the Netherlands, bottom ash has also been used as aggregate in asphalt and concrete.

Also shown in Table 5.14 are the composition of flyash and air pollution control (APC) residues from a dry/semi-dry control system and a wet control system (IAWG 1997). The chemical characteristics of flyash are greatly influenced by the additives used in the air pollution control system. For example, where additives such as lime and activated carbon are added, the air pollution control residue would comprise flyash, lime, activated carbon

Table 5.14 *Typical composition of bottom ash, flyash and air pollution control (APC) residues from a dry/semi-dry system and a wet control system*

Polluting substance	Bottom ash	Flyash	APC residue dry/semi-dry	APC residue wet
Aluminium (g kg^{-1})	22–73	49–90	12–83	21–39
Silicon (g kg^{-1})	91–310	95–210	36–120	78
Calcium (g kg^{-1})	37–120	74–130	110–350	87–200
Potassium (g kg^{-1})	0.7–16	22–62	5.9–40	0.8–8.6
Magnesium (g kg^{-1})	0.4–26	11–19	5.1–14	19–170
Sodium (g kg^{-1})	2.9–42	15–57	7.6–29	0.7–3.4
Iron (g kg^{-1})	4.1–150	12–44	2.6–71	20–97
Chlorine (g kg^{-1})	0.8–4.2	29–210	62–380	17–51
Cadmium (mg kg^{-1})	0.3–71	50–450	140–300	150–1400
Chromium (mg kg^{-1})	23–3200	140–1100	73–570	80–560
Copper (mg kg^{-1})	190–8200	600–3200	16–1700	440–2400
Mercury (mg kg^{-1})	0.02–7.8	0.7–30	0.1–51	2.2–2300
Nickel (mg kg^{-1})	7–4300	60–260	19–710	20–310
Lead (mg kg^{-1})	98–14 000	5300–26 000	2500–10 000	3300–22 000
Zinc (mg kg^{-1})	610–7800	9000–70 000	7000–20 000	8100–53 000
PCDD (ng g^{-1})	0.25–0.48	115–1040	0.7–32	—
PCDF (ng g^{-1})	0.54–0.102	48–280	1.4–73	—
I-TEQ (ng g^{-1})	0.0018–0.002	1.5–2.5	0.8–2.0	—

Source: IAWG 1997.

and the associated adsorbed pollutants. In some cases, the lime may comprise up to 50 wt% of the APC residue. Flyash is characterised by spherical particles associated with aggregates of polycrystalline, amorphous and glassy material. Attempts to identify possible mineral phases in the flyash show that, apart from the amorphous and glassy material, various chemical compounds and mineral species can also be identified. The spherical particles common in flyash are composed of complex calcium, sodium and potassium aluminosilicates, whilst the associated amorphous and crystalline material is enriched in the more volatile elements (Eighmy et al 1995). Compounds and minerals identified in flyash include Pb_3SiO_5, $Pb_3O_2SO_4$, $Pb_3Sb_2O_7$, $PbSiO_4$, $Cd_5(AsO_4)_3Cl$, $CdSO_4$, K_2ZnCl_4, $ZnCl_2$, $ZnSO_4$, Fe_3O_4, Fe_2O_3, SiO_2, $CaSiO_3$, Al_2SiO_5, $Ca_3Si_3O_9$, $CaAl_2SiO_6$, $Ca_3Al_6Si_2O_{16}$, $NaAlSi_3O_8$, and $KAlSi_3O_8$ (Eighmy et al 1995; Kirby and Rimstidt 1993; Hundesrugge et al 1989; Evans and Williams 2000).

The presence of PCDD and PCDF has been shown in flyashes derived from the incineration of solid waste – Table 5.15 shows total PCDD and PCDF and certain congeners found in flyashes from German, Canadian and Dutch incinerators (Hagenmaier et al 1987; Olie et al 1982; Shaub 1989). The German data represented the average of 52 flyash samples from 10 incinerators and the Canadian data an average of 8 samples. The significant concentrations of PCDD and PCDF found in flyash samples illustrates the de-novo synthesis route to their formation, a route catalysed by the flyash itself.

The contamination of flyash with heavy metals, PCDD and PCDF, results in the ash attaining the status of hazardous waste and consequently requiring special permits for landfill disposal to hazardous waste landfill sites (EC Waste Landfill Directive 1999;

Table 5.15   *Concentration of PCDD and PCDF in municipal solid waste incinerator flyash*

PCDD/PCDF	Reference [1] (ng/g)	Reference [2] (ng/g)	Reference [3] (ng/g)
PCDD			
TCDD	11	13	93
PeCDD	34	23	254
HxCDD	50	26	604
HpCDD	57	15	760
OCDD	65	6	345
Total PCDD	217	83	2056
PCDF			
TCDF	72	—	173
PeCDF	95	—	312
HxCDF	82	—	459
HpCDF	56	—	314
OCDF	13	—	51
Total PCDF	318	—	1309

Sources: [1] Hagenmaier et al 1987; [2] Olie et al 1982; [3] Shaub 1989.

Mulder 1996; Sakai et al 1996). The high heavy metal concentrations present in the ash residues from incineration become of more significance when they are placed in landfill sites, where leaching of the pollutants may be a source of groundwater contamination. Generally, the flyash is more readily leached than the clinker fraction since the heavy metals largely occur in the smallest-size fraction of less than 10 μm and are concentrated at or near the surface of the particles (Buekens and Schoeters 1984). In addition, the high chlorine content of the waste, results in the majority of the metal species being present as the metal chlorides which are generally more soluble in water than other species (Brunner and Monch 1986; Denison and Silbergeld 1988). It has been shown that up to 32.5% of the available zinc, 1.75% of lead, 5.7% of manganese and 94% of the available cadmium can be leached from flyash (Buekens and Schoeters 1984). Whilst water in contact with flyash produces alkaline solutions rather than acidic; copper, lead, zinc and cadmium show increased solubilities at high alkalinities, that is, they are amphoteric in nature, showing significant solubilities at both low and high pH values (Denison and Silbergeld 1988).

Apart from landfilling in hazardous waste landfill sites, other treatment methods used or under investigation to stabilise incinerator flyash are solidification, chemical stabilisation, ash melting or vitrification and extraction/recovery processes (Sakai et al 1996; European Commission 2004). Solidification involves the use of, for example, cement to produce a low permeability, low porosity product, which inhibits the leaching of pollutants. Thermal treatment methods include vitrification at temperatures of 1300–1500 °C, followed by a rapid quenching process to obtain an amorphous glassy material. Such thermal processes produce a homogeneous, dense product with low leaching characteristics. Acid extraction and chemical stabilisation of heavy metals has also been carried out.

Table 5.16 *Emissions to air, wastewater and solid residues from the incineration of one tonne of municipal solid waste, and yearly total for a typical 250 000 tonnes per year incinerator*

Parameter	Emission (g/tonne incinerated)	Yearly emission (kg/year)
Dust	165	41 250
Hydrogen chloride	70	17 500
Hydrogen fluoride	2.2	550
Sulphur dioxide	129	32 250
Nitrogen oxides	2141	535 250
Carbon monoxide	126	31 500
TOC	19	4 750
Mercury	0.048	12
Cadmium + thallium	0.095	24
Heavy metals (Sb, As, Pb, Cr, Co, Cu, Mn, Ni, V, Sn)	1.737	434
PCDD + PCDF	0.000000250	0.0000625
Wastewater (flue gas treatment)	0.15–0.3 (m³/tonne)	37 500–75 000 (m³/year)
Bottom ash	200–350 (kg/tonne incinerated)	50 000–87 500 (tonnes/year)
Boiler ash + flue gas treatment flyash	20–40 (kg/tonne incinerated)	5000–10 000 (tonnes/year)

Source: European Commission 2004.

The total emissions from the incineration of each tonne of municipal solid waste can be estimated. Table 5.16 shows an estimate for a typical European municipal solid waste incinerator of the emissions from a representative 250 000 tonne per year incinerator (European Commission 2004).

Dispersion of emissions from the chimney stack   A serious consideration in relation to incinerators of all types is the height of the chimney stack and the dispersion of the plume of exhaust gases. The dispersion of the plume involves an initial rise, followed by a horizontal spreading about the plume centre line (Figure 5.16, Clarke 1986; Clarke 1999). The dispersion is assumed to proceed from an imaginary point source. The concentration profiles in the vertical and horizontal directions are assumed to approach a Gaussian distribution form. The dispersion of the plume will influence the down-wind ground concentrations of pollutants. Such considerations are vitally important in the assessment of the impact of an incinerator on the local environment. The rate of dispersion, and hence ground concentrations, are influenced by meteorological considerations, wind speed and rate of plume emission. Unstable conditions promote dispersion, whilst stable conditions, such as fog, result in very slow spreading plumes. The ground level concentration close to the stack is zero. However, eventually the plume spreads out and reaches ground level a point known as the radius of maximum effect. For incinerators, this point could be several kilometres from the incinerator. At further distances from the stack, the ground concentration becomes reduced as the plume becomes more diluted in the atmosphere. Figure 5.16 shows plume dispersal in stable, unstable and neutral meteorological conditions. Unstable conditions give the highest ground level concentration closer to the stack than for stable

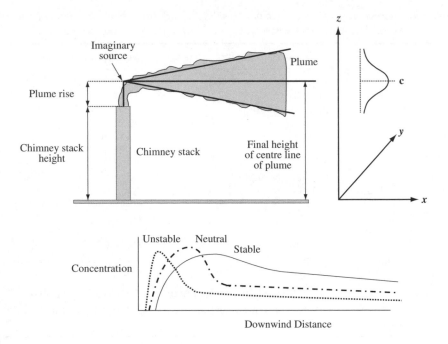

Figure 5.16   *Plume dispersion from chimney stacks and ground level concentration in relation to atmospheric stability. Sources: Clarke 1986; Clarke 1999.*

conditions. There is a range of computer software to calculate the dispersion from the chimney stack. Such calculations involve input data in relation to a range of different parameters, for example, the rate of pollutant emissions, stack height, volumetric emissions rate, gas temperature, surrounding geographical terrain, surrounding building heights, wind speed, wind direction, atmospheric stability, any atmospheric chemical reactions, etc.

Dispersion is also influenced by the height of the stack, as higher stacks promote increased rates of dispersion. The use of dispersion models and guidelines give recommended stack heights for incinerators.

5.3.1.5   *Energy Recovery via District Heating, Electricity Generation and Combined Heat and Power*

The modern municipal waste incinerator relies on the production of steam for electricity generation or district heating to ensure the cost-effectiveness of the process. Some schemes may incorporate both electricity generation and district heating as combined heat and power (CHP) systems. Reported thermal efficiencies for electricity generation-only systems are between 25 and 30%, district heating systems are between 80 and 90% and combined heat and power systems are between 70 and 85% (European Commission 2004). Electricity is generated from the steam produced in the boilers via a steam-condensing turbine. The high-pressure, high-temperature steam enters the turbine and passes through the various stages of the turbine and, as it does so, it expands and reaches high velocity, turning the blades of the turbine and hence the turbine shaft, which generates the electricity (Gilpin 1982; Porteous 1992). In general, about 0.3–0.7 MWh of electricity can be generated in a municipal solid waste incinerator from one tonne of municipal solid waste, depending on the plant size, steam parameters, steam utilisation efficiency and the calorific value of the waste (European Commission 2004). Where district heating is the objective, the high-temperature, high-pressure steam passes through heat exchangers which generate hot water under pressure for distribution to homes, offices and institutions. The water is often superheated (European Commission 2004). CHP systems would use a different type of steam turbine which would generate a lower amount of electricity, but the steam effluent from the turbine would be at a higher temperature, enabling district heating to be incorporated (Porteous 1992).

Evaluation of the best option is very much a site-specific issue. Contracts to sell the electricity or heat supply to the local district should be secured. In addition, contracting for waste to fuel the plant on a long-term basis should also be secured. District heating schemes rely on a market for the heat which, in the case of domestic and commercial premises, may be seasonal. The demand for heat will be different from summer to winter, but this is not a problem for electricity generation. In addition, the incinerator plant itself will require energy for hot water, steam and power. Figure 5.17 shows the energy production from municipal solid waste incineration for various European countries (European Commission 2004).

There has been some concern that recycling schemes may reduce the calorific value of municipal solid waste and adversely affect the performance and energy recovery potential of large-scale municipal solid waste incineration. However, it has been shown that source separation of waste does not significantly affect the calorific value of the derived waste used in the incinerator (Atkinson et al 1996). The study examined a range of 'bring' and 'kerbside' collection systems for recovering recyclable materials. In each case the level of recovery for the four types of recycled materials examined: newspapers and magazines;

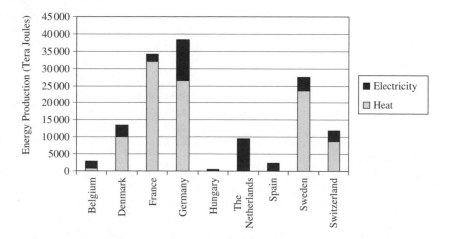

Figure 5.17 *Energy production from the incineration of municipal solid waste in various European countries (tera (10^{12}) Joules). Source: European Commission 2004.*

glass bottles and jars; metal cans; and plastic bottles, was estimated to be between 4.4 and 13% of the household waste stream. However, the reduction in the calorific value of the waste was estimated at only between 5 and 10%. This was due to the fact that non-combustible materials such as metal cans and glass are removed, but also due to a decrease in moisture and ash content of the derived waste. It was concluded that such variations in calorific value were not significant and would make little difference to the energy recovery performance of a municipal solid waste incinerator.

5.3.2   Other Types of Incineration

There are a wide variety of incineration types used to incinerate a wide variety of wastes. In this section a number of different design types will be discussed in relation to their technologies and application to different types of waste:

- fluidised bed incinerators;
- starved air incinerators;
- rotary kiln incinerators and cement kilns;
- liquid and gaseous waste incinerators.

5.3.2.1   Fluidised Bed Incinerators

Fluidised bed incinerators have been used for a wide variety of wastes including municipal solid waste, sewage sludge, hazardous waste, liquid and gaseous wastes and those wastes with difficult combustion properties (Buekens and Patrick 1985; Neissen 1978; Waste Incineration 1996; European Commission 2004). Fluidised beds are mainly of the bubbling, turbulent or circulating bed type, although some pressurised fluidised beds have been built for coal combustion for power generation. Figure 5.18 shows schematic diagrams of bubbling bed, turbulent and circulating fluidised beds (Buekens and Schoeters 1984; European Commission 2004).

   Fluidised beds consist of a bed of sand particles contained in a vertical refractory-lined chamber through which the primary combustion air is blown from below; the sand particles are hence fluidised by adjusting the air flow. Increasing the air flow produces a turbulent flow of solids, and to prevent elutriation of the bed material out of the free-board, cyclones are placed within the freeboard to re-circulate the solids back into the bed. Further increase in air flow produces a circulating fluidised bed where, intentionally, the solids are elutriated out of the bed into a cyclone and the material is re-circulated back to the bed. However, in the circulating fluidised bed, combustion also takes place in the cyclone. Such beds are much longer and produce longer residence times of the solid particles of waste in the hot zone, resulting in higher burn-out of the products of combustion and reduced organic emissions.

   The bed of sand is heated by pre-heated air or gas or oil burners to raise the temperature, such that incoming waste will ignite and combust efficiently. The processed waste, in the form of shredded municipal solid waste or refuse-derived fuel pelts, or indeed other wastes such as sewage sludge or industrial waste, is fed continuously into the hot sand bed. The waste combusts and start-up fuel is then no longer required. The fluidised bed reactor promotes the dispersion of incoming waste, with rapid heating to ignition temperature and promotes sufficient residence time in the reactor for their complete combustion. In

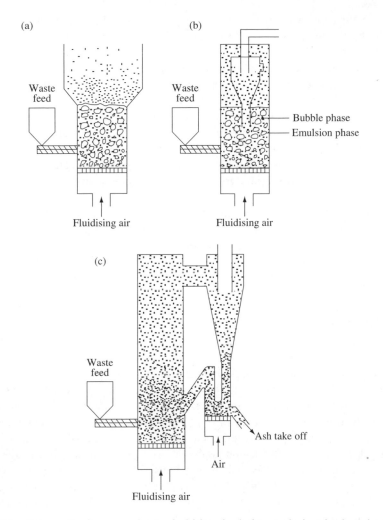

Figure 5.18 *Schematic diagram of: (a) a bubbling bed; (b) a turbulent bed; and (c) a circulating fluidised bed.*

the fluidised bed, drying, devolatilisation, ignition and combustion all take place within the bed (European Commission 2004). Secondary functions include the uniform heating of excess air, good heat transfer for heat exchange surfaces within the bed, and the ability to reduce gaseous emissions by control of temperature, or the addition of additives directly to the bed to adsorb pollutants, for example, the addition of lime to reduce sulphur dioxide emissions (Probert et al 1987; Waste Incineration 1996). A further feature of fluidised beds is their lower operating temperatures, typically around 850–950 °C maximum combustion temperatures, which therefore produce lower levels of thermal NO_x (European Commission 2004). The fluidised bed reactor greatly increases the burning rate of waste, since the rate of pyrolysis of the solid waste material is increased by direct contact with the hot, inert bed material. Gases in the bed are continuously mixed by the bed material, thus enhancing the flow of gases to and from the burning solid surface and enhancing the completeness and rate of the gas-phase combustion reaction, this factor becomes more pronounced with circulating fluidised beds which have longer residence times in the hot zone. In addition, the charred surface of the burning solid material is continuously abraded by the bed material, enhancing the rate of new char formation and the rate of char oxidation. Fluidised beds are compact and have high heat storage with fast dynamic response to throughput or demand. They also have high heat transfer rates and thus enable faster ignition of low combustible waste. Because of the high heat transfer rates found in fluidised beds, they are very good for heat-recovery processes and the heat-transfer surfaces may be placed within the bed (Buekens and Schoeters 1984; European Commission 2004). The ash residue from waste combustion is usually removed from the bottom of the furnace.

Fluidised beds, by the nature of their combustion, are able to incinerate a range of wastes. Municipal solid waste may be incinerated in a fluidised bed incinerator but is best achieved by some form of pre-screening and shredding, or the production of refuse derived fuel (RDF) pellets (Ballantyne et al 1980). The solid waste is then fed by screw gravity feeding or pneumatically. Processing routes have involved the production of refuse-derived fuel in shredded or pelletised form. Large particles in fluidised beds can cause problems due to agglomeration, which prevents fluidisation, and the bed then consequently slumps. The agglomerated masses will then have to be removed. Fusion of ash particles in very hot zones in the bed can also cause agglomerates to form. Fluidised beds for municipal solid waste tend to be in the range 12 000–200 000 tonnes per year, whereas large-scale municipal solid waste incinerators range from 50 000 to 800 000 tonnes per year (Whiting 2003). Fluidised bed incinerators for processed municipal solid waste are common in the USA, Japan and Sweden (DTI OSTEMS Mission 1996). Fluidised beds in the USA are often linked to integrated materials recycling and energy recovery facilities, which pre-sort and recover recyclable materials and then shred the waste to produce a combustible fraction for the fluidised bed. Japan also has a large fluidised bed industry to incinerate MSW, where the incinerators are fed by waste which has already been pre-sorted by the householder to produce a combustible fraction (Whiting 2003; Waste Management in Japan 1995). Consequently, there is a lower requirement for pre-processing, only shredding being generally required.

Fluidised beds have been successfully developed for the incineration of sewage sludge (Frost 1992; Waste Incineration 1996). Sewage sludge has a high water content, typically 96% water. The dried solids have a relatively high calorific value of about 20–24 MJ/kg but a high ash content of between 20 and 50% (Bruce et al 1989; Frost 1992). De-watering the sludge is expensive so that a balance is struck between de-watering and raising the calorific value of the wet sludge sufficiently to enable combustion to take place. Lower levels of de-watering can be allowed if a supplementary fuel is used with the sludge. Normally, the sludge entering the fluidised bed would have been thickened and de-watered to some extent, using mechanical de-watering. The limiting dry solids content of a sludge which, if fed to the furnace, would require no supplementary fuel, is termed the 'autothermic solids content' (Frost 1992). A typical bubbling fluidised bed for sewage sludge inciner-ation is shown in Figure 5.19 (Frost 1992; Oppelt 1987). Figure 5.20 shows two sewage sludge fluidised bed incinerator designs (Frost 1992). Figure 5.20 (a) shows a simple mechanical de-watering system to produce 24% of dry solids in the sludge. Such a sludge

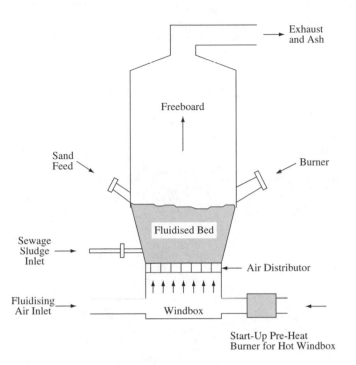

Exhaust
and Ash

Freeboard

Sand
Feed

Burner

Fluidised Bed

Sewage
Sludge
Inlet

Air Distributor

Fluidising
Air Inlet

Windbox

Start-Up Pre-Heat
Burner for Hot Windbox

Figure 5.19 *Schematic diagram of a typical bubbling fluidised bed incinerator for sewage sludge. Sources: Frost 1992; Oppelt 1987.*

would not be autothermic and therefore supplementary fuel is required. Figure 5.20 (b) shows that, after the mechanical dewatering stage, flue gas heat from the combustion process is used to dry the sludge and increase the solids content to 45%. Supplementary fuel is therefore not required, except in the initial start-up of the process. The hot flue gases may be passed to a boiler system for energy recovery. Typical throughputs of sewage sludge are between 15 000 and 25 000 tonnes of dry solids per year.

Sewage sludge incinerators of the multiple-hearth type have also been used extensively (Figure 5.21, Oppelt 1987; Brunner 1991). The incinerator has between 5 and 12 hearths and is designed to handle wastes with high moisture contents. The sludge, which requires dewatering to at least 15% dry solids content, is fed to the top of the incinerator. The sludge moves down through the hearths by movement of the rabble arms which move the sludge alternatively through the centre and edge of the hearths. Flue gases pass up through the furnace. The upper hearths act as drying hearths utilising the hot flue gases from the burning sludge in the middle and lower hearths. Furnace temperatures are up to 900 °C. Ash residues exit at the base of the incinerator (Bruce et al 1989; Frost 1992; Hall 1992).

5.3.2.2 Starved Air Incinerators

Starved air or pyrolytic incinerators are two-stage combustion-type incinerators which are widely used for clinical waste incineration and also for some industrial wastes. The

(a)

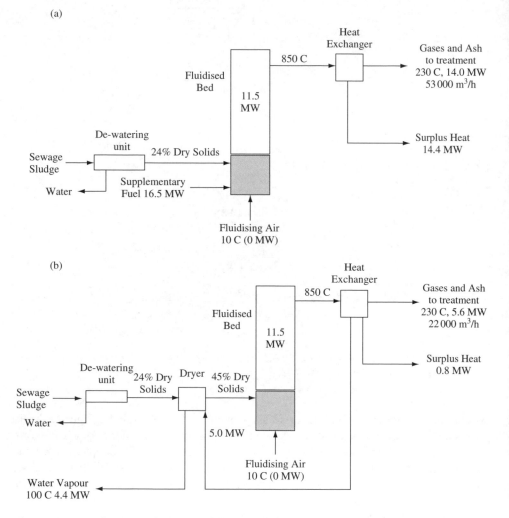

(b)

Figure 5.20 *Schematic diagrams of bubbling fluidised bed designs for incineration of sewage sludge. Source: Frost 1992.*

two stages consist of a pyrolytic stage and a combustion stage. Typical throughputs of waste are between 400 and 25 000 tonnes per year (Whiting 2003). The system is used mainly for solid waste. A typical two-stage solid waste incinerator with waste heat recovery system is shown in Figure 5.22 (Oppelt 1987). The advantages of the starved air incinerator are a more controlled combustion process leading to lower releases of volatile organic compounds and carbon monoxide (Waste Incineration 1996). In addition, the low combustion air flow results in low entrainment of particulate in the flue gases, which also reduces other particulate-borne pollutants, such as heavy metals, dioxins and furans.

Pyrolysis is defined as the chemical decomposition of the waste by the action of heat. Heating the waste in an inert atmosphere produces a gas which, when ignited, is

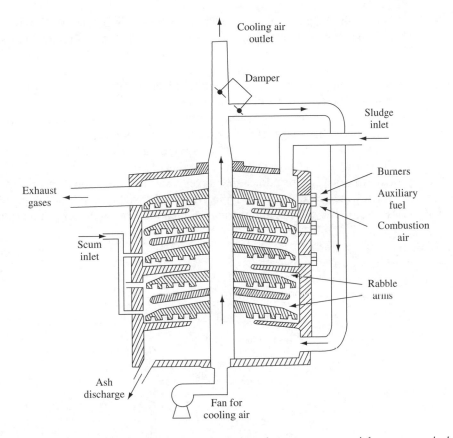

Figure 5.21   *Schematic diagram of a multiple-hearth incinerator used for sewage sludge. Sources: Oppelt 1987; Brunner 1991.*

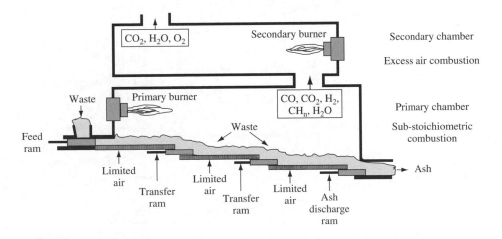

Figure 5.22   *Schematic diagram of a starved air two-stage incinerator. Source: Oppelt 1987.*

self-supporting in air. In practice the two-stage combustor relies on semi-pyrolysis, where the heat for the thermal decomposition or gasification of the waste is produced by sub-stoichiometric combustion of the waste. The waste is combusted under sub-stoichiometric conditions, i.e., where there is insufficient air to provide complete combustion and therefore there is a high proportion of the products of incomplete combustion which pass through to the second stage (Priest 1985; Brunner 1991).

The two-stage process ensures that gas velocities are relatively low and particulate matter is largely retained in the first stage. However, with the stringent legislative requirements on emissions, the full range of gas clean-up systems are required to control other emissions such as acid gases, heavy metals and dioxins and furans.

The pyrolytic/gasification reactions that take place are numerous and complex. Figure 5.23 shows the reactions for a typical hydrocarbon of chemical composition $(CH_2)n$ (Priest 1985). The sub-stoichiometric conditions produce a reducing atmosphere

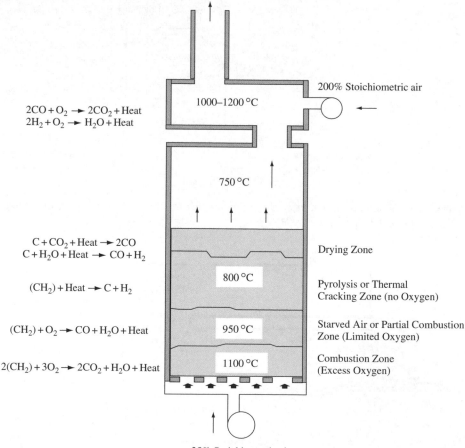

Figure 5.23 *Combustion reactions for a typical hydrocarbon in a starved air, two-stage inciner-ator. Source: Priest 1985. This material has been reproduced from an Institution of Mechan-ical Engineers' Seminar Volume, Energy Recovery from Refuse Incineration, 1985, Figure 1, Page 42 by GM Priest by permission of the Council of the Institution of Mechanical Engineers.*

within the primary chamber, and the heat generated breaks down the hydrocarbon in the pyrolysis zone into carbon and hydrogen. The carbon reacts with the CO_2 and H_2O, generated earlier, to give CO and H_2 which passes to the second-stage combustion zone where complete combustion takes place.

The temperature of the gases leaving the pyrolytic section are of the order of 700–800 °C since a high proportion of the heat generated is used in the endothermic pyrolytic process. These gases will then pass to the secondary section, where secondary excess air, approximately 200% stoichiometric, is added to give a temperature of 1000–1200 °C, which completes the combustion process, combusting the hydrogen, carbon monoxide and hydrocarbons. The two-stage combustion process inhibits the formation of NO_x (Waste Incineration 1996).

The relatively long residence time, within the secondary chamber, plus the high temperature of over 1000 °C, will destroy any dioxins and PCBs that are contained in the secondary gases. This does not preclude any de-novo formation of dioxins and consequently clean-up procedures of, for example, additive activated carbon and/or lime plus a fabric filter, are also required. In practical terms, the smaller two-stage incinerator of up to 0.75 tonnes/h capacity tend to be vertical units operating on a batch basis. The larger units 0.75–5 tonnes/h are designed as horizontal units and have automatic feeding and de-ashing (Priest 1985).

In most cases it is not necessary to pre-treat the waste prior to loading onto the furnace. Charging is normally via a hydraulic ram system which pushes the waste, via a refractory-lined guillotine door, to the pyrolysis chamber. The primary chamber process consists of combustion, partial combustion, pyrolysis and drying. Primary air is fed through the grate to give even combustion at the base of the waste. The residence time of the waste in the primary chamber will be dependent on the hearth area and the characteristics of the waste, and should produce an almost carbon-free ash, the time may be anything from 6–12 h. The ash is finally discharged at the rear of the incinerator into a water trough.

The gases entering the secondary chamber have a sufficient calorific value to be self-sustaining in combustion. Secondary air is introduced to provide the excess air conditions with a high degree of turbulence to create sufficient mixing in order to sustain combustion without the use of support fuel at typical operating temperatures of between 1000 and 1200 °C. Long residence times in the secondary chamber also enable complete burn-out of the combustible gases, vapours, tars and soot. Auxiliary burners are also employed for the initial start-up and then combustion is sustained by the gases from the pyrolytic stage, when the burner may be switched off (Priest 1985; Brunner 1991).

5.3.2.3 Rotary Kiln Incinerators

The rotary kiln is a two-stage incineration type, but the first stage is usually operated in the oxidative mode, i.e., with about 50–200% excess air, rather than the semi-pyrolytic mode found in starved air incinerators. Typical throughputs of waste in rotary kiln incinerators are of the order of 4000–50 000 tonnes per year (Whiting 2003). Rotary kilns have been used for a wide variety of wastes, including municipal solid waste, sewage sludge, industrial waste and hazardous waste and for clean-up of contaminated soils (Waste Incineration 1996). However, they are most common for the treatment of hazardous, clinical and industrial wastes, where in some cases whole drums of waste are fed to the rotary kiln to be completely destroyed (Waste Incineration 1996; European Commission 2004). Figure 5.24 shows a schematic diagram of a rotary kiln incinerator (Oppelt 1987; European Commission 2004).

The rotary kiln is the primary chamber, consisting of an inclined cylinder lined with ceramic material which rotates on rollers at rates which can vary between two revolutions per minute to six revolutions per hour, depending on the type of waste and type of rotary kiln. The size of the rotary kiln can be 1–6 m in diameter and 4–20 m in length. The kiln is rotated by a series of rollers on which the kiln is located. The waste is fed to the front end and ignited by a burner, the combusting wastes are tumbled and agitated by the rotation of the kiln and move down the kiln to reach the end as ash. Residence times of the waste in the rotary kiln are generally more than 30 minutes (European Commission 2004). Internal baffles may be used to increase the mixing and turning of the waste. The kiln typically operates at temperatures around 1200 °C when incinerating hazardous wastes (European Commission 2004). The 'slagging type' of rotary kiln operates at temperatures up to 1500 °C. The high temperature of the 'slagging type' rotary kiln allows the formation of a molten slag of the ash and whole drums of waste can be incinerated as the metal drums will melt (Waste Incineration 1996). The presence of the molten slag absorbs particulate matter, including heavy metals. The ash or molten slag exits the kiln into a quench pit. The molten ash forms a glass-like material, which is less susceptible to leaching of the dissolved metals (Brunner 1991).

The gases from the primary rotary kiln pass to the secondary chamber where excess air conditions with auxiliary burners serve to completely burn out the combustible gases, vapours, tars and soot. A secondary chamber is particularly necessary for hazardous wastes where the time, temperature and turbulence may be insufficient to guarantee the complete combustion of all the organic components of the waste in the primary chamber

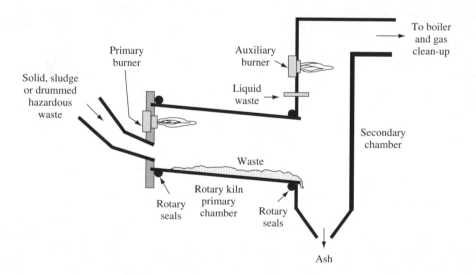

Figure 5.24 *Schematic diagram of a rotary kiln incinerator. Sources: Oppelt 1987; European Commission 2004.*

(Oppelt 1987). Typical temperatures in the secondary chamber would be up to 1400 °C, with residence times of between 1 and 3 s and up to 200% excess air levels. The rotary kiln combines a long residence time plus a high temperature, which enables the complete combustion of complex hazardous wastes (Oppelt 1987). To protect the rotary kiln and secondary chamber from the high temperatures of combustion, the walls are lined with refractory ceramic material.

Cement kilns used for the production of cement are used in some countries to dispose of a variety of wastes including municipal solid waste, industrial waste, tyres and hazardous wastes. Substitute liquid fuel or Secondary liquid fuel (SLF) used in cement kilns is produced by blending organic wastes. The main constituents of SLF include solvents, working fluids (oils, lubricants, etc.) contaminated fuels, organic sludge (e.g. food industry wastes) and other organic chemical products. The wastes are derived as solvent wastes produced by the chemical industry, but also include some aqueous wastes and wastes containing high concentrations of halogen or metal contents. The waste is blended with other fuels in cement kilns which utilise rotary kiln technology. The length of the rotary kiln cyclinder for cement manufacture is exceedingly long, typically up to 250 m in length and 4 m diameter and is lined with alumina bricks. Normally coal or oil is used as the fuel but has been supplemented with waste. Chalk or limestone, plus clay or shale are mixed with water to form a slurry which is passed through the high-temperature furnace to form cement clinker. After processing through the kiln, the cement clinker is ground and gypsum is added to produce cement. The combustion temperatures within the cement kiln are very high in the cement kiln, typically more than 1400 °C (Gilpin 1982). The process is very energy intensive and the use of waste material offsets the costs of fuel. The high temperatures and long residence times used in the process serve to destroy the waste. In addition, when chlorinated or fluorinated wastes are combusted, the large mass of alkaline clinker from the process absorbs and neutralises the acidic stack gases (Holmes 1995; Benestad 1989).

The EC Waste Incineration Directive (2000) covers not only the incineration of waste but also the co-incineration of waste. A co-incineration plant is any plant whose main purpose is the generation of energy or production of material products and which uses wastes as a regular or additional fuel, or in which waste is thermally treated for the purpose of disposal. The emission-limit values to air for co-incineration of waste are set down in the Directive. The Directive states that the co-incineration of waste in plants not primarily intended to incinerate waste should not be allowed to cause higher emissions of polluting substances in that part of the exhaust-gas volume resulting from such co-incineration, than those plants permitted for dedicated incineration. For the co-incineration of waste, the air emission-limit value is determined by a mixing rule formula.

The use of tyres in cement kilns has been shown to reduce the emissions of NO_x, through the formation of reducing zones when the tyres are being burned (Environment Agency PGN 1996). It has also been proposed that the use of Secondary Liquid Fuel (SLF), at a fuel input level of 40%, reduced NO_x emission levels by 50% (House of Lords 1999). Similar reductions in NO_x have been reported when plastic waste has been used in cement kilns (Tokheim et al 2001). The influence on the emissions of other pollutants from cement kilns using waste has shown, for example, no significant difference in SO_2 emissions whether coal or coal plus SLF was combusted in the cement kiln (House of Commons 1997). The influence of using waste in cement kilns on heavy metal emissions is not clear. It has been suggested that the efficiency of the cement kiln in retaining the heavy metals present in the waste, binds the metals to the cement or cement kiln dust (House of Commons 1997). Consequently, no significant increase in heavy metal emissions may be expected. However, Sarofim et al (1994) and Guo and Eckert (1996) report that, when waste-derived fuels are used in cement kilns, there can be an increase in the emissions of certain metals. Because of the high operational temperature and long residence times of cement kilns, their destruction efficiency for organic compounds including PCDD, PCDF and PCBs present in the fuel, waste or raw feed material is high, of the order of >99.995% (Eduljee 1999(b)). The emissions of PCDD/PCDF from cement kilns using conventional fuel and waste fuels such as tyres, refuse-derived fuel and solvent-derived fuels, have been reviewed (Eduljee 1999(b)). In general it was concluded that the ranges of PCDD/PCDF emission concentration resulting from the use of conventional fuel, such as coal and petroleum coke, overlap with the ranges obtained with the use of secondary waste-derived fuels and raw materials, regardless of the type of secondary fuel. It was also shown that, irrespective of which fuel was used in the cement kiln, emissions were below the target emission standard of $0.1\,\mathrm{ng/m^3}$ generally applied throughout Europe, to regulate emissions of dioxins and furans from incinerators. Examination of cement kiln emissions, when using waste tyres as fuel, in relation to health-impact assessment, ambient monitoring, soil sampling and air-quality modelling, have concluded that, in general, the use of tyres as substitute fuel does not increase environmental impacts from the cement-making process (Environment Agency 2003).

5.3.2.4   *Liquid and Gaseous Waste Incinerators*

Liquid and gaseous waste incinerators pass the waste into a burner which mixes the (combustible) waste with air to form a flame zone which burns the waste. Figure 5.25 shows a typical liquid waste burner (Oppelt 1987; European Commission 2004). Supplementary fuel may be required, depending on the calorific value of the waste, or else the liquid

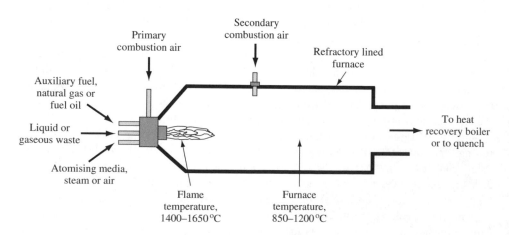

Figure 5.25   *Typical liquid and gaseous waste incinerator. Sources: Oppelt 1987; European Commission 2004.*

waste is pumped directly into a flame generated by the burner, fired by a conventional fuel. The conventional fuel, such as natural gas or fuel oil, helps to maintain steady combustion conditions. The flame is fired into a ceramically lined combustion chamber which radiates heat back into the exhaust gases, thus providing an extended hot zone to completely burn out the products from the combustion of the waste. Very high temperatures occur in the flame, of the order of 1400–1650 °C and furnace chamber temperatures are between 820 and 1200 °C. The chamber may be horizontal or vertical. Liquid waste incinerators are used extensively for the combustion of hazardous wastes. The key section of the incinerator is the burner, which essentially serves to atomise the waste to form a fine spray of droplets, and vapour which ignites to form the flame. Several different designs of burner nozzle exist to cope with the wide range of properties found with liquid wastes and sludges (European Commission 2003(b); Brunner 1991). The liquid waste is pumped under high pressure through the burner nozzle, which produces a fine spray of atomised droplets of size typically between 10 and 150 µm. The smaller the droplet size, the easier vaporisation becomes and consequently the burn-out of each droplet takes place in a much shorter time (Williams 1990(a))

Gaseous waste incinerators operate on a similar system to liquid waste incinerators but the difficulties of producing a fine spray or vapour for combustion are already overcome. Gaseous wastes usually consist of organic hydrocarbons which are combustible. The gases or vapours may be of low concentration and consequently are not autothermic, and therefore the gases or vapours are passed into a burner with either the supplementary fuel gas or combustion air, or may be passed directly into the flame zone. The combustion chamber provides a long residence time for complete burn-out of the gaseous waste (Brunner 1991).

References

Ahad E., Adrados M.A., Caixach J., Fabrellas B. and Rivera J. 2000. Chemosphere 40, 1143–1147.

Adams B., Goemans M. and Joannes J. 2001. 3rd International Symposium on Incineration and Flue Gas Treatment Technologies. Brussels, 2–4 July.

Ahlborg U.G. and Victor K. 1987. In Waste Management and Research, 5, 203.

Alcock R.E., Sweetman A.J. and Jones K.C. 2001. Chemosphere, 43, 183–194.

Allsopp M., Costner P. and Johnston P. 2001. Incineration and human health: state of knowledge of the impact of waste incinerators on human health. Greenpeace Research Laboratories, Exeter University, UK.

Atkins G. 1996. Energy from waste; the best practicable environmental option. In, University of Leeds Short Course – Incineration of municipal waste with energy recovery. Department of Fuel and Energy, The University of Leeds. SELCHP Ltd, London.

Atkinson W., New R., Papworth R., Pearson J., Poll J. and Scott D. 1996. The impact of recycling household waste on down stream energy recovery systems. ETSU Report B/RI/00286/REP. Energy Technology Support Unit, Department of Trade and Industry, London.

Baker P.G. 1981. Analytical Processes. November, 478–480.

Ballantyne W.E., Huffman W.J., Curran L.M. and Stewart D.H. 1980. Energy recovery from municipal solid waste and sewage sludge using multisolid fluidised bed combustion technology. In, Jones J.L. and Radding S.B. (Eds.), Thermal Conversion of Solid Wastes and Biomass, ACS Symposium Series 130. American Chemical Society, Washinghton DC.

Barfnecht T.R., Andon B.M., Thilley W.G. and Hites R.A. 1980. In, Proceedings of the fifth international symposium on PAH. Columbus, OH.

Barton R.G., Clark W.D. and Seeker W.R. 1990. Combustion Science and Technology, 74, 327–342.

Benestad C. 1989. Incineration of hazardous waste in cement kilns, Waste Management and Research, 7, 351–361.

Bergstrom J.G.T. 1986. Waste Management and Research, 4, 57–64.

Bertazzi P.A., Pesatori A.C., Consonni D., Tironi A., Landi M.T. and Zocchetti C. 1993. Epidemiology, 4, 398–406.

Bonte J.L., Fritsky K.J., Plinke M.A. and Wilken M. 2002. Waste Management, 22, 421–426.

Bouscaren B. 1988. In, Brown A., Evemy P. and Ferrero G.L. (Eds.), Energy Recovery Through Waste Combustion. Elsevier Applied Science, Essex.

British Medical Association. 1991. Hazardous Waste and Human Health. Oxford University Press, Oxford.

Brna T.G. 1990. Combustion Science and Technology, 74, 83.

Bruce A.M., Colin F. and Newman P.J. 1989. Treatment of Sewage Sludge. Elsevier Applied Science, London.

Brunner C.R. 1985. Hazardous Air Emissions from Incineration. Chapman & Hall, New York.

Brunner C.R. 1991. Handbook of Incineration Systems. McGraw-Hill Inc., New York.

Brunner P.H. and Monch H. 1986. Waste Management and Research, 4, 105–119.

Brzuzy L.P. and Hites R.A. 1996. Environmental Science and Technology, 30, 1797–1804.

Buekens A. and Patrick P.K. 1985. Incineration. In, Solid Waste Management: Selected Topics. Suess M.J., World Health Organization, Copenhagen.

Buekens A. and Schoeters J. 1984. Thermal methods in waste disposal. Study performed for the EEC under contract EC1 1011/B721/83/B, Buekens A. Free University of Brussels.

Bumb R.R., Crummett W.B., Cutie S.S., Gledhill J.R., Hummel R.H., Kagel R.O., Lamparski L.L., Luomoa E.V., Miller D.L., Nestrick L.A., Shadoff L.A., Stehl R.H. and Woods J.S. 1980. Science, 210, 385.

Buser H. 1979. Chemosphere, 8, 157.

Buser H.R. and Rappe C. 1984. Analytical Chemistry 56. 3. 442–448.

Cains P.W. and Eduljee G.H. 1997. Chemosphere, 34, 51–69.

CEC. 2001. Community strategy for dioxins, furans and polychlorimnated biphenyls. Commission of the European Communities, COM(2001)593 Final, Brussels, Belgium.

CEC. 2004. Implementation of the community strategy for dioxins, furans and polychlorinated biphenyls. Commission of the European Communities, COM(2004) 240 Final, Brussels, Belgium.

Chang M.B. and Lin J.J. 2001. Chemosphere, 45, 1151–1156.

Chow W., Miller M.J. and Torrens D. 1994. Fuel Processing Technology, 39, 5–20.

Clarke, A.G. 1986. The Air. In, Hester R.E. (Ed.), Understanding our Environment. The Royal Society of Chemistry, London.

Clarke A.G. 1998. Industrial Air Pollution Monitoring. Chapman & Hall, London.

Clarke A.G. 1999. Chimney Design. In, University of Leeds Short Course – Incineration of municipal waste with energy recovery. Department of Fuel and Energy, The University of Leeds.

Clayton P., Coleman P., Leonard A., Loader A., Marlowe I., Mitchell D., Richardson S., Scott D. and Woodfield M. 1991. Review of solid waste incineration in the UK. Warren Spring Laboratory (National Environmental Technology Centre) Report No., LR776 (PA). Department of Trade and Industry, London.

Colmsjo A.C., Zebuhr Y.U. and Ostman C.E. 1986. Atmospheric Environment, 20, 2279–2282.

Commoner B., Shapiro K. and Webster T. 1987. Waste Management and Research, 5, 327.

Council Directive 89/369/EEC. 1989. Reduction of air pollution from new municipal waste incinerators. Official Journal of the European Communities, Brussels.

Council Directive 89/429/EEC. 1989. Reduction of air pollution from existing municipal waste incinerators. Official Journal of the European Communities, Brussels.

Darley P. 2003. Energy Recovery. In, University of Leeds Short Course – Incineration of Municipal Waste. The University of Leeds. Darley & Associates.

Davies I.W., Harrison R.M., Perry R., Ratnayaka D. and Wellings R.A. 1976. Environmental Science and Technology, 10, 451–453.

Davis S.B., Gale T.K., Wendt J.O.L. and Linak W.P. 1998. 27th, Symposium (International) on Combustion. The Combustion Institute, Pittsburgh, 1785–1791.

Denison R.A. and Silbergeld E.K. 1988. Risk Analysis, 8, 343–355.

Dioxin Risk: Are we Sure Yet? 1995. EPA draft report review. Environmental Science and Technology, 29, 24A–35A.

Donaldson K., Stone V., Clouter A., Renwick L. and MacNee W. 2001. Occupational Environmental Medicine, 58, 211–216.

DTI OSTEMS Mission. 1996. Energy from waste, Waste Incineration and Anaerobic Digestion with Energy Recovery. Institution of Civil Engineers, London.

Dyke P.H., Foan C. and Fiedler H. 2003. Chemosphere, 50, 469–480.

EC Waste Incineration Directive. 2000. Directive 2000/76/EC of the European Parliament and of the Council on the Incineration of waste. European Commission, Brussels.

EC Waste Landfill Directive. 1999. Council Directive 1999/31/EC, L 182/1, 26.4.1999, Landfill of Waste. Official Journal of the European Communities, Brussels.

Eduljee G. 1999(a). Environmental and Waste Management, 2, 45.

Eduljee G. 1999(b). Environmental and Waste Management, 2, 45–54.

Eighmy T.T., Eusden J.D., Krzanowski J.E., Domingo D.S., Stampfli D., Martin J.R. and Erikson P.M. 1995. Environmental Science and Technology, 29, 629–646.

Energy from Waste: Best Practice Guide. 1996. Department of Trade and Industry, London.

Environment Agency. 1996. Risk assessment of dioxin releases from municipal waste incineration processes. Environment Agency (HMIP), Department of the Environment, HMSO, London.

Environment Agency PGN. 1996. Process Guidance Note S2 3.01, Cement Manufacture, Lime Manufacture and Associated Processes. Environment Agency, HMSO, London.

Environment Agency. 2003. Review of tyre burning in cement kilns, Framework contract 10499 (produced by AEA Technology, Culham, Abingdon). Environment Agency, HMSO, London.

EPAQS. 2001. Airborne Particles, Department of the Environment, Expert Panel on Air Quality Standards. HMSO, London.

Europa. 2003. The European Union on-line. European Union WEB information, Brussels, Belgium.

European Commission. 2003(a). Waste Generated and Treated in Europe. European Commission, Office for Official Publications of the European Communities, Luxembourg.

European Commission. 2003(b). Integrated Pollution Prevention and Control; Draft reference document on best available techniques for the waste treatments industries. European Commission, Brussels.

European Commission. 2004. Integrated pollution prevention and control. Draft reference document on the best available techniques for waste incineration. European Commission, Brussels.

European Commission DG Environment. 1999. Compilation of EU dioxin exposure and health data. European Commission Directorate General Environment, European Commission, Brussels.

Evans J. and Williams P.T. 2000. Trans.I.Chem.E., 78, Part B, 40–46.

Frost R. 1992. Incineration of Sewage Sludge, University of Leeds Short Course 'Incineration and Energy from Waste'. The University of Leeds.

Gelboin H.V. and Tso P.O.P. (Ed.) 1978. Polycyclic Hydrocarbons and Cancer. Academic Press, New York.

Gilpin A. 1982. Dictionary of Energy Technology. Butterworth Scientific and Ann Arbor Science, London.

Giugliano M., Cernuschi S., Grosso M., Aloigi E. and Miglio R. 2001. Chemosphere, 43, 743–750.

Giugliano M., Cernuschi S., Grosso M., Miglio R. and Aloigi E. 2002. Chemosphere, 46, 1321–1328.

Greenberg R.R., Zoller W.H. and Gordon G.E. 1978. Environmental Science and Technology, 12, 566–573.

Guo Q. and Eckert J.O. 1996. Journal of Hazardous Materials, 51, 47–65.

Hagenmaier H., Brunner H., Haag, R., Kraft M. and Lutzke K. 1987. Waste Management and Research, 5, 239.

Hagenmaier H., Kraft M., Haag R. and Brunner H. 1988. In, Brown A., Evemy P. and Ferrero G.L. (Eds.), Energy Recovery through Waste Combustion. Elsevier Applied Science, Essex.

Hall J.E. 1992. Treatment and use of sewage sludge. In (The Treatment and Handling of Wastes) Ed., Bradshaw A.D., Southwood R. and Warner F. Chapman & Hall, London.

Hall G.S. and Knowles M.J.D. 1985. Good design and operation reaps benefits on UK refuse incineration boilers over the last ten years. In, Energy Recovery from Refuse Incineration. Institution of Mechanical Engineers, London.

Ho T.C., Chen C., Hopper J.R. and Oberacker D.A. 1992. Combustion Science Technology, 85, 101–116.

Holmes J. 1995. The UK Waste Management Industry, 1995. Institute of Waste Management, Northampton.

Horne P.A. and Williams P.T. 1997. 1st International Symposium on Incineration and Flue Gas Treatment Technologies, Sheffield, July.

House of Commons. 1997. House of Commons Environment Committee, Third Report, The Environmental Impact of Cement Manufacture. The Stationary Office, London.

House of Lords. 1999. Waste Incineration, House of Lords Select Committee on the European Communities, Session 1998–99, 11th Report Waste Incineration, HL Paper 71. HMSO, London.

Hu S.W. and Shy C.M. 2001. Journal of the Air and Waste Management Association, 51, 1100–1109.

Huang H. and Buekens A. 2001. Chemosphere, 44, 1505–1510.

Hundesrugge T., Nitsch K.H., Rammensee W. and Schoner P. 1989. Mull Abfall, 6, 318–324.

Hunsinger H., Kreisz S. and Seifert T. 1998. Chemosphere, 37, 2293–2297.

Hutzinger O., Frei R.W., Meriam E. and Pocchiari F. 1982. Chlorinated Dioxins and Related Compounds. Pergamon Press, New York.

IAWG. 1997. International Ash Working Group, Municipal Solid Waste Incinerator Residues, International Ash Working Group. Elsevier Publications, London.

IEH. 1997. Health effects of waste combustion products. Institute for Environmental Health, Medical Research Council, University of Leicester.

Kirby C.S. and Rimstidt J.D. 1993. Environmental Science Technology, 27, 652–660.

Kogevinas M. 2001. Human Reproduction Update, 7, 331–339.

Krause H.H. 1991. In, Bryers R.W. (Ed.), Incinerating Municipal and Industrial Waste. Hemisphere Publishing Corporation, New York.

Kreisz S., Hunsinger H. and Vogg H. 1996. Chemosphere, 32, 73–78.

Lee M.L., Novotny M. and Bartle K.D. 1981. Analytical Chemistry of Polycyclic Aromatic Compounds. Academic Press, New York.

Liljelind P., Unsworth J., Maaskant O. and Marklund S. 2001. Chemosphere, 42, 615–623.

Lindqvist O. 1986. Waste Management and Research, 4, 35–44.

Longwell J.P. 1983. In, Lahaye J.L. and Prado G. (Eds.), Soot in Combustion Systems and its Toxic Properties. Plenum Press, New York.

Love L.B. and Seskin E.P. 1970. Science, 169, 723.

Lustenhouwer J.W.A., Olie K. and Hutzinger O. 1980. Chemosphere, 9, 501.

Maaskant O.L. 2001. 3rd International Symposium on Incineration and Flue Gas Treatment Technologies. Brussels, 2–4 July.

Marklund S., Rappe C., Tysklind M. and Egeback K.E. 1987. Chemosphere, 16, 29.

McCahey S., McMullan J.T. and Williams B.C. 1999. Fuel, Vol. 78, 1771–1778.

Morf L.S., Brunner P.H. and Spaun S. 2000. Waste Management Research, 18, 4–15.

Mulder E. 1996. Waste Management, 16, 181–184.

Mulholland J.A. and Sarofim A.F. 1991(a). Environmental Science Technology, 25, 268.

Mulholland J.A. and Sarofim A.F. 1991(b). Environmental Progress, 10, 83.

Nchida S. and Kamo H. 1983. Industrial Engineering Chemistry. Proc. Des. Dev., 22, 144.

Nielsen K.K., Moeller J.T. and Rasmussen S. 1986. Chemosphere, 15, 1247.

Niessen W.R. 1978. Combustion and Incineration Processes. Marcel Dekker Inc., New York.

Oakland D. 1988. Filtration and Separation, January/February, 39.

Ogura I., Masunaga S. and Nakanishi J. 2001. Chemosphere, 45, 173–183.

Olie K., Lustenhouwer J.W.A. and Hutzinger O. 1982. In, Hutzinger O., Frei R.W., Merian E. and Pocchiari F. (Eds.), Chlorinated Dioxins and Related Compounds; Impact on the Environment. Pergamon Press, Oxford.

Oppelt H.M. 1987. Journal of the Air Pollution Control Association, 37, 558–586.

Ozvacic V., Wong G., Tosine H., Clement R.E. and Osborne J. 1985. Journal of the Air Pollution Control Association, 35, 849.

Parker C. and Roberts T. 1985. Energy from Waste: an Evaluation of Conservation Technologies. Elsevier Applied Sciences, London.

Pfranf-Stotz G. and Reichelt J. 1997. Characterisation and valuation of municipal solid waste incineration bottom ashes. In, 1st International Symposium on Incineration and flue gas treatment technologies. Sheffield, July.

Pollution Paper 27. 1989. Dioxins in the Environment. Department of the Environment, HMSO, London.

Porteous A. 1992. Dictionary of Environmental Science and Technology. John Wiley & Sons, Chichester.

Pranghofer G.G. and Fritsky K.J. 2001. 3rd International Symposium on Incineration and Flue Gas Treatment Technologies. Brussels, 2–4 July.

Prestbo W., Miller M.J. and Torrens I.M. 1995. J. Air, Water and Soil Poll.Assoc., 80, 145–158.

Priest G.M. 1985. Producing steam from the combustion of solid waste. In, Energy Recovery from Refuse Incineration, Seminar, Institution of Mechanical Engineers, Strathclyde, August.

Probert S.D., Kerr K. and Brown J. 1987. Harnessing energy from domestic, municipal and industrial refuse. Applied Energy, 27, 89–168.

Quab U., Fermann M. and Broker G. 2000. Chemosphere, 40, 1125–1129.

Rabl A. and Spadaro J.V. 2002. Issues in Environmental Science and Technology, 18, 171–193.

Reimann D.O. 1987. Waste Management and Research, 5, 147.

Richter H. and Howard J.B. 2000. Progress in Energy and Combustion Science, 26, 565–608.

Royal Commission on Environmental Pollution. 1993. 17th Report, Incineration of Waste, HMSO, London.

Sakai S., Sawell S.E., Chandler A.J., Eighmy T.T., Kosson D.S., Vehlow J., van der Sllot H.A., Hartlen J. and Hjelmar O. 1996. World trends in municipal solid waste management. Waste Management, 16, 341–350.

Saracci R., Kogevinas M., Bertazzi P.-A., Mesquita B.H., Coggon D., Green L.M., Kauppinen T., L'Abbe K.A., Littorin M., Lynge E., Mathews J.D., Neuberger M., Osman J., Pearce N. and Winkelmann R. 1991. The Lancet, 338, 8774, 1027–1032.

Sarofim A.F., Pershing D.W., Dellinger B., Heap M.P. and Owens W.D. 1994. Hazardous Waste and Hazardous Materials, 11, 169–192.

Seeker W.R. 1990. Waste Combustion, 23rd Symposium (International) on Combustion. The Combustion Institute, Pittsburgh, 867–885.

Shaub W.M. 1989. An overview of what is known about dioxin and furan formation, destruction and control during incineration of MSW. Coalition on Resource, Recovery and the Environment (CORRE), US EPA, MSW Technology Conference, San Diego, CA, January–February.

Steisel N., Morris R. and Clarke M.J. 1987. Waste Management and Research, 5, 381.

Tiernan T.O. 1983. In, Choudhary G., Keith L.M. and Rappe C. (Eds.), Chlorinated Dioxins and Dibenzofurans in the Total Environment. Butterworth, London.

Tokheim L.-A., Gautestad T., Axelsen E.P. and Bjerketvedt D. 2001. Energy recovery from wastes: Experience with solid alternative fuels combustion in a precalciner cement kiln. 3rd International Symposium on Incineration and flue gas treatment technologies. Brussels, 2–4 July.

Tosine H. 1983. In, Choudhary G., Keith L.M. and Rappe C. (Eds.), Chlorinated Dioxins and Dibenzofurans in the Total Environment. Butterworth, London.

Tuppurainen K., Halonen I., Ruokojärvi P., Tarhanen J. and Ruuskanen J. 1998. Chemosphere, 36, 1493–1511.

Turner B. 1996. Refractory materials for incinerators. In, University of Leeds Short Course – Incineration of municipal waste with energy recovery. Department of Fuel and Energy, The University of Leeds. Hepworth Refractories Ltd, Sheffield.

Uberoi M. and Shadman F. 1990. AIChE Journal, 36, 307–309.

Uchida S., Kamo H. and Kubota H. 1988. Industrial Engineering Chemical Research, 27, 2188.

Visalli J.R. 1987. Journal of the Air Pollution Control Association, 37, 1451.

Vogg H., Braun H., Metzger M. and Scneider J. 1986. Waste Management and Research, 4, 65–74.

Vogg H., Metzger M. and Steiglitz L. 1987. Waste Management and Research, 5, 285.

Wade J. 2003. Flue Gas Emissions Control. In, University of Leeds Short Course – Incineration of municipal waste. The University of Leeds.

Walsh D.C., Chillrud S.N., Simpson H.J. and Bopp R.F. 2001. Environmental Science and Technology, 35, 2441–2447, 2001.

Waste Management Paper 28. 1992. Recycling. Department of the Environment, HMSO, London.

Waste Incineration. 1996. IPC Guidance Note, S2 5.01, Processes subject to integrated pollution control; waste Incineration. Environment Agency, Bristol.

Waste Management in Japan. 1995. Department of Trade and Industry, Overseas Science and Technology Expert Mission Visit Report. Institute of Wastes Management, Northampton.

Weber R., Iino F., Imagawa T., Takeuchi M., Saurai T. and Sadakata M. 2001. Chemosphere, 1429–1438.

Wevers M. and De Fre R. 1998. Organohalogen Compounds, 36, 342–347.

Wey M.Y., Su J.L. and Chen J.C. 1999. Journal of the Air and waste Management Association, 49, 444–453.

Whiting K. 2003. In, University of Leeds Short Course – Incineration of Municipal Waste. Department of Fuel and Energy, The University of Leeds.

Wilhelm J., Stieglitz L., Dinjus E. and Will R. 2001. Mechanistic studies on the role of PAHs and related compounds in PCDD/F formation on model fly ashes. Chemosphere, 42, 797–802.

Williams A. 1990(a). Combustion of Liquid Fuel Sprays. Butterworths, Ltd, London.

Williams P.T. 1994. Pollutants from Incineration: An Overview. In, Waste Incineration and the Environment, Hester R.E. and Harrision R.M. (Eds.), Issues in environmental science and technology. The Royal Society of Chemistry, London.

Williams P.T. 1990(b). Journal of the Institute of Energy, 63, 22–30.

World Health Organization. 1990. Heavy metal and PAH compounds from municipal incinerators. Environmental health Series No. 32. World Health Organization, Copenhagen.

Yasuda K. and Takahashi M. 1998. Journal of the Air and Waste Management Association, 20, 2279–2287.

York, D. 2003. In, University of Leeds Short Course – Incineration of Municipal Waste. The University of Leeds.

Zhang F.S., Yamasaki S., Nanzyo M. and Kimura K. 2001. Environmental Pollution, 115, 253–260.

Zimmerman R., Blumenstock M., Schramm K.W. and Kettrup A. 2000. Proceedings Dioxin 2000, V46, 78.

6

Other Waste Treatment Technologies: Pyrolysis, Gasification, Combined Pyrolysis–Gasification, Composting, Anaerobic Digestion

Summary

This chapter discusses other options for waste treatment and disposal. Pyrolysis of waste, the types of products formed during pyrolysis and their utilisation, and the different pyrolysis technologies are discussed. Gasification of waste, gasification technologies and utilisation of the product gas are described. Composting of waste is described, including the composting process and the different types of composter. Anaerobic digestion of waste, the degradation process and operation and technology for anaerobic digestion are discussed. Current examples of the different types of pyrolysis, gasification, composting and anaerobic digestion are described throughout.

6.1 Introduction

The hierarchy of waste management and the concept of sustainable waste management have led to the development of alternative waste treatment and disposal options rather than the traditional reliance on the options of landfill and incineration. Alternatives which have a minimal environmental impact, with a view to recycling or energy recovery with low pollution, have received particular attention. Amongst such technologies are

Waste Treatment and Disposal, Second Edition   Paul T. Williams
© 2005 John Wiley & Sons, Ltd   ISBNs: 0-470-84912-6 (HB); 0-470-84913-4 (PB)

pyrolysis, gasification, combined pyrolysis–gasification, composting and anaerobic digestion.

The thermal treatment options of pyrolysis, gasification and combined pyrolysis/gasification systems, are generating increasing interest as viable alternative environmental and economic options for waste processing. These options have a number of advantages over conventional incineration or landfilling of waste. Depending on the technology, the waste can be processed to produce not only energy, but also gas or oil products for use as petrochemical feedstocks and/or a carbonaceous char for use in applications such as effluent treatment or for gasification feedstock. The production of storable end products such as a gas, oil or char, enables the possibility of de-coupling the end use of that product, either for energy production or petrochemical use from the waste treatment process. The EC Waste Incineration Directive (2000), regulates the emissions to air, land and water from incineration and also details the operational, emissions monitoring, process conditions, etc., of the incineration plant. Incineration is defined in the Directive as any thermal process dedicated to the thermal treatment of wastes, with or without energy recovery. In addition, the Directive specifically includes the thermal treatment processes of pyrolysis and gasification processes insofar as the substances resulting from the treatment are subsequently incinerated.

Composting of the biodegradable fraction of waste, particularly municipal solid waste, results in diversion of that waste away from landfill. Biodegradable wastes are defined as wastes such as food and garden waste, paper, cardboard, textiles wood, etc., that are degraded over long periods of time by various aerobic and anaerobic bacteria to produce a liquid leachate and landfill gas. The proportion of biodegradable waste in municipal solid waste in Europe varies from 66% to more than 90%, depending on country (European Environment Agency 2002). The EC Waste Landfill Directive (1999) seeks to reduce the amount of biodegradable waste sent to landfill to 75% of the 1995 levels by 2006, 50% of 1995 levels by 2009 and 35% of 1995 levels by 2016 (EC Waste Landfill Directive 1999). The main aim of the Directive is to reduce the amount of emissions of landfill gas (which is composed of the greenhouse gases, carbon dioxide and methane) being emitted to the atmosphere, and also to encourage more recycling of waste. Composting satisfies both of these criteria and, in addition, generates a product useful in agricultural and horticultural applications.

Anaerobic digestion of waste is also an attractive option for the treatment of municipal solid waste and other wastes such as sewage sludge, agricultural waste and animal manure. Anaerobic digestion involves the same substances that generate landfill gas: carbon dioxide and methane, in a waste landfill site, but in a controlled, closed environment. The generated gas, again composed of carbon dioxide and methane, can be used directly as fuel, or upgraded to a higher quality gaseous fuel or chemical feedstock. In addition, the residue can be used as a soil conditioner.

6.2   Pyrolysis

Pyrolysis is the thermal degradation of organic waste in the absence of oxygen to produce a carbonaceous char, oil and combustible gases. How much of each product is produced is dependent on the process conditions, particularly temperature and heating rate. Figure 6.1

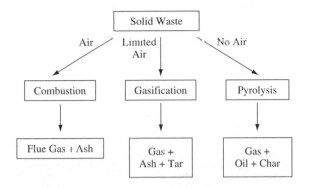

Figure 6.1 *Process characterisation of incineration, gasification and pyrolysis.*

characterises the main differences between pyrolysis, gasification and incineration. The key difference is the amount of oxygen supplied to the thermal reactor. For pyrolysis there is an absence of oxygen, and for gasification there is a limited supply of oxygen, such that complete combustion does not take place, instead the combustible gases; carbon monoxide and hydrogen are produced. The oxygen for gasification is supplied in the form of air, steam or pure oxygen. Incineration involves the complete oxidation of the waste in an excess supply of oxygen to produce carbon dioxide, water and ash, plus some other products such as metals, trace hydrocarbons, acid gases, etc.

Waste materials are composed of complex chemical compounds, for example, municipal solid waste contains paper and cardboard which are composed of large, complex polymereric, organic molecular chains such as cellulose, hemicellulose and lignin. Similarly, wastes such as forestry wastes and biomass are also mainly composed of cellulose, hemicellulose and lignin polymeric molecules. Plastics are also composed of large polymer chains. The process of thermal degradation or pyrolysis of such materials, in the absence of oxygen, results in the long polymer chains breaking to produce shorter molecular weight chains and molecules. These shorter molecules result in the formation of the oils and gases characteristic of pyrolysis of waste. The exact mechanisms of thermal degradation of waste are not clear. Figures 6.2 and 6.3 show examples of the large polymer chains of cellulose, hemicellulose, lignin, and several plastics and rubbers found in waste materials (Graham et al 1984; Menachem and Goklestein 1984).

Relatively low temperatures are used for pyrolysis, in the range 400–800 °C. The application of pyrolysis to waste materials is a relatively recent development. In particular, the production of oils from the pyrolysis of waste has been investigated, with the aim of using the oils directly in fuel applications or, after upgrading, to produce refined fuels. The pyrolysis oils derived from a variety of wastes have also been shown to be complex in composition and contain a wide variety of chemicals which may be used as chemical feedstock. The oil has a higher energy density, that is a higher energy content per unit weight, than the raw waste. The solid char can be used as a solid fuel or as a char–oil, char–water slurry for fuel. Alternatively the char can be used as carbon black or upgraded to activated carbon. The gases generated have medium to high calorific values and may contain sufficient energy to supply the energy requirements of a pyrolysis plant (Bridgwater and Bridge 1991).

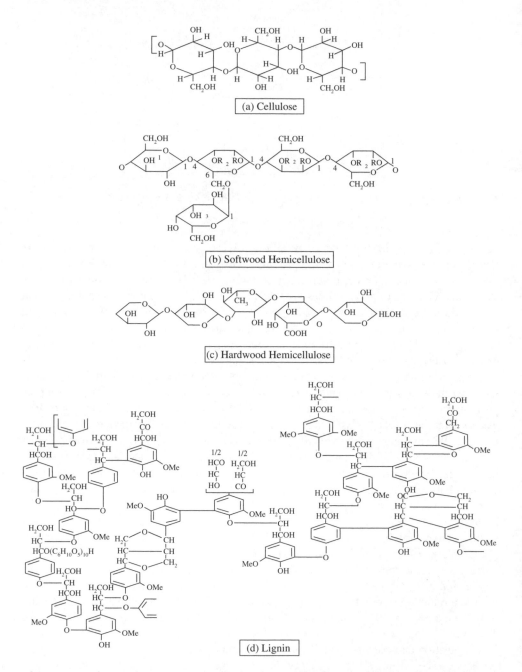

Figure 6.2 *Chemical structures of cellulose, hemicellulose and lignin. Sources: Menachem and Goklestein 1984; Graham et al 1984. Figure (d) copyright 1984, with permission from Elsevier.*

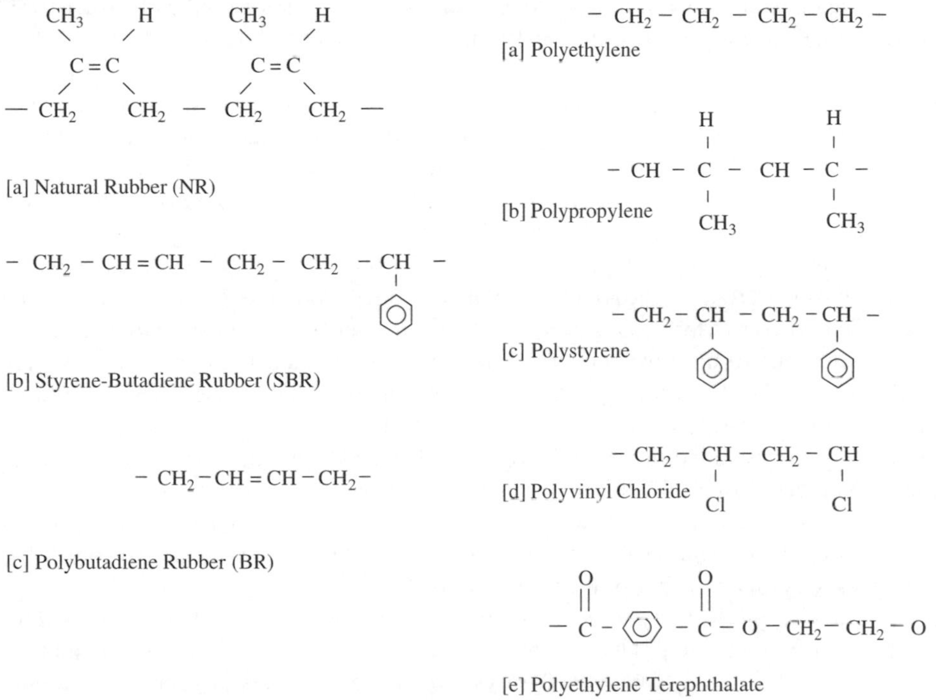

CH₃   H   CH₃   H
  \   /   \   /
   C = C   C = C
  /   \   /   \
— CH₂   CH₂ — CH₂   CH₂ —

[a] Natural Rubber (NR)

— CH₂ — CH = CH — CH₂ — CH₂ — CH —

[b] Styrene-Butadiene Rubber (SBR)

— CH₂ — CH = CH — CH₂ —

[c] Polybutadiene Rubber (BR)

— CH₂ — CH₂ — CH₂ — CH₂ —
[a] Polyethylene

                H   H
                | |
— CH — C — CH — C —
                | |
[b] Polypropylene   CH₃   CH₃

— CH₂ — CH — CH₂ — CH —
[c] Polystyrene

— CH₂ — CH — CH₂ — CH
              | |
[d] Polyvinyl Chloride   Cl   Cl

      O   O
      || ||
— C — 〈◯〉 — C — O — CH₂ — CH₂ — O

[e] Polyethylene Terephthalate

Figure 6.3   *Chemical structures of rubbers and plastics.*

The process conditions are altered to produce the desired char, gas or oil end product, with the pyrolysis temperature and heating rate having the most influence on the product distribution. The heat is supplied by indirect heating, such as the combustion of the gases or oil, or directly by hot gas transfer. Pyrolysis has the advantage that the gases or oil product derived from the waste can be used to provide the fuel for the pyrolysis process itself. Pyrolysis systems for municipal solid waste, tyres, plastics, composite plastics, sewage sludge, textile waste and biomass have been investigated (Williams et al 1995; Bridgwater and Evans 1993; Lee 1995; Roy 1994; European Commission 2004; Williams 2001, 2003; Cunliffe et al 2003; Cunliffe and Williams 1998, Williams and Reed 2003, 2004; Reed and Williams 2004).

Very slow heating rates coupled with a low final maximum temperature, maximises the yield of char, because the production of char from wood in the form of charcoal involves a very slow heating rate to moderate temperatures. The process of carbonisation of waste results in reduced concentrations of oil/tar and gas product and these are regarded as by-products of the main charcoal-forming process. Moderate heating rates in the range of about 20°C/min to 100°C/min and maximum temperatures of 600°C give an approximately equal distribution of oils, char and gases. This is referred to as conventional pyrolysis or slow pyrolysis. Because of the slow heating rates and generally slow removal of the products of pyrolysis from the hot pyrolysis reactor, secondary reactions of the products can take place. Generally, a more complex product slate is found.

Very high heating rates of about 100 °C/s to 1000 °C/s at temperatures below 650 °C and with rapid quenching, lead to the formation of a mainly liquid product, which is referred to as fast or flash pyrolysis. Liquid yields up to 70% have been reported for biomass feedstocks using flash pyrolysis. In addition, the carbonaceous char and gas production are minimised. The primary liquid products of pyrolysis are rapidly quenched and this prevents breakdown of the products to gases in the hot reactor. The high reaction rates also cause char-forming reactions from the oil products to be minimised (Bridgwater and Bridge 1991; Bridgwater and Evans 1993).

At high heating rates and high temperatures the oil products quickly breakdown to yield a mainly gas product. The typical yield of gas from the original feedstock hydrocarbon is 70%. This process differs from gasification which is a series of reactions involving carbon and oxygen in the form of oxygen gas, air or steam, to produce a gas product consisting mainly of CO, CO_2, H_2 and CH_4. Table 6.1 shows the typical characteristics of different types of pyrolysis (Bridgwater and Bridge 1991; Bridgwater and Evans 1993).

Pyrolysis process conditions can be optimised to produce either a solid char, gas or liquid/oil product. Table 6.2 shows the yields of char, oil/liquid and gas from various waste feedstocks (Williams and Besler 1996; Horne and Williams 1996; Williams et al 1995; Kaminsky and Sinn 1980; Williams and Besler 1992; Rampling and Hickey 1988; Williams and Williams 1997; Williams and Reed 2003).

The solid char product from carbonisation or slow pyrolysis of wood has been used for centuries as the process to produce charcoal for use as fuel and charcoal product yields of between 30 and 40% are common. Pyrolysis of waste materials also produces a char

Table 6.1 *Typical characteristics of different types of pyrolysis*

Pyrolysis	Residence time	Heating rate	Reaction environment	Pressure (bar)	Temperature (°C)	Major product
Carbonisation	hrs–days	Very low	Combustion products	1	400	Charcoal
Conventional	10 s–10 min	Low–moderate	Primary/ secondary products	1	<600	Gas, char liquid
Flash–liquid	<1 s	High	Primary products	1	<600	Liquid
Flash–gas	<1 s	High	Primary products	1	>700	Gas
Ultra	<0.5 s	Very high	Primary products	1	1000	Gas, chemicals
Other pyrolysis types:						
Vacuum	2–30 s	Medium	Vacuum	<0.1	400	Liquid
Hydropyrolysis	<10 s	High	H_2+primary	~20	<500	Liquid, chemicals
Methanolysis	0.5–1.5 s	High	CH_4+primary products	~3	1050	Benzene, toluene, xylene+ alkenes

Sources: Bridgwater and Bridge 1991; Bridgwater and Evans 1993.

Table 6.2   *Product yields from the pyrolysis of waste*

Waste	Pyrolysis process	Temperature (°C)	Heating rate	Char (%)	Liquid (%)	Gas (%)
Wood[1]	Moderate (batch)	600	20 °C/min	22.6	50.4	27.0
Wood[2]	Fast (fluidised bed)	550	~300 °C/s	17.3	67.0	14.9
Tyre[3]	Moderate (batch)	600	20 °C/min	39.2	54.0	6.8
Tyre[3]	Slow/moderate (batch)	850	~5 °C/min	49.5	32.5	18.0
Tyre[4]	Fast (fluidised bed)	640	—	38	40	18
RDF[5]*	Moderate (batch)	600	20 °C/min	35.2	49.2	18.8
RDF[6]*	Moderate (batch)	700	—	30	49	22
Plastic[7] (mixed)	Moderate (batch)	700	25 °C/min	2.9	75.1	9.6
Textile flax[8]	Slow (batch)	450	2 °C/min	25.0	52.5	22.5

* Refuse derived fuel from municipal solid waste.

Sources: [1] Williams and Besler 1996; [2] Horne and Williams 1996; [3] Williams et al 1995; [4] Kaminsky and Sinn 1980; [5] Williams and Besler 1992; [6] Rampling and Hickey 1988; [7] Williams and Williams 1997; [8] Williams and Reed 2003.

product, the percentage production depending on process conditions. Pyrolysis of municipal solid waste produces a 35% char product, which has a high ash content of up to 37%, and tyre pyrolysis under slow heating rate conditions produces a char of up to 50% with an ash content of about 10%. The chars may be used directly as fuels, briquetted to produce solid fuels, used as adsorptive materials such as activated carbon, upgraded to produce a higher grade activated carbon, or crushed and mixed with the pyrolysis oil product to produce a slurry for combustion.

The calorific value of the chars are relatively high, for example, char derived from municipal solid waste has a calorific value of about 19 MJ/kg, tyre char about 29 MJ/kg and wood waste produces a char of calorific value about 33 MJ/kg (Williams and Besler 1996; Williams et al 1995; Williams and Besler 1992; Rampling and Hickey 1988). These figures compare with a typical bituminous coal of calorific value 30 MJ/kg. As such, the chars could be used as a medium grade solid fuel.

The significance of a high ash content in the chars means that the value of the char as a fuel is reduced. In addition, the use of pyrolysis chars as substitutes for activated carbon are greatly diminished if they have a high ash content. Chars from wood have very low ash contents, typically less than 2%, wheras the ash content of tyre-derived pyrolysis chars are over 10% (Cunliffe and Williams 1998). The upgrading of pyrolysis chars to activated carbon for biomass-derived pyrolysis chars, has been achieved using steam activation. However, the upgrading of tyre chars to activated carbon requires an additional processing step of de-ashing to make the product acceptable to the activated carbon industry. In addition, the specifications of activated carbon derived from traditional routes such as coconut shell are well established, and as with most new products, it is difficult for an alternative product to break into an established market. Even though the waste-derived chars may be cheaper, the specifications, quality and maintenance of quality have to be guaranteed. Commercially used activated carbons have surface areas typically in the range of 500–2000 m²/g and pore sizes which can be manipulated by the process conditions or source feedstock to produce the desired pore-size distribution for a particular application. Activated carbons may be produced by either physical or chemical activation. Chemical activation involves impregnation with a chemical, such as zinc chloride, followed by carbonisation using the pyrolysis process. Physical activation involves pyrolysis of the source material to produce a char, followed by steam or carbon dioxide gasification. Such techniques have been applied to waste materials to enhance the properties of the derived char in order to produce an activated carbon with properties similar to those produced commercially (Williams and Reed 2003, 2004; Reed and Williams 2004; Cunliffe and Williams 1998). For example, char derived from tyres has an initial surface area of about $60\,m^2/g$, but activation of the carbonaceous char with steam at temperatures above $800\,°C$ produces an activated carbon with a surface area of over $650\,m^2/g$ (Cunliffe and Williams 1998). The action of the steam is to react with the carbon to produce carbon monoxide, carbon dioxide and hydrogen, opening up pores and increasing the surface area. Surface areas of flax textile waste are less than $5\,m^2/g$, but steam activation can produce activated carbons of over $900\,m^2/g$ (Williams and Reed 2003). The chemical activation of waste, i.e., where an absorbent waste, such as textile flax and hemp biomass waste absorbs a reacting chemical, such as zinc chloride or potassium hydroxide, followed by pyrolysis at $500\,°C$ has produced very high surface area activated carbons of over $2000\,m^2/g$ (Williams and Reed 2004).

The product oil from pyrolysis of waste has the advantage of being able to be used in conventional electricity-generating systems, such as diesel engines and gas turbines. However, the properties of the pyrolysis oil fuel may not match the specifications of a petroleum-derived fuel and may require modifications to the power plant or upgrading of the fuel. In some cases the oil product is described as a liquid but, depending on the feedstock and the pyrolysis process conditions, it may represent either a true oil, an oil/aqueous phase, separated oil and aqueous phases or, for some waste feedstocks, a waxy material. The advantages of producing an oil product from waste are that the oil can be transported away from the pyrolysis process plant and therefore de-couples the processing of the waste from the product utilisation. The oil may be used directly as a fuel, added to petroleum refinery stocks, upgraded using catalysts to a premium grade fuel, or used as a chemical feedstock. The composition of the oil is dependent on the chemical composition of the feedstock and the processing conditions. For example, oils derived from biomass have a high oxygen content, of the order of 35% by weight, due to the content of cellulose, hemicellulose and lignin in the biomass. These are large polymeric structures containing mainly carbon, hydrogen and oxygen. Similarly, oils derived from municipal solid waste have a high oxygen content due to the presence of cellulosic components in the waste such as paper, cardboard and wood. Biomass and municipal solid waste pyrolysis oils derived from flash pyrolysis processes, tend to have a lower viscosity and consist of a single water/oil phase. The oils are therefore high in water, which markedly reduces their calorific value. Slow pyrolysis produces liquid products with higher viscosities which tend to have two phases due to the more extensive degree of secondary reactions which occur. Oils derived from scrap tyre pyrolysis and plastics, on the other hand, are composed of mainly carbon and hydrogen.

The oils have significant calorific values ranging from 25 MK/kg for oils derived from municipal solid waste to 42 MJ/kg for oils derived from scrap tyres, compared with a typical petroleum-derived fuel oil at 46 MJ/kg (Table 6.3). Table 6.3 shows the properties of oils derived from the pyrolysis of tyres, municipal solid waste and wood (Williams et al 1995; Pober and Bauer 1977; Rick and Vix 1991). Comparison with petroleum-derived diesel

Table 6.3 *Typical fuel properties of waste derived pyrolysis oils*

Parameter	Tyre oil	MSW oil	Biomass oil	Diesel oil
Carbon residue (%)	0.7	—	—	<0.35
Mid B.Pt. (°C)	230	—	—	300
Viscosity (cSt)	2.12 (60 °C)	—	17 (100 °C)	1.3 (60 °C)
	3.50 (40 °C)	—	90 (50 °C)	3.3 (40 °C)
Density (kg/m^3)	0.91	1.3	1.2	0.78
API gravity	20.41	—	—	31
Flash Point (°C)	24	56	110–120	75
Hydrogen (%)	9.98	7.6	7–8	12.8
Carbon (%)	87.0	57.5	50–67	—
Nitrogen (%)	0.4	0.9	0.8–1	—
Oxygen (%)	0.7	33.4	15–25	—
Initial B.Pt. (°C)	80	—	—	180
10% B.Pt. (°C)	140	—	—	—
50% B.Pt. (°C)	230	—	—	300
90% B.Pt. (°C)	340	—	—	—
CV (MJ/kg)	42.0	24.4	24.7 (lower)	46.0
Sulphur (%)	1.5	0.1–0.3	<0.01	0.9

Sources: Williams et al 1995; Pober and Bauer 1977; Rick and Vix 1991.

fuel shows that in many respects the oils derived from waste are quite similar. However, the direct use of such fuels in combustion systems designed and optimised on fuels refined from petroleum, may be difficult. For example, biomass and municipal solid waste pyrolysis oils can be viscous, highly acidic, due to the organic acids present in the oils, and can readily polymerise. In addition, pyrolysis oils may contain solid char particles due to carry-over from the pyrolysis reactor. Consequently, their use in liquid spray or atomisation combustion systems such as diesel engines, furnaces and boilers, may result in the spray or atomisation system becoming blocked and/or corroded.

Performance guarantees for the use of non-standard fuels in combustion systems may invalidate the manufacturers warranties, which would be based around standard, i.e., petroleum-refined fuels. Emission limits from the combustion system, set at National and European level, would also have to be met irrespective of the fuel being used. However, the fuels derived from waste materials such as tyres, wood and municipal solid waste, have been successfully combusted in a variety of systems (Bridgwater and Evans 1993).

The oils derived from the pyrolysis of waste materials tend to be chemically very complex, due to the polymeric nature of the wastes and the range of potential primary and secondary reactions. Biomass and municipal solid waste pyrolysis oils contain hundreds of different chemical compounds including organic acids, phenols, alcohols, aldehydes, ketones, furans, etc. (Desbene et al 1991). Tyre pyrolysis oils consist mainly of alkanes, alkenes and monoaromatic and polycyclic aromatic compounds (Williams et al 1995). Oils derived from mixed plastic waste at typical pyrolysis temperatures of 500 °C are highly viscous and consist largely of alkanes, alkenes and aromatic compounds. Where single plastics are pyrolysed, the wax or oil is similar to the basic structure from which the plastic was formed. Consequently, for example, polyethylene and polypropylene will produce mainly alkane and alkene waxes, whilst polystyrene will produce an oil consisting of alkane, alkene and aromatic product (Williams and Williams 1997). Polyvinyl chloride also produces an aromatic oil product when pyrolysed, in addition to hydrocarbon gases and hydrogen chloride gas.

Because of the range of compounds found in pyrolysis oils, there is some interest in using the oils as chemical feedstocks for speciality chemicals. For example, wood pyrolysis oils contain oxygenated compounds such as methylphenols (cresol), methyoxyphenol (guaiacol), furaldehyde (fufural) and methoxypropenylphenol (isoeugenol) which have applications in the pharmaceutical, food and paint industries (Bridgwater and Evans 1993; Stoikos 1991). Tyre oil contains dl-limonene used in the formulation of industrial solvents, resins and adhesives and as a replacement for chlorofluorocarbon for cleaning electronic circuit boards (Pakdel et al 1991). The oils also contain significant concentrations of benzene, xylenes, styrene and toluene used extensively in the chemical and pharma-ceutical industries (Williams et al 1995). The wax/oil-like product derived from the pyrolysis of mixed plastic waste has been successfully re-processed in petroleum refineries, using catalyst cracking to produce gasoline or plastics (Lee 1995).

To overcome some of the problems of high oxygen content, high viscosity, acidity and polymerisation associated with the oils derived from waste materials containing high oxygen contents, e.g., biomass and municipal solid waste, research has been undertaken to upgrade the oils (Stoikos 1991). The research has concentrated on the use of catalysts to produce a premium quality fuel or high value chemical feedstock. Two main routes to catalytic upgrading have been investigated: high pressure catalytic hydrotreatment, and low pressure catalysis using shape-selective catalysts of the zeolite type. Catalytic hydro-treatment of the oils with hydrogen or hydrogen and carbon monoxide under high pressure and/or in the presence of hydrogen donor solvents using transition metal catalysts, has produced oils similar in composition to gasoline and diesel. The upgrading takes place through deoxygenation and hydrocracking of the heavy fractions in the oil. Zeolite ZSM-5 catalysts have a strong acidity, high activities and shape selectivities, which convert the oxygenated oil to a light hydrocarbon mixture in the C_1–C_{10} range by dehydration and deoxygenation reactions. The oxygen in the oxygenated compounds of biomass pyrolysis oils is converted largely to CO, CO_2 and H_2O, and the resultant oil is highly aromatic with a dominance of single-ring aromatic compounds, and is similar in composition to gasoline (Bridgwater and Bridge 1991; Williams and Horne 1994).

Catalytic upgrading of pyrolysis oils has also been undertaken for tyres and plastics, also using Zeolite-type catalysts. The application of catalysts to the processing of scrap tyres and waste plastics is to enhance the oil product from the pyrolysis of such wastes to produce a premium-grade fuel or chemical feedstock. The derived oils from pyrolysis–catalysis of waste plastics have been shown to be very aromatic (Bagri and Williams 2002; Williams and Bagri 2004). Combined pyrolysis–catalysis of tyres with Zeolite-type catalysts produces an oil very high in concentrations of the high-value chemicals benzene, xylenes and toluene, such that it has the potential to be used as a chemical feedstock rather than a liquid fuel (Williams and Brindle 2002, 2003).

The gases produced from municipal solid waste and biomass waste pyrolysis are mainly carbon dioxide, carbon monoxide, hydrogen, methane and lower concentrations of other hydrocarbon gases. The high concentration of carbon dioxide and carbon monoxide is derived from the oxygenated structures in the original material, such as cellulose, hemicellulose and lignin. In addition, the gas contains a significant proportion of uncondensed pyrolysis oils. The pyrolysis of scrap tyre and mixed plastics waste produces higher concentrations of hydrogen, methane and other hydrocarbon gases, since the waste material is high in carbon and hydrogen compounds and has less oxygenated compounds. The gases have a significant calorific value, for example, the gas produced from the conventional pyrolysis of municipal solid waste has a calorific value of the order of $18 \, MJ/m^3$ and wood waste produces a gas of calorific value $16 \, MJ/m^3$ (Williams and Besler 1992, 1996). Tyre pyrolysis produces a gas of much higher calorific value, of about $40 \, MJ/m^3$ depending on the process conditions (Williams et al 1995). The high calorific value is due to the high concentrations of hydrogen and other hydrocarbons. By comparison, the calorific value of natural gas is about $37 \, MJ/m^3$. The high calorific value of pyrolysis gases means that the gas could be used to provide the energy requirements for the pyrolysis process plant.

The gases are produced from the thermal degradation reactions of the waste constituents as they breakdown, and also through secondary cracking reactions of the primary products. Consequently, higher gas yields are found where the products of pyrolysis spend a relatively longer time in the hot zone of the reactor rather than rapid quenching which produces higher oil yields and lower char yields. Also, higher gas yields are found at pyrolysis temperatures above about $750\,°C$, where accelerated cracking of the pyrolysis products occurs. At such high-temperature conditions, the main product is gas and some process descriptions term this process as gasification. However, pyrolysis implies the absence of any oxygen, whereas gasification implies that limited oxygen is supplied to the process as air, steam or pure oxygen, to gasify the waste.

A wide variety of pyrolysis technologies have been investigated for the pyrolysis of waste materials. Examples of technologies which have been used for waste pyrolysis include fluidised beds, fixed-bed reactors, ablative pyrolysis at hot surfaces, rotary kilns, entrained flow reactors and vacuum pyrolysis (Bridgwater and Evans 1993). The design is dictated by the type of pyrolysis being undertaken, for example, fast or slow heating rates to produce the targeted end product. Many are still at the pilot-scale stage, whilst others are at the commercial or near commercial stage. Boxes 6.1 and 6.2 show examples of pyrolysis systems for wastes.

Box 6.1
Ensyn Engineering Associates Inc., Canada, Biomass Pyrolysis System

The Ensyn Engineering Associates Inc., Canada, biomass pyrolysis system is a rapid heating fast pyrolysis system to produce an oil product for use as a chemical feedstock and fuel oil. The materials pyrolysed include wood waste, agricultural waste, heavy petroleum oils and tyre crumb. The system utilises a solid heat carrier of sand with a carrier gas, such as nitrogen, to carry heat into a turbulent vertical reactor. The sand material used to transfer heat to the feedstock is heated by external furnaces. Very rapid interaction occurs with the waste feedstock to produce fast pyrolysis primary products, which are then rapidly quenched to produce a liquid product. Temperature ranges from 400 to 950°C are possible and residence times from 50 to 1500 ms. Maximum feedstock throughputs are proposed in the range of 25 tonne/day. High liquid yields

Continued on page 336

Continued from page 335

up to 67% of the original feedstock have been recorded depending on the feedstock material. The products from the pyrolysis are quenched with water. The solid char is removed in a drop-out vessel which separates the char and sand and the sand is recycled back to the reactor. A condenser and filter/cyclone system and electrostatic precipitator is used to separate the liquids from the gaseous products.

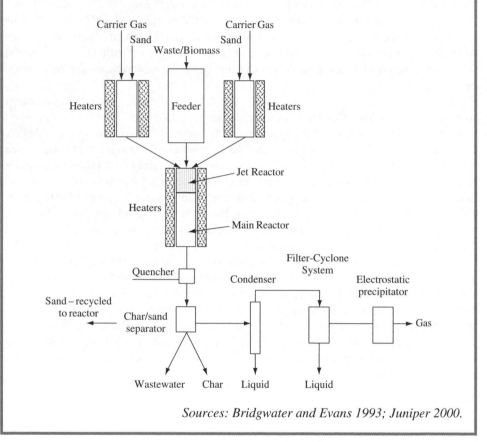

Sources: Bridgwater and Evans 1993; Juniper 2000.

Box 6.2

Pyrovac International Inc., Canada, Waste Pyrolysis System

The Pyrovac system developed in Canada is based on vacuum pyrolysis, where the pyrolysis process takes place under vacuum, such that the pyrolysis gases are rapidly removed from the hot zone of the reactor. The process is able to pyrolyse a wide range of wastes including municipal solid waste, biomass wastes, sewage sludge, scrap tyres, biomedical wastes, automotive shredder residue and petroleum residues. The process involves a drying stage, followed by pyrolysis under vacuum at 15 kPa and 500 °C. The wastes are conveyed through the reactor over horizontal plates heated by

Continued on page 337

> *Continued from page 336*
>
> molten salts to the process temperature of 500 °C. The molten salts are a mixture of KNO_3, $NaNO_2$ and $NaNO_3$ and are heated by the combustion of product pyrolysis gas. The product gases are removed by a vacuum pump and condensed to produce an oil product. The non-condensable product gases are combusted to produce energy. The oil product has been used as a fuel and as a chemical feedstock. Oil production from the Pyrovac process has produced 50 wt% oil from scrap tyres, 46 wt% from municipal solid waste, 45 wt% from forestry residues and 20 wt% from automotive shredder residue.
>
>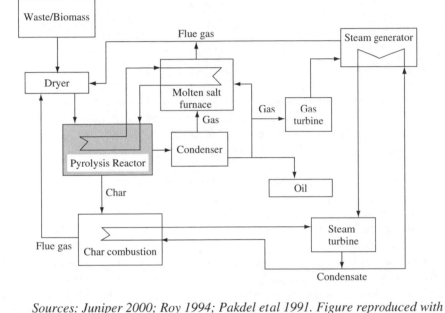
>
> *Sources: Juniper 2000; Roy 1994; Pakdel et al 1991. Figure reproduced with the permission of Juniper Consultancy Services Ltd, Uley, UK.*

6.3 Gasification

Gasification differs from pyrolysis in that oxygen in the form of air, steam or pure oxygen is reacted at high temperature with the available carbon in the waste to produce a gas product, ash and a tar product. Partial combustion occurs to produce heat and the reaction proceeds exothermically to produce a low to medium calorific value fuel gas. The operating temperatures are relatively high compared to pyrolysis, at 800–1100 °C with air gasification, and 1000–1400 °C with oxygen. Calorific values of the product gas are low for air gasification, in the region of 4–6 MJ/m³, and medium, about 10–15 MJ/m³ for oxygen gasification (Bridgwater and Evans 1993). Steam gasification is endothermic for the main char–steam reaction and consequently steam is usually added as a supplement to oxygen gasification to control the temperature. Steam gasification under pressure is, however, exothermic and steam gasification at pressures up to 20 bar and temperatures of between

700 and 900 °C produces a fuel gas of medium calorific value, approximately 15–20 MJ/m^3 (Bridgwater and Evans 1993; Rampling 1993). The product calorific values can be compared with natural gas at about 37 MJ/m^3.

The principle reactions occurring during gasification of waste in air are (Rampling 1993; Whiting 2003; Francis and Peters 1980)

$$C + O_2 \Rightarrow CO_2 \qquad \text{Oxidation – exothermic}$$
$$C + CO_2 \Rightarrow 2CO \qquad \text{Boudouard reaction – endothermic}$$

Overall

$$2C + O_2 \Rightarrow 2CO \qquad \text{Exothermic}$$

Steam gasification

$$C + H_2O \Rightarrow CO + H_2 \qquad \text{Carbon–steam reaction – endothermic}$$
$$C + 2H_2O \Rightarrow CO_2 + 2H_2 \qquad \text{Carbon–steam reaction – endothermic}$$
$$CO + H_2O \Rightarrow CO_2 + H_2 \qquad \text{Water–gas shift reaction – exothermic}$$
$$C + 2H_2 \Rightarrow CH_4 \qquad \text{Hydrogenation – exothermic}$$

In high pressure steam gasification, additional reactions include

$$CO + 3H_2 \Rightarrow CH_4 + H_2O \qquad \text{Hydrogenation – exothermic}$$
$$CO_2 + 4H_2 \Rightarrow CH_4 + 2H_2O \qquad \text{Hydrogenation – exothermic}$$

In practice there is usually some moisture present with the air which produces some hydrogen. In addition, the heating of the waste produces pyrolytic reactions and methane, and higher molecular weight hydrocarbons or tar are formed. When air is used, the non-combustible nitrogen in the air inevitably reduces the calorific value of the product gas by dilution. Therefore, the major components of the product gas from waste gasification are carbon monoxide, carbon dioxide, hydrogen and methane and, where air gasification is used, nitrogen will also occur as a major component.

For wastes and biomass, the development of gasification technologies has been via air or oxygen/steam gasification (Bridgwater and Evans 1993). Table 6.4 shows examples of the systems used for waste gasification (Rampling 1993; Whiting 2003). The characteristics of the gasifier system, the waste composition and operational conditions can give rise to tars, hydrocarbon gases and char; these are products of the incomplete gasification of the waste. The characteristics of the gasifier have most influence on the quality of the product gas, for example, down-draft gasifiers have all the products of gasification passing through a high temperature zone and with high turbulence. This arrangement results in a high conversion of the pyrolysis intermediates and a gas with a low tar content, whereas, the up-draft gasifier produces gas which is hot and, when passing up through the down-flowing waste, produces pyrolysis reactions and a higher concentration of tar in the final product gas. Fluidised bed reactors produce intermediate pyrolysis tar products which are passed out of the fluid bed into the

Table 6.4   *The main types of waste gasifier reactor systems*

Updraft gasification
Air flows up from the base of the reactor with the waste flowing down counter-current to the air flow. Gasification takes place in a slowing moving 'fixed' bed. Because the moisture, tar and gases generated do not pass through a hot bed of char there is less thermal breakdown of the tars and heavy hydrocarbons, therefore the product gas is relatively high in tar. The tars may be condensed and recycled to increase thermal breakdown of the tars.

Downdraft gasification
The air and the waste flow co-currently down the reactor. Gasification takes place in a slowing moving 'fixed' bed. There is an increased level of thermal breakdown of the tars and heavy hydrocarbons as they are drawn through the high-temperature oxidation zone, producing increased concentrations of hydrogen and light hydrocarbons. The air/steam or oxygen is introduced just above a 'throat' or narrow section in the reactor, which influences the degree of tar cracking.

Fluidised bed gasification
Waste is fed into the fluidised bed at high temperature. The fluidised bed may be a bubbling bed where the solids are retained in the bed through the gasification process. Alternatively, circulating beds may be used with high fluidising velocities; the solids are elutriated, separated and recycled to the reactor in a high solids/gas ratio resulting in increased reaction. Twin fluidised-bed reactors may be used where the first bed is used to gasify the waste, and the char is passed to a separation unit and then to a second fluidised bed where combustion of the char occurs to provide heat for the gasifier reactor.

Entrained flow gasification
A widely used technology, where the gasification reactions take place in suspension in an entrained flow of gas. The waste feedstock is introduced into a vertical reactor with steam and oxygen. The residence time is very short and the gasification takes place at high temperature and pressure. The waste can be in liquid or solid form, but where solids are used, the particle size must be small. The entrained flow gasifier tends to produce a high conversion of the waste to produce a low tar content gas.

Rotary kiln gasification
Rotary kilns involve a slowly rotating, inclined, ceramic-lined cylinder, which slowly moves the waste down the cylinder, whilst the waste is gasified. Gasification is with air, steam or oxygen. The residence time is much longer than for fluidised bed and entrained flow gasification reactors.

Sources: Rampling 1993; Whiting 2003; Juniper 2000.

freeboard by the fluidising gas. As the tars pass up through the hot freeboard of the fluidised bed, some thermal cracking of the tars to gases may occur, but an overall gas tar content similar to that of an up-draft gasifier is usually found (Rampling 1993). The entrained flow gasifier and fluidised bed gasifier designs require the waste to be processed to produce fine granules to enable efficient feeding of the waste to the gasifier. For wastes such as sewage sludge, this is not such a problem, but for municipal solid waste and forestry residues, pre-treatment of the waste is required. For rotary kilns, no such pre-treatment and size reduction of the waste is required. Table 6.5

Table 6.5   *Product gas characteristics from different gasifier types*

Gasifier type	Calorific value of the product gas (MJ/m^3)	Gas Quality[2]	Efficiency (%)
Downdraft–air	4.0–6.0	****	70–90
Downdraft–O$_2$	9–11	****	60–80[1]
Updraft–air	4.0–6.0	***	75–95
Updraft–O$_2$	8–14	***	65–85[1]
Fluidised bed–air	4–6	***	70–90
Fluidised bed–O$_2$	8–14	***	60–75[1]
Fluidised bed–steam	12–18	***	70–80
Circulating fluidised bed–air	5–6.5	**	75–95
Circulating fluidised bed–O$_2$	10–13	***	70–80[1]
Twin fluid bed	13–20	***	65–75
Cross flow–air	4.0–6.0	*	75–95
Horizontal moving bed–air	4.0–6.0	**	60–70
Rotary kiln–air	4.0–6.0	**	70–85
Multiple hearth	4.0–6.0	**	60–80

[1] Oxygen system efficiencies include a notional energy used for oxygen production.
[2] Gas quality is a relative assessment in terms of tars and particulates in raw gas: *, worst; *****, best.
Source: Bridgwater and Evans 1993.

shows the characteristics of gasification product gas from different gasifier types (Bridgwater and Evans 1993).

Utilisation of the gaseous product is often by direct combustion in a boiler or furnace. The heat energy is used for process heat or to produce steam for electricity generation. However, the raw gas will contain tar, char and hydrocarbon gases and therefore the boiler or furnace burner system must be able to tolerate these contaminants and not be susceptible to fouling or clogging. In addition, gasification of heterogeneous waste, such as municipal solid waste, produces a gas which can vary in composition, and consequently, the burner system of the boiler or furnace should be able to handle a range of gas compositions and calorific values. The advantages of direct combustion systems are that the gas does not have to be cleaned to any great extent before combustion and that the gases are used hot, maintaining the sensible heat in the system.

Where the utilisation of the product gas is into gas turbines or internal combustion engines to generate power or electricity, then the gas has to be cleaned to a higher specification than in direct combustion systems. Piping of the gas to the combustion unit requires that it be cooled and cleaned before utilisation to prevent pipe corrosion and deposition of tars and water. Removal of particulate material is by cyclones and bag filters, and tar removal is by secondary cracking at high temperature or catalyst cracking at lower temperatures. Gas turbines have been suggested as a suitable utilisation system for electricity generation, particularly for pressurised waste gasifiers. However, the fuel gas specifications for gas turbines are very stringent (Bridgwater and Evans 1993). Boxes 6.3 and 6.4 show examples of gasification systems for wastes (Bridgwater and Evans 1993; Juniper 2000).

Box 6.3
The Lurgi Circulating Fluidised Bed Waste Gasifier

The Lurgi circulating fluidised bed gasifier has been used for the gasification of processed municipal solid waste in the form of refuse-derived fuel, wood bark and waste wood, coal and petroleum coke. The full throughput capacity of the system is approximately 6500 kg/h and uses air as the reactant gas, although oxygen or steam can be used. The product gas is of low calorific value, between 4 and 6 MJ/m^3, which has been used as fuel for firing a lime kiln. The figure shows a schematic diagram of the commercial system used to gasify forestry residues in the form of wood bark. The wet bark is dried to approximately 10% moisture content in a rotary drier utilising the derived gases from the gasifier as fuel to provide the heat. The dried bark is fed to the gasifier at an operating temperature of 800 °C and atmospheric pressure. Solids removed from the gasifier are captured by the cyclones and are returned to the fluidised bed. Ash formed at the base of the gasifier is cooled and removed for disposal. The product gas is then used directly to fuel a lime kiln using a multifuel burner. The product gas composition for wood bark gasification is: hydrogen, 20.2%; carbon monoxide, 19.6%; carbon dioxide, 13.5%; hydrocarbons, 3.8%; nitrogen, 42.9%. Nitrogen content is high due to the use of air as the source of oxygen, and consequently the product gas has a low calorific value.

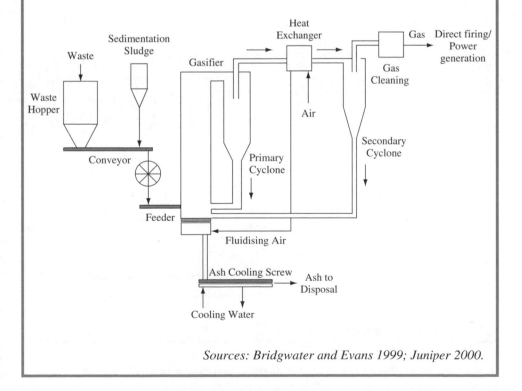

Sources: Bridgwater and Evans 1999; Juniper 2000.

Box 6.4

The ALSTOM Power TwinRec System

The acceptable particle size for feeding is 300mm which, for wastes such as automotive shredder residues, means no pre-treatment is required. Waste is gasified in an air-blown, internally circulating, fluidised bed gasifier, operating at temperatures of between 500 and 600 °C. The design of the distributor within the fluidised bed provides more turbulence and improved combustion efficiency, but at lower fluidisng velocity than a circulating fluidised bed. The somewhat low gasification temperature provides for a slower, more easily controllable process. The high specific thermal load enables a small gasifier cross-section. The synthesis gas and entrained carbonaceous char are passed to a cyclonic combustion chamber, where combustion takes place with added secondary air at between 1350 and 1450 °C. The high temperatures of the cyclonic combustor melt the flyash to form a molten slag which is then quenched to produce a non-leachable granulate. Energy is recovered as steam for either electricity generation or district heating. Bottom ash from waste gasification is recovered from the base of the fluidised bed where recovery of metals, such as iron, copper and aluminium, takes place. The technology produces a reduced mass flow of the flue gas, allowing for a reduction in the size of the steam boiler and emissions control system, located after the combustor.

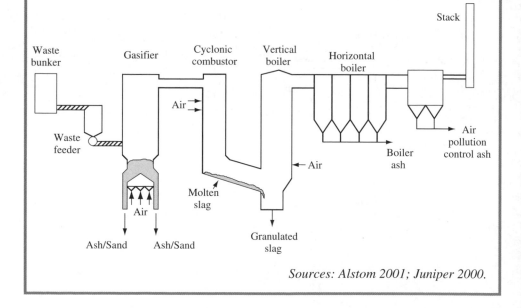

Sources: Alstom 2001; Juniper 2000.

6.4   Combined Pyrolysis–Gasification

Some modern developments in thermochemical processing of waste have utilised both pyrolysis and gasification in combined technologies, which may then involve a further combustion step to combust the gases produced in the first two stages. Such pyrolysis/

gasification/combustion technologies are, in effect, incinerators, but each step is separated into a separate temperature and pressure controlled reactor rather than in an incinerator, where the three thermal degradation steps are combined in a one-step grate combustion system. The de-coupling of the thermal degradation steps has the advantage of flexibility in determining which targeted end product is best suited to each application. Further advantages include the option that the product gas may be cleaned to remove acid gases prior to the combustion of the gas for energy recovery. This results in reduced high-temperature corrosion within the energy recovery system. Also, pyrolysis/gasification systems produce significantly reduced gas volumes for clean-up compared with a conventional waste incinerator, resulting in scale-down of the gas cleaning system and a consequent reduction in cost. In addition, as an alternative, the gas product may be further cleaned and used as a chemical feedstock (Williams 2001; European Commission 2004; Whiting 2001, 2003).

More than 100 pyrolysis/gasification process technologies have been identified worldwide, of which 60 have been technologically and economically evaluated in detail (Whiting 2001; Juniper 2000). Figure 6.4 shows the different combinations of different processes involved under the category of pyrolysis, gasification and combined pyrolysis–gasification systems. Several pyrolysis, gasification and combined systems for processing wastes have been described (Juniper 2000; European Commission 2004). The commercial

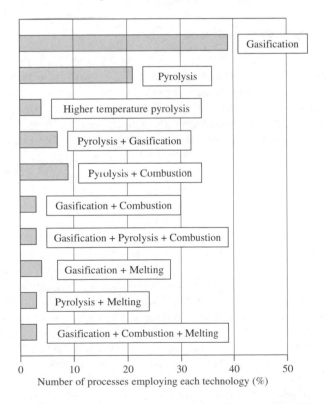

Figure 6.4 *Types of technology combinations employed in combined thermal processing systems (60 types analysed). Sources: Whiting 2003; Juniper 2000. Reproduced with the permission of Juniper Consultancy Services Ltd, Uley, UK.*

development of the various systems ranges from pilot scale through to commercialisation. Clearly, the range and complexity of such systems and the lack of a proven long-term track record, inhibits the full-scale commercial development of pyrolysis–gasification systems as an alternative to mass burn incineration. It has been suggested that the increasing complexity of combined pyrolysis–gasification systems, which may then also include combustion and ash melting, have developed because of the need to handle more complex and less homogeneous waste streams (Juniper 2000). The processes that include ash melting, aim to ensure the stability of the solid residue output and thereby maximise the recycling potential of the waste in the process. Systems where the second stage consists of combustion, such as pyrolysis–combustion or gasification–combustion aim to maximise the energy recovery without the need to clean the product gas to any extent. The complex systems such as pyrolysis–gasification–ash melting, followed by full gas cleaning, aim at producing a clean product gas with low levels of particulate, tars and acid gases, suitable for use in combined-cycle gas turbine power generation (Juniper 2000).

Box 6.5 shows the combined pyrolysis–gasification Noell process, developed in Germany (Leipnitz 1995; Juniper 2000). Box 6.6 shows the combined pyrolysis–gasification Thermoselect process (Thermoselect 2003; Juniper 2000; Drost and Kaiser 2001)

Box 6.5
The Noell Waste Treatment Process

The Noell waste treatment process is based on a combination of pyrolysis and entrained flow gasification. The system is designed to treat domestic waste, sewage sludge, hazardous waste and biomass with throughputs of up to 100 000 tonnes per year. The figure shows a schematic diagram of the system. The pyrolysis section consists of an indirectly heated, gas fired rotary kiln, operated under an inert gas atmosphere, the gas being derived from the waste treatment process. Shredded waste is fed into the pyrolysis reactor at approximately 550 °C and solids retention times in the kiln are of about 1 h. The char product is separated, and ferrous and non-ferrous metals are separated from the char. The char is then ground and passed to the entrained flow gasifier. The pyrolysis gases and oil, water and dust carry-over are quenched. The condensed oil, dust carry-over and gas, plus the ground char, are passed to the gasifier. The gasifier is an entrained flow type where an inert solid material of particle size <1 mm and high loading of about 350 kg/m^3 of solids is fed, with the pyrolysis products and oxygen, into a burner operating a sub-stoichiometric conditions. High temperatures of the order of 1400 °C are produced in the gasifier. The gasifier reaction under partial oxygenation conditions, i.e., sub-stoichiometric, generates a gas composed of over 80% carbon monoxide and hydrogen. Any solid inert material is converted to an ash slag, because of the high temperatures involved, and is quenched and granulated. The resultant gas is cooled, scrubbed and utilised for energy recovery. The high gasifier temperatures completely destroy toxic hydrocarbon compounds and, because the operating conditions, are reducing the de-novo synthesis of dioxins and furans is eliminated, thus reducing the costs of gas clean-up.

Continued on page 345

Continued from page 344

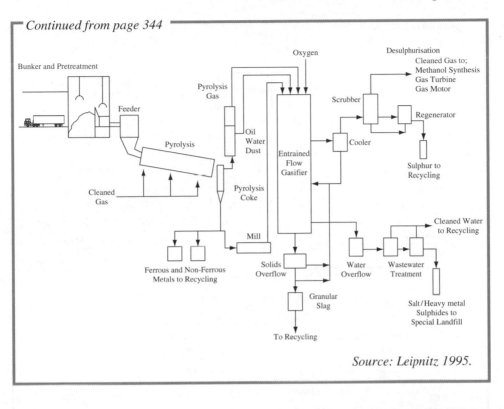

Source: Leipnitz 1995.

Box 6.6
The Thermoselect Process

Untreated waste is compacted at pressures of over 1000 tonnes to 10% of its original volume to form plugs of waste, devoid of most of the air. These are then fed to the pyrolysis reactor, which is heated indirectly at 600 °C. The resultant organic pyrolysis gases, vapours and char are fed to a high-temperature gasification chamber operated at approximately 1200 °C, with oxygen as the gasifying agent. The product synthesis gas is shock-quenched to 90 °C and undergoes several cleaning steps to produce a clean gas suitable as a chemical feedstock or for energy recovery applications. At the base of the gasification reactor, temperatures of 2000 °C produce melting of the metal and mineral components of the waste. The liquid melt flows to a homogenisation chamber at 1600 °C, where sufficient residence time allows the separation of two phases, a metal alloy and a mineral phase. Rapid quenching of the melt produces a granulate mineral material for use in road building, construction and aggregates, and a metal alloy for recovery of metals. Other products recovered are sulphur, from the gas cleaning process, and zinc and lead concentrate, sodium chloride and purified water from the water treatment process. The high temperatures involved in the process and rapid cooling of the gases ensures that very low levels of dioxins and furans are reported from the process emissions.

Continued on page 346

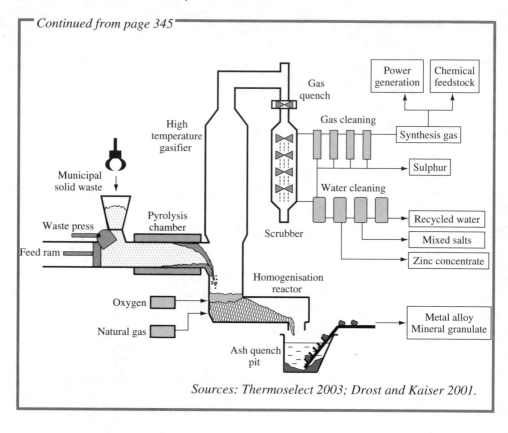

Continued from page 345

Sources: Thermoselect 2003; Drost and Kaiser 2001.

6.5   Composting

Composting is the aerobic, rather than anaerobic, biological degradation of biodegradable organic waste such as garden and food waste. Composting is a relatively fast biodegradation process, taking typically about 4–6 weeks to reach a stabilised product. Composting is practised on a small scale at the individual household level, and on a large-scale via composting schemes, where the organic waste collected from parks, household garden waste collected from civic amenity sites and garden and food waste collected directly from households in separate kerbside collections, is composted at large central facilities. The degraded product is a stabilized product which is added to soil to improve soil structure, especially for clay soils, or which acts as a fertilizer improving the nutrient content, or as a much being used to retain moisture in the soil. Compost is also used for land restoration and landscaping, where it is used as a mulch (McLanaghan 2002).

Composting removes a large part of the organic biodegradable waste from the waste stream and in this way helps to fulfil the obligations placed on member States of the EU in meeting the requirements of the EC Waste Landfill Directive (1999). The Directive seeks to reduce the amount of biodegradable waste sent to landfill to 75% of the 1995 levels by 2006, 50% of 1995 levels by 2009 and 35% of 1995 levels by 2016 (EC Waste Landfill Directive 1999). Consequently the biodegradation process operating in landfills will be reduced, resulting in less generation of the landfill gases, carbon dioxide and methane which are both 'greenhouse gases'.

The organic waste used for large-scale composting schemes can be variable and may also contain contaminants due to incorrect sorting. This in turn may lead to a variable product composition, and quality may be difficult to guarantee for the end user. Acceptance of the product in the market place relies on a wide range of criteria including price, quality and consistency of the product, and guarantees that the product will be free of contaminants such as heavy metals, glass and other inert materials and also free from plant and animal pathogens. The impact on consumer confidence which would occur if a contaminant such as glass or a hypodermic syringe were found in composted municipal solid waste, would be extremely adverse (Border 1995).

A high proportion of municipal solid waste in Europe is characterised as biodegradable, representing from 66% to more than 90%, depending on country (European Environment Agency 2002). The biodegradable fraction of municipal solid waste includes food and garden waste, paper and board, wood and some textiles and therefore has the potential to be composted. Figure 6.5 shows the fraction of municipal solid waste that is collected as a separate compostable fraction and also the fraction of biodegradable waste which is collected as a separate fraction for several countries within Europe (Hogg et al 2002). Whilst the percentage of municipal solid waste that is collected for composting is relatively small for many countries, for example, 21% for Austria, 5.3% for Italy, and 2% for the UK, the fraction of the biodegradable/compostable waste that is collected is much higher. For example, in Austria, 75% of the compostable waste present in municipal solid waste is collected as a separate fraction; for Germany, the figure is 78%; Sweden, 27% and for the UK it is 6% (Hogg et al 2002). It should also be considered that, while home composting is a major activity in many European countries, it is very variable across Europe. For example, Figure 6.6 shows the percentage of households undertaking home composting for several countries across Europe (Hannequart and Radermaker 2003).

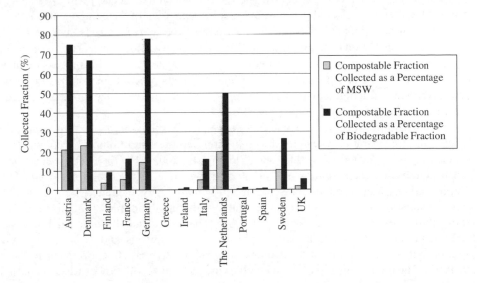

Figure 6.5   *The fraction of municipal solid waste and biodegradable waste collected as a separate compostable fraction. Source: Hogg et al 2002.*

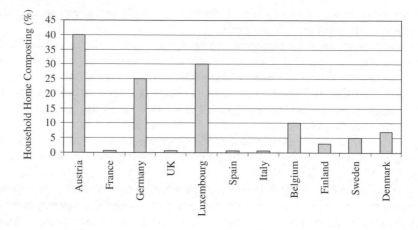

Figure 6.6 *Household home composting in selected European countries. Source: Hannequart and Radermaker 2003.*

In terms of the management of municipal solid waste and other biodegradable wastes, it is the large-scale centralised composting facility which is viewed as an alternative to waste landfill and incineration. The composting process for municipal solid waste involves a number of stages which are shown in Figure 6.7 (McLanaghan 2002). The initial stage involves collection of the waste as source-segregated waste by the householder, where segregation of the biodegradable fraction is made through kerbside collection or 'bring' schemes. Pre-processing of 'green waste', i.e., garden waste from parks and civic amenity sites may only require shredding or pulverisation. Source-segregated organic waste, segregated by the householder and collected in kerbside or 'bring' systems would require a greater degree of pre-processing to remove contaminants and poorly segregated wastes. Mixed municipal solid waste would require separation of the component waste on the scale of a materials recovery facility, to remove the inert materials such as glass, ferrous and non-ferrous metals, etc. The input biodegradable waste is delivered to a reception and storage area prior to shredding and homogenisation. This process reduces the size of the waste and produces a more homogeneous product for composting. The separated organic fraction may be shredded or pulverised to give a size range of between 1 and 10 cm^2 depending on the type of waste.

The composting process is aerobic and consequently relies on a plentiful supply of oxygen. Regular aeration is required to maintain aerobic conditions. The composting process may be characterised by three stages (Swan et al 2002). The first stage is characterised by increasing temperatures and involves a high rate of microbiological activity. Simple carbohydrates and proteins are readily biologically degraded by mesophilic micro-organisms, followed by thermotolerant and thermophilic micro-organisms as the temperature rises above 45 °C (Swan et al 2002). The second stabilisation stage involves biodegradation of the waste by thermophilic micro-organisms and is an exothermic process so that temperatures in the compost pile can reach up to 70 °C (Warmer Bulletin 29, 1991). The high-temperature stage involves the thermal destruction of weed seeds and

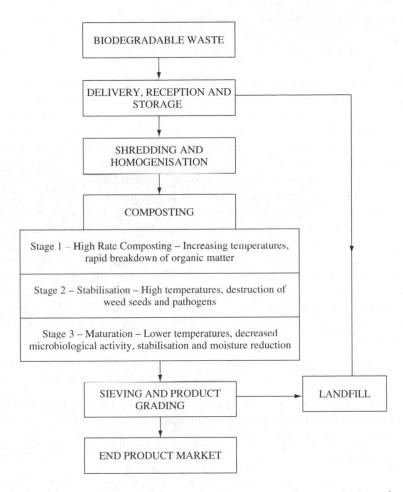

Figure 6.7 Schematic diagram of the composting process. Sources: McLanaghan 2002; Swan et al 2002.

pathogenic micro-organisms. The compost also includes a third maturation stage and is characterised by lower temperatures. The maturation stage involves further biodegradation of the intermediate compounds and may take several weeks for completion (Warmer Bulletin 29, 1991). Box 6.7 shows the biological processes operating during composting (de Bertoldi et al 1983). The final stages of composting would be processes such as sieving and grading to remove un-composted materials and contaminants such as glass, plastics and metals and then size reduction and screening. Whilst there is great interest in composting organic waste from domestic sources and also parks and garden waste deposited at civic amenity sites, there are other major sources of organic waste suitable for composting. For example, agricultural waste, sewage sludge, forestry waste and food waste (Border 1995).

Box 6.7
The Biological Processes Operating During Composting

The table below shows the organic fraction composition of MSW. The simple carbon compounds, such as soluble sugars and organic acids, are easily metabolised and mineralised by heterotrophic and heterogeneous micro-organisms. High metabolic activity and exothermic processes produce an increased temperature in the compost heap which, because of the low thermal conductivity, cannot dissipate the heat and consequently temperatures rise. This rise in temperature allows only the thermophilic micro-organisms to be active. Cellulose, pectin, starch and lignin are degraded later by fungi and actinomycetes; the decrease in temperature and also moisture and pH, increases the activity of the fungi and actinomycete micro-organisms. Cellulose decomposition is intense throughout the process, but particularly during the final stages, mostly through degradation by eumycete micro-organisms. The degradation of lignin is restricted to the basidiomycete group of fungi. The nitrogen content is decreased via ammonia formation and volatilisation during the composting process. However, the loss of carbon dioxide and water from carbon- and hydrogen-containing constituents in the waste, results in an overall decrease in the C:N ratio, representing a relative (to C) increase in nitrogen overall. In addition, the nitrogen content later slightly increases by nitrogen fixation from the micro-organisms.

The main factors influencing composting are listed below.

1. Suitable oxygen content to maintain aerobic conditions: a minimum oxygen content in the compost of 18% is recommended.

2. Temperature: maximum micro-organism activity is observed in the temperature range, 30–35 °C.

3. Moisture content: below a minimum 40% moisture content, biodegradation is significantly reduced; high moisture contents are also to be avoided since they occupy intraparticle spaces and thereby produce anaerobic conditions.

4. pH range of the waste material: optimal composting is achieved in the pH range 5.5–8. Bacteria prefer a near-neutral pH, whereas fungi develop better in a slightly acidic environment.

5. C:N ratio of the waste material: optimal C:N ratio in the starting waste material is about 25, higher values resulting in a slow rate of decomposition, and lower ratios resulting in nitrogen loss. The organic fraction of MSW has a C:N ratio between 26 and 45 and for raw sewage sludge it is 7–12.

6. Size range of waste material: shredding of the starting waste material increases the surface area and results in enhanced rates of composting.

Continued on page 351

Continued from page 350

Constituent	%
Volatile matter	70–90
Protein	2–8
Lipids	5–10
Total sugar	5
Cellulose	35–55
Starch	2–8
Lignin	3–8
Phosphorus	0.4–0.7
Potassium	0.7–1.7
Crude fibre	35–40

Source: de Bertoldi et al 1983.

The composting stage involves the biodegradation of the sample under aerobic conditions and therefore requires aeration of the waste. Aeration is achieved by several methods and as these increase in sophistication and control of the process, they also increase in cost. Such processes include the 'windrow' system, aerated static piles, in-vessel systems and vermicomposting.

The 'windrow' system involves the biodegradable waste being piled into elongated conical heaps about 2 m high, about 3–4 m in width and 50 m in length (Figure 6.8). The waste is turned periodically by mechanical turning (Swan et al 2002; Warmer Bulletin 29 1991; Diaz et al 1993). Turning rates vary from one turn/day in the early stages of composting to one turn/five days towards the end of the process. The turning process serves to introduce fresh air and release trapped heat, moisture and stale air. The windrows are usually placed on a gravel bed to aid the collection of any leachate that may be formed. Windrows would normally be arranged in rows and the mechanical turning vehicle moves up and down each row turning the waste, thereby periodically fully aerating the pile.

Forced aeration systems involve air being blown or sucked through the pile of composting waste by a fan. The compost pile is located on an aeration block and remains undisturbed. Air is distributed via a perforated pipe covered with a porous base material, which is usually finished compost which acts as a filter and even distributor of the air. The waste pile for composting is constructed over the filter and perforated pipe. Typical forced aeration systems have pile heights of 2–3 m, or 2–6 m in width and up to 30 m in length (Diaz et al 1993). The air is passed through the pile either continuously or periodically. If the air is drawn down through the pile, the odours from the compost are contained in the system, allowing for control and treatment if required. Where the air is blown up through the compost pile, this serves to transfer the heat from the inner pile to the outer regions. Figure 6.8 shows a schematic diagram of the windrow and forced aeration systems.

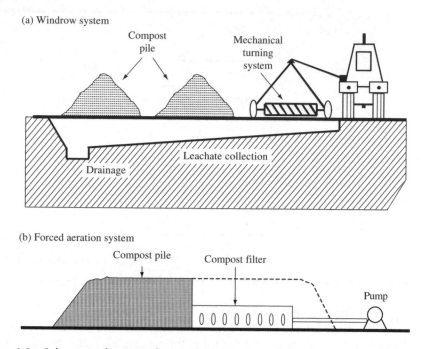

Figure 6.8 *Schematic diagram of a windrow system and forced air system for aeration of compost. Reproduced by permission of R.C. Strange.*

In-vessel systems allow closer control of temperature, moisture, aeration and waste-mixing rates (Diaz et al 1993). These include containers, silos or towers, enclosed halls, tunnels, rotating drums or reactor tank systems (Swan et al 2002). Containers are usually small-scale systems for food processing, and catering wastes operating on a batch system. Silos or tower systems are vertical units operating on a continuous basis, where the biodegradable waste is fed to the top of the silo and composting takes place as the waste moves down the tower. The compost is collected at the bottom of the tower after several days' duration, followed by a maturation stage. Composting in enclosed halls involves large floor areas where composting takes place inside a purpose-built composting building. Tunnel composting systems are large-scale systems which may be either batch or continuous and may also involve some form of mechanical agitation. Rotating drum systems involve the waste being placed in a long rotating drum, typically 3–4 m in diameter and 50 m long, combined with forced aeration (Swan et al 2002). The air passes through the rotating drum while the waste is continuously stirred and tumbled. Residence times in the drum are typically only several days and would require a subsequent maturation stage to reach completion of the composting process. Figure 6.9 (a) shows a typical rotating drum composter (Diaz et al 1993). Composting tank reactors are similarly more sophisticated composting systems, and one example, a reactor system, is shown in Figure 6.9 (b) (Diaz et al 1993). The waste is stirred by a series of augers which are perforated and allow air to be blown into the composting waste pile. The augers are located on a stirring arm.

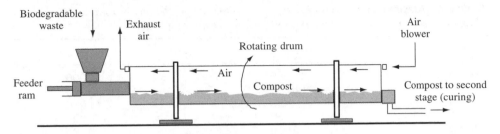

(a) Rotating drum composting system

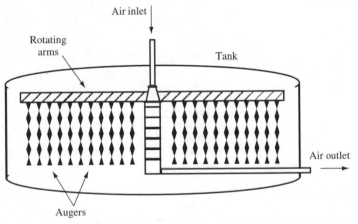

(b) Reactor tank composting system

Figure 6.9 *Schematic diagrams of the rotating drum composting system and the reactor tank composting system. Source: Diaz et al 1993. Reprinted with permission of CRC Press, LLC.*

Vermicomposting is the composting of biodegradable waste using selected species of earthworms. The process takes place in long troughs in which the temperature is kept below 35 °C. The vermicomposting process relies on the earthworms to mix, aerate and fragment the waste, combined with the biodegradation process of the micro-organisms.

Since the composting process is a biodegradation process it consequently leads to the formation not only of compost, but other products which may require control and treatment. Leachate may form in cases of high moisture content. The leachate will have many of the properties and composition of leachate generated in the early stages of landfill. The leachate is allowed to collect in channels and is discharged to sewer or is treated on site depending on the level of leachate generated. Gaseous emissions from the composting process consist mainly of volatile organic compounds. The emissions are often malodorous and potentially toxic.

Table 6.6 shows some gaseous emissions from the composting of municipal solid organic waste using a variety of composting systems from windrows, forced aeration, to reactor systems (Eitzer 1995). The results represent the maximum observed concentrations and in most cases represent localised high concentrations, since gas samples were

Table 6.6   Maximum emissions of selected volatile organic chemicals from direct sampling of waste and compost from municipal solid waste composting facilities

Component	Maximum emission ($\mu g/m^3$)	Limit value[1] ($\mu g/m^3$)
Trichlorofluoromethane	915 000	5 620 000
Acetone	166 000	1 800 000
Carbon disulphide	150	31 000
Methylene chloride	260	174 000
2-Butanone	320 000	590 000
Chloroform	54	49 000
1,1,1-Trichloroethane	15 000	1 900 000
Carbon tetrachloride	290	31 000
Benzene	700	32 000
Trichloroethene	1300	270 000
2-Hexanone	6600	20 000
Toluene	66 000	188 000
Tetrachloroethene	5600	339 000
Chlorobenzene	29	46 000
Ethylbenzene	178 000	434 000
m, o-Xylene	15 000	434 000[2]
p-Xylene	6900	434 000[2]
Styrene	6100	213 000
Isopropyl benzene	370	246 000
1,3,5-Trimethylbenzene	2200	123 000[2]
1,2,4-Trimethylbenzene	1000	123 000[2]
1,4-Dichlorobenzene	90	451 000
Naphthalene	1400	52 000

[1] American Conference of Governmental Industrial Hygienists, threshold limit values-time weighted average for workplace air.
[2] Value for the sum of all the isomers of each compound.

Source: Eitzer 1995. Copyright 1995. Reproduced by permission of the American Chemical Society.

taken directly from the waste and compost piles. However, the results show that, whilst a wide range of emissions were detected, they were below permissible workplace exposure limits.

A further aspect of airborne emissions from the composting process is the likely presence of microbiological organisms. The putrescible-rich fraction of municipal solid waste has been composted and airborne emissions monitored for bacteria, fungi and actinomycetes (Newport et al 1993). Swan et al (2002) also report that the main organisms associated with composting sites are bacteria, fungi and actinomycetes. Before composting, moulds and enteric bacteria were found to be the most common microbial species present in the air emissions from mechanically separated waste. During the composting process, the main organisms were fungi and actinomycetes, both of which produce large numbers of spores and can lead to acquired allergic responses. During processing after composting, airborne particles increased markedly, with identified species including Aspergillus fumigatus and Penicillilium spores being identified. Both of which are known to cause allergies. Swan et al (2002) have comprehensively reviewed the health effects associated with composting various wastes. They suggest that the potential ill-health effects of composting sites are exacerbated by poorly managed sites. In addition, they report that the concentration of bioaerosols reach background levels at a distance of approximately

100–500 m from the composting site. Evaluation of the data associated with the potential health impacts of composting sites have reported that composting facilities do not pose any unique endangerment to the health and welfare of the general public (Milner et al 1994). However, Swan et al (2002) suggest that the exposure of the bioaerosols generated from the composting process may represent a risk, and therefore more research is required.

A key element of the production of compost from biodegradable waste is the issue of the quality of the final end product in relation to compost derived from traditional non-waste sources which are well established in the market place. The proposed EC Biological Treatment of Biowaste Directive (2001) includes a proposal for the establishment of standards for two classes of compost and one for stabilised biowaste (Hogg et al 2002). The compost standards relating to quality and the stabilised biowaste relating to a lower quality product are only suitable for those applications not involving food and fodder production, such as landscaping, road construction, golf courses and football pitches. There are also restrictions on the amount and frequency of application of such stabilised biowaste. The proposed standards are set out in Table 6.7. The standards may be compared with the concentrations of heavy metals in compost derived from different waste sources shown in Table 6.8 (Hogg et al 2002). Table 6.9 shows examples of large-scale

Table 6.7 *Compositional limit values for heavy metals and other components of compost and stabilised biowaste from the proposed EC Directive on the Biological Treatment of Biowaste (mg/kg dry matter)*

Component	Compost class I	Compost class II	Stabilised biowaste
Impurities >2 mm	<0.5%	<0.5%	<3.0%
Gravel and stones >5 mm	<5%	<5%	—
Cadmium	0.7	1.5	5
Chromium	100	150	600
Copper	100	150	600
Mercury	0.5	1	5
Nickel	50	75	150
Lead	100	150	500
Zinc	200	400	1500
Polychlorinated biphenyls	—	—	0.4
Polyaromatic hydrocarbons	—	—	3

Sources: EC Biological Treatment of Biowaste Directive 2001; Hogg et al 2002.

Table 6.8 *Heavy metal composition of compost from different waste sources (mg/kg dry matter)*

Component	MSW compost	Biowaste compost	Green waste compost
Cadmium	4.5	0.9	1.4
Chromium	122.0	28.5	45.6
Copper	161.8	95.9	50.8
Mercury	1.6	0.6	0.5
Nickel	59.8	23.8	22.4
Lead	318.1	85.5	87.3
Zinc	541.5	288.5	186.4

Source: Hogg et al 2002.

Table 6.9   *Examples of large-scale separate collection and composting schemes in Europe*

Country	Scheme name	Households (1000s)	Inhabitants (1000s)	Biodegradable waste collected (tonnes/year)	Compost produced (tonnes/year)
Spain	Baix Camp	8	25	4000	360
	Barcelona	55	137	10 700	1900
	Montejurra	23	52	10 000	2000
France	Gironde	20	—	36 000	24 000
	Niort	12	—	8500	4500
Ireland	Limerick	2.8	—	950	450
Italy	Monza	—	119	10 000	—
	Padova	—	205	16 500	—
UK	Castle Morpeth	20.4	—	5000	3000

Source: Directorate General Environment 2000.

collection and centralised composting schemes for various countries of Europe (Directorate General Environment 2000). It has been estimated that more than 7.5 million tonnes of composting capacity is available in Europe (De Baere 2000). Box 6.8 presents examples of large-scale separate collection and composting schemes (Directorate General Environment 2000).

Box 6.8

Examples of Large-scale Separate Collection and Composting Schemes

Gironde Composting Scheme, France

The Gironde composting scheme in France in centred on the Sud Basin district in the department of Gironde. The scheme involves 20 000 households. Biodegradable kitchen and garden waste are collected separately and composted centrally. Householders can also take garden waste to waste collection centres. The composting facility has a total capacity of 40 000 tonnes per year and produces between 24 000 and 36 000 tonnes of compost per year. The facility covers an area of 20 hectares with the composting unit covering 14 000m^2. The process involves a reception, storage and crushing area. The composting is achieved in open air windrow systems over a two–three-week period, involving watering, turning, ventilation and humidifcation. The product is sieved using two decreasing mesh sizes to remove contaminants and oversize particles, followed by a final maturation stage of between five and six weeks.

Barcelona Composting Scheme, Spain

The Barcelona composting scheme in Spain involves approximately 137 000 inhabitants in 55 000 households over four municipalities covering the southern part of Barcelona. The scheme consists of a separate household kerbside collection of biodegradable domestic waste using 10 litre bins, which are deposited in dedicated kerbside containers and are then collected by the local municipalities. In addition, separate collection of

Continued on page 357

Continued from page 356

biodegradable waste from large-scale producers, such as food markets, is carried out using compactor lorries. Approximately 10 700 tonnes of such waste is treated each year, generating 1700 tonnes of compost in a centralised facility. Lorries deliver the biodegradable waste to the composting plant where the waste is mixed with sewage sludge and wood waste. The composting process is via three composting tunnels. The tunnels take 2–3 days to fill, and the mix spends between 10 and 14 days composting. The composting conditions are controlled in terms of air flow, temperature, oxygen content, etc., by means of sampling probes and air ventilators. The mix is matured in piles in the open for two months with regular turning and finally the compost is sieved using screens to produce two grain sizes of compost.

Padova Composting Scheme, Italy

The scheme covers the Bacino Padova district of northern Italy covering 26 local municipalities, involving 205 000 inhabitants. A door-to-door collection scheme collects the biodegradable waste fraction from each household. In addition, there are centralised 'bring' schemes for garden waste, and more than 35% of households are involved in home composting. Approximately 16 500 tonnes of biodegradable waste is collected each year. The composting process involves a pre-treatment stage of open-air shredding of the waste and mixing with sewage sludge. Composting takes place in a composting hall where the biodegradable waste is placed in piles over an aerated floor where air is supplied via pipes from a blower to the piles. The piles are turned and mixed every 3–4 days for one month, to promote rapid thermophilic decomposition. The compost is then transferred to outside windrows where the compost is turned every 8–10 days in a maturation stage. The compost is finally sieved using screens to provide two grades of compost, and is stored under a roofed area.

Source: Directorate General Environment 2000.

6.6 Anaerobic Digestion

The anaerobic degradation processes found in landfills which lead to the formation of methane and carbon dioxide from organic waste are utilised in anaerobic digestion but in an enclosed, controlled reactor. The better control of the process means that all of the gas is collected for utilisation unlike landfills where collection efficiencies are relatively low at 50% or less. In addition, the process in a waste landfill typically takes many years to anaerobically degrade the biodegradable waste, but using an anaerobic digestion system, the process is complete within a period of weeks (McLanaghan 2002). The solid residue arising from anaerobic digestion can also be cured and used as a fertiliser. The main aim of the process is to produce a product gas, rich in methane, which can be used to provide a fuel or act as a chemical feedstock. Anaerobic digestion has been used to treat sewage sludge and agricultural wastes for many years and has also been developed for municipal solid wastes and industrial wastes (Verstraete and Vandeviere 1999). Figure 6.10 shows the main steps in the anaerobic digestion of biodegradable waste (IEA Bioenergy 1996; McLanaghan 2002). The biodegradable fraction of the waste requires separation from the

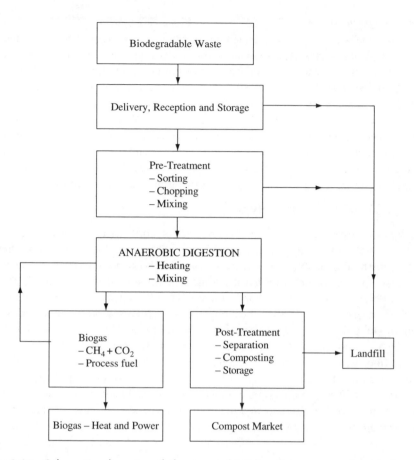

Figure 6.10 *Schematic diagram of the anaerobic digestion process for biodegradable waste. Sources: McLanaghan 2002; IEA Bioenergy 1996.*

other components of the waste. Source separation using kerbside or bring systems to civic amenity sites, or mechanical separation, may be used. In some cases biodegradable sewage sludge or agricultural wastes are processed using anaerobic digestion either as separate systems or via co-digestion with the biodegradable fraction from municipal solid waste (Verstraete and Vandeviere 1999). The biodegradable waste fraction is delivered to a reception area and stored prior to processing. Pre-treatment involves the removal of contaminants and homogenisation of the waste to aid efficient anaerobic digestion, which also protects the down-stream processes. The main stage of anaerobic digestion involves heating and mixing the waste and generates a biogas consisting of methane and carbon dioxide. The methane is combusted to produce energy which provides heat for the anaerobic digestion process and also for export to provide heat and power. The anaerobic process also serves to stabilise the waste and also to disinfect and deodorise it. Post-treatment involves removal of further contaminants, such as glass and plastics, and further stabilisation takes place through composting of the residue to produce a composted product (IEA Bioenergy 1996).

The organic components of waste can be classified into broad biological groups represented by proteins, carbohydrates and lipids or fats. Carbohydrates are by far the major component of biodegradable wastes and include cellulose, starch and sugars. Proteins are large complex organic materials composed of hundreds or thousands of amino acids groups. Lipids or fats are materials containing fatty acids. The degradation of the organic components within the anaerobic digester takes place largely by biological processes but also involves inter-related physical and chemical processes (Wheatley 1990; Diaz et al 1993).

The initial stages of the decomposition involve the hydrolysis and fermentation of the cellulosic, protein and lipid compounds in the waste by micro-organisms, the facultative anaerobes, which can tolerate reduced oxygen conditions. Carbohydrates, proteins and lipids are hydrolysed to sugars which are then further decomposed to carbon dioxide, hydrogen, ammonia and organic acids. Proteins decompose via de-aminisation to form ammonia and also carboxylic acids and carbon dioxide. Gas concentrations may rise to levels of up to 80% carbon dioxide and 20% hydrogen. The second stage of anaerobic digestion is the acid stage where organic acids formed in the hydrolysis and fermentation stage are converted by acetogen micro-organisms to acetic acid, acetic acid derivatives, carbon dioxide and hydrogen. Other organisms convert carbohydrates directly to acetic acid in the presence of carbon dioxide and hydrogen. Hydrogen and carbon dioxide levels then begin to decrease. The final stage of anaerobic digestion is the main methane gas forming stage. Low hydrogen levels promote the methane-generating micro-organisms, the methanogens, which generate methane and carbon dioxide from the organic acids and their derivatives generated in the earlier stages. There are two classes of micro-organisms which are active in the methanogenic stage, the mesophilic bacteria which are active in the temperature range 30–35 °C and the thermophilic bacteria active in the range 45–65 °C. The methanogenesis stage is the main gas generation stage of anaerobic digestion with the gas composition generated at approximately 60% methane and 40% carbon dioxide. Ideal conditions for the methanogenic micro-organisms are a pH range from 6.8 to 7.5 (Diaz et al 1993; Wheatley 1990; IEA Bioenergy 1996).

Anaerobic digestion takes place in an enclosed, closely controlled reactor. There are a range of systems available, as shown in Table 6.10. Biodegradtion takes place in a slurry of waste and micro-organisms. The yield of gas from anaerobic digestion depends on the composition and biodegradability of the waste. The rate of decomposition of the waste

Table 6.10   *Types of anaerobic digestion systems*

Dry continuous digestion
The waste is fed continuously to a digestion reactor with a digestate dry matter content of 20–40%. Both completely mixed and plug-flow systems are available, with plug-flow systems relying on external recycling of a proportion of the outgoing digested waste to be mixed with the incoming waste feedstock in order to initiate digestion.

Dry batch digestion
The waste is fed to the reactor with digested material from another reactor. The reactor is then sealed and left to digest naturally. Leachate derived from the biodegradation process is collected from the bottom of the reactor and recirculated to maintain a uniform moisture content and redistribute nutrients and micro-organisms. When digestion is complete the reactor is opened, unloaded and refilled to start the batch process again.

Table 6.10   *Continued*

Leach-bed or sequencing batch process
The process is similar to the dry batch process but the leachate derived from the biodegradation of the waste is exchanged between established and new batches of waste to facilitate start-up of the biodegradation process. After methanogenesis has become established in the waste, the reactor is uncoupled and reconnected to fresh solid waste in a second reactor.

Wet continuous digestion
The waste is slurried with a large proportion of water to provide a dilute (10% dry solids) waste feedstock that can be fed to a conventional, completely mixed digester. Effective removal of glass and stones is required to prevent accumulation in the bottom of the reactor. When used for municipal solid waste alone, filter pressing of the wet digestate to recover liquor to recycle for feed preparation is required, to avoid generating an excessive volume of diluted digestate for disposal. Alternatively, the process can be used for co-digestion with dilute wastes such as sewage sludge.

Multi-stage wet digestion
There are also a range of multi-stage wet digestion processes where municipal solid waste is slurried with water or recycled liquor and fermented by hydrolytic and fermentative micro-organisms to release volatile fatty acids, which are then converted to gas in a specialist high-rate industrial anaerobic digester.

Sources: IEA Bioenergy 1996; IEA Bioenergy 2001.

depends on the micro-organism population and temperature. The slurry consists mainly of water, and sewage sludge, agricultural wastes and some industrial wastes might have over 90% moisture content, while the biodegradable fraction of municipal solid waste might have about 60% moisture content. Lower moisture contents are preferred since they reduce the liquid effluents from the plant (Wheatley 1990; IEA Bioenergy 1996).

Operational parameters for the reactor would be controlled to give anaerobic conditions and maximum gas yield. For maximum gas production a temperature in the range 30–35 °C for mesophilic bacteria and 45–65 °C for thermophilic bacteria, would be used. Heating of the anaerobic reactor vessel is via the use of the process biogas, methane and carbon dioxide, produced from the anaerobic waste digestion process. Whilst higher gas production is found at the higher temperature range where the thermophilic bacteria are active, there is a balance against the costs of the higher energy input. Depending on the type of waste, some form of shredding may be required to increase the surface area for reaction, whereas for sewage sludge, agricultural slurries and industrial liquids/sludges, no pre-processing size reduction will be required. However, other factors such as the pH of the waste, the nutrient content and the C:N ratio are important factors which require control. For example, a high C:N ratio produces a high acid content and low methane production. The biodegradable fraction of municipal solid waste can have a high C:N ratio of above 50, whereas sewage sludge has a C:N ratio below 10. Therefore the co-digestion of biodegradable municipal solid waste and sewage sludge is common. The optimum pH in the reactor would be 7, although high gas production rates are observed between pH 6.8 and 7.5. Low pH, that is acidic conditions, inhibit the activity of the methanogenic micro-organisms, but a certain level of organic acids are required as nutrients for the methanogens (Wheatley 1990; Diaz et al 1993).

Mixing of the waste slurry within the digester is important in maintaining a high rate of anaerobic biodegradation and consequently a high production of gas. The mixing process serves to disperse the incoming waste within the actively digesting sludge, improving contact with the micro-organisms and replacing the previously degraded products with fresh nutrients in the form of the waste. Mixing also reduces any stratification in the digester. For example, different temperatures may result in differential biological activity or sedimentation, where gravity separation results in a liquid top layer with low biodegradation activity, an active biodegradation layer and a lower degraded product sludge layer. The methods of mixing include internal mechanical mixing using rotating paddles, using the product gas to recirculate through the digester, or recirculation of the slurry (Wheatley 1990).

The methanogenic micro-organisms are the centre of the process and their continued biodegradable activity is essential to the continued operation of the digester. Consequently, the pH range, temperature, nutrient level, C:N ratio, mixing, etc., would be controlled to maintain maximum activity. The initiation of the whole process is by introduction of the waste and a starter sludge of micro-organisms, for example, taken from an already operating digester or anaerobically biodegrading organic waste from other sources (Wheatley 1990; IEA Bioenergy 1996; Diaz et al 1993).

Gas production rates from anaerobic digestion depend on the starting waste materials and the operational characteristics. Table 6.11 shows the production of gas represented by methane and carbon dioxide in relation to different types of waste and operational characteristics (Diaz et al 1993). Other gases present in low concentration would be hydrogen, hydrogen sulphide, nitrogen and hydrocarbon gases at trace levels. Anaerobic digestion produces a gas which is similar in composition to landfill gas and the potential uses are similar to those of landfill gas. The gas has a calorific value of between 20 and 25 MJ/m^3 (IEA Bioenergy 1996). The gas may be used directly as replacement fuel for kilns, boilers and furnaces located close to the anaerobic digestion site. If the gas is to be used in power generation, a greater degree of gas clean-up is required, for example, to remove corrosive trace gases, moisture and vapours from the gas stream. Such possible pre-treatments may include further filtration, gas chilling to condense certain constituents, absorption and adsorption systems to scrub the gases, and other gas clean-up systems such as membranes and molecular sieves to remove trace contaminants (IEA Bioenergy 1996, 2001).

Box 6.9 describes the Valorga, France, anaerobic digestion plant for municipal solid waste (Cayrol et al 1990; de Laclos et al 1997). Box 6.10 describes the Dranco anaerobic digestion system for organic wastes (De Baere 1998).

Table 6.11 *Production of gas from the anaerobic digestion of various wastes*

Waste material	Gas production (m^3/kg dry solids)	Temperature (°C)	Methane content (%)	Retention time (Days)
Cattle manure	0.20–0.33	11.1–31.1	—	—
Poultry manure	0.31–0.56	32.6–50.6	58–60	9–30
Pig manure	0.49–0.76	32.6–32.9	58–61	10–15
Sheep manure	0.37–0.61	—	64	20
Municipal refuse[1]	0.31–0.35	35–40	55–60	15–30

[1] US refuse, estimated yield per kilogram of organic solids.

Source: Diaz et al 1993.

Box 6.9

The Valorga Anaerobic Digestion Plant for Municipal Solid Waste

The Valorga company of France has developed an anaerobic digestion plant for the treatment of municipal solid waste. The commercial plant at Amiens, north of Paris, is designed to treat 109000 tonnes of municipal solid waste per year, which includes 104000 tonnes/year domestic waste and 5000 tonnes/year of industrial effluent waste. These tonnages represent approximately 52000 tonnes of biodegradable waste processed each year. The process is a semi-continuous, high-solids, one-step, plug-flow-type process. The mixed waste is delivered to the plant and a grab transfers the waste to a crusher. The large combustible components in the waste are screened through a trommel, and ferrous metals are removed by a magnetic belt, and glass and other inert materials are removed in a densimetric sorting apparatus. The large combustible materials are combusted in a two-stage pyrolytic/combustion unit and energy is recovered via a steam boiler. The sorted organic material from the pre-processing stage is passed to the anaerobic digestion stage (methanisation unit). The waste is digested in two reactors of $3300\,m^3$ capacity each. The waste is mixed with liquid effluent from the digestor and passed into the reactor as a slurry of between 30 and 35% solids content. Operation of the reactor is under mesophilic (35–40 °C) or thermophilic (45–60 °C), conditions. The product gas produced contains between 55 and 60% methane. The gas is decarbonised to remove carbon dioxide, desulphurised to remove traces of hydrogen sulphide (3000–4000 ppm), dried and compressed, and then sold for direct use or to the national gas company of France, for injection into the public network. The residual material from the digester is pressed to reduce the moisture content to about 40% and is used as a soil conditioner. The liquid from the press is recirculated with the incoming organic fraction of the waste back into the digester. The plant is estimated to produce about $6550\,m^3$ product gas/day.

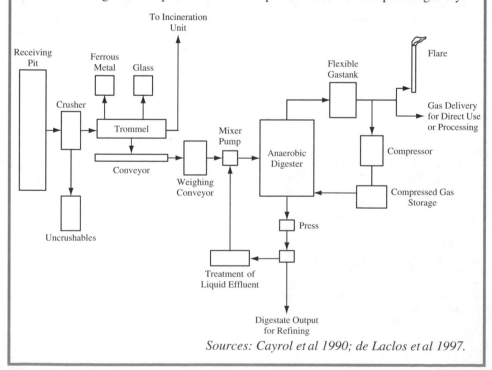

Sources: Cayrol et al 1990; de Laclos et al 1997.

Box 6.10

The Dranco Anaerobic Digestion Plant for Biodegradable Waste

The Dranco anaerobic digestion plant has been used to process biowaste, including biodegradable waste from municipal solid waste. The system involves a pre-treatment stage and a single-stage thermophilic, anaerobic-digestion step, followed by a short aerobic composting stage. Pre-treatment involves particle-size reduction to less than 40 mm and also removal of inert material. Anaerobic digestion takes place in an upright, closed reactor which is fed from the top. The digested residual material is extracted from the bottom of the reactor. The input waste is mixed with the digester residue prior to injection into the reactor to promote the anaerobic digestion process. During the mixing process, steam is added to the mixing unit to raise the temperature of the mixture to about 55 °C which promotes the thermophilic micro-organism degradation. Biogas, consisting of methane and carbon dioxide, is produced, cleaned and used for heat and power production. The digested residual material is de-watered to about 50% dry matter content and composted under aerobic conditions for a period of about two weeks. The final product is used as a soil conditioner.

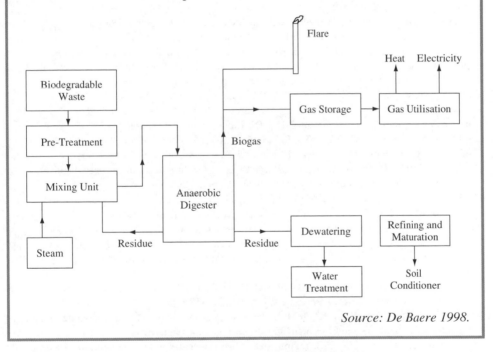

Source: De Baere 1998.

The end products from anaerobic digestion depend on the type of digestion process. The main product from the dry digestion of waste is a solid residue or digestate which can be matured into a compost product. Small quantities of surplus liquor are also produced which are generally similar to a dilute digested sewage sludge. Products from the wet digestion process are generally similar to a concentrated digested sewage sludge. They can be spread directly onto farmland or de-watered to provide separate liquid fertiliser and solid compost products.

Anaerobic digestion of sewage sludge and agricultural waste is a well established technology, although less so for the biodegradable fraction of municipal solid waste.

However, it has been estimated that worldwide there are more than 125 anaerobic digestion plants treating biodegradable municipal solid waste (IEA Bioenergy 2001). In Europe, Denmark, Germany, Italy, the Netherlands, Sweden and Switzerland have fully developed anaerobic digestion plants for handling organic wastes. For example, Denmark treats manure, organic industrial waste and biowaste totalling 1.1 million tonnes/year, Germany treats sewage sludge, manure, biowaste and organic industrial waste totalling 360 000 tonnes/year and Italy handles manure, mixed waste, organic industrial waste, and manure totalling 660 000 tonnes/year (IEA Bioenergy 1996). The application of anaerobic digestion for biodegradable municipal solid waste is also a growing process in Europe, with an estimate of more than 1 million tonnes processed in more than 50 plants across Europe (De Baere 2000).

References

Alstom 2001. Alstom Power Ltd. company literature, Alstom Power Ltd, Baden, Switzerland.

Bagri R. and Williams P.T. 2002. Journal of Analytical and Applied Pyrolysis, 63, 29–41.

Border D. 1995. The Waste Manager, 12–15 March.

Bridgwater A.V. and Bridge S.A. 1991. In, A.V. Bridgwater and G. Grassi (Eds.), Biomass Pyrolysis Liquids, Upgrading and Utilisation. Elsevier Applied Science, London.

Bridgwater A.V. and Evans G.D. 1993. An assessment of thermochemical conversion systems for processing biomass and refuse. Energy Technology Support Unit, Harwell, Report ETSU/T1/00207/REP, Department of Trade and Industry.

Cayrol F., Peillex J.-P. and Desbois S. 1990. In, Biomass for energy and Industry, 5th EC Conference, Grassi G., Gosse G. and dos Santos G. Elsevier Applied Science, London.

Cunliffe A.M., Jones N. and Williams P.T. 2003. Environmental Technology, 24, 653–663.

Cunliffe A.M. and Williams P.T. 1998. Environmental Technology, 19, 1177–1190.

Cunliffe A.M. and Williams P.T. 1999. Energy and Fuels, 13, 166–175, 1999.

De Baere L. 1998. Development of anaerobic digestion technologies in Europe: Applications to different types of waste. Organic Waste Systems, Gent, Belgium.

De Baere L. 2000. Water Science and Technology, 41, 283–290.

de Bertoldi M., Vallini G. and Pera A. 1983. Waste Management and Research, 1, 157–176.

de Laclos H.F., Desbois S. and Saint-Joly C. 1997. Water Science and Technology, 36, 457–462.

Desbene P.-L., Essayegh M., Desmazieres B. and Basselier J.-J. 1991. In, A.V. Bridgwater and G. Grassi (Eds.), Biomass Pyrolysis Liquids, Upgrading and Utilisation. Elsevier Applied Science, London.

Diaz L.F., Savage G.M., Eggerth L.L. and Golueke C.G. 1993. Composting and Recycling of Municipal Solid Waste. Lewis Publishers, Boca Raton, USA, 1993.

Directorate General Environment. 2000. Success stories on composting and separate collection. Directorate General for the Environment, European Commission, Brussels.

Drost U. and Kaiser H. 2001. Proceedings of the 3rd International Symposium on Waste Treatment Technologies, Brussels. Institution of Chemical Engineers Publications, Rugby, Vol. 1, p. 10.

EC Biological Treatment of Biowaste Directive. 2001. Proposed EC Directive. Brussels, Belgium.

EC Waste Incineration Directive. 2000. Directive 2000/76/EC of the European Parliament and of the Council on the Incineration of Waste. European Commission, Brussels.

EC Waste Landfill Directive. 1999. Council Directive 1999/31/EC, L 182/1, 26.4.1999, Landfill of Waste Official Journal of the European Communities, Brussels, Belgium.

Eitzer B.D. 1995. Emissions of volatile organic chemicals from municipal solid waste composting facilities. Environmental Science and Technology, 29, 896–902.

European Commission. 2004. Integrated pollution prevention and control. Draft reference document on the best available techniques for waste incineration. European Commission, Brussels.

European Environment Agency. 2002. Biodegradable municipal waste management in Europe, Part 1 Strategies and instruments. European Environment Agency, Copenhagen.

Francis W. and Peters M.C. 1980. Fuels and Fuel Technology. Pergamon Press, Oxford.

Graham R.G., Bergougnou M.A. and Overend R.P. 1984. Journal of Analytical and Applied Pyrolysis, 6, 95–135.

Hannequart J.P. and Radermaker F. 2003. Methods, costs and financing of waste collection in Europe: General review and comparison of various national policies. Association of Cities and Regions for Recycling (ACRR) in Europe, Brussels, Belgium.

Hogg D., Barth J., Favoino E., Centemero M., Caimi V., Amlinger F., Devliegher W., Brinton W. and Antler S. 2002. Comparison of compost standards within the EU, North America and Australasia. Waste and Resources Action Programme, Banbury, UK.

Horne P.A. and Williams P.T. 1996. Fuel, 75, 1051–1059.

IEA Bioenergy. 1996. Bioenergy, biogas from municipal solid waste: Overview of systems and markets for anaerobic digestion of MSW. IEA Bioenergy, Energy Recovery from MSW Task Anaerobic Digestion Activity. National renewable energy Laboratory, Harwell, 1996.

IEA Bioenergy. 2001. Bioenergy, biogas and more: Systems and markets overview of anaerobic digestion. IEA Bioenergy. AEA Technology, Culham, Oxfordshire, UK.

Juniper. 2000 Pyrolysis and Gasification of Waste. Juniper Consultancy Services Ltd, Uley, Gloucestershire, UK.

Kaminsky W. and Sinn H. 1980. In, Jones J.L. and Radding S.B. (Eds.), Thermal conversion of solid wastes and biomass., ACS Symposium Series 130. American Chemical Society Publishers, Washington DC.

Lee M. 1995. Chemistry in Britain, July 1995, 515–516.

Leipnitz Y. 1995. Combined pyrolysis and gasification, Noell Abfall Und Energietechnik GMBH, Germany. In, The Future of European thermal Waste Treatment. Librairie des Enterprises, Conference 7/8 September, Paris.

McLanaghan S.R.B. 2002. Delivering the Landfill Directive: The role of new and emerging technologies. Associates in Industrial Ecology, Penrith, UK.

Menachem L. and Goklestein I. 1984. International Fibre Science, Vol 11, Wood Structure and Composition. Marcel Dekker Inc., New York.

Milner P.D., Olenchock S.A., Epstein E., Rylander R., Haines J., Walker J., Ooi B.L., Horne E. and Maritato M. 1994. Compost Science and Utilisation, 2, 6057.

Newport H.A., Bardos R.P., Hensler K., Goss E., Willett S. and King L. 1993. Municipal Waste Composting, Report No. CWM/074/93. DoE, HMSO, London.

Pakdel H., Roy C., Aubin H., Jean G. and Coulombe S. 1991. Environ. Sci. Technol. 25, 1646–1649.

Pober K.W. and Bauer H.F. 1977. The nature of pyrolytic oil from municipal solid waste. In, Anderson and Tillman (Eds.), Fuels from Waste. Academic Press, London.

Rampling T.W.A. 1993. Downdraft fixed bed gasification, Paper W93058, Presented to the Department of Trade and Industry, Renewable Energy Programme Advanced Workshop. Natural Resources Institute, Chatham, Kent, December.

Rampling T.W. and Hickey T.J. 1988. The laboratory characterisation of refuse-derived fuel. Warren Spring Laboratory Report, LR643. Department of Trade and Industry, HMSO, London.

Reed A.R. and Williams P.T. 2004. International Journal of Energy Research, 28, 131–145.

Rick F. and Vix U. 1991. In, Bridgewater A.V. and Grassi G., Biomass Pyrolysis Liquids Upgrading and Utilisation. Elsevier Applied Science, London.

Roy C. 1994. In, Wallonia, International Meeting, Thermolysis: a technology for recycling and depollution, 24–25 March. ISSeP, Liege, Belgium.

Stoikos T. 1991. In, A.V. Bridgwater and G. Grassi (Eds.), Biomass Pyrolysis Liquids, Upgrading and Utilisation. Elsevier Applied Science, London.

Swan J.R.M., Crook B. and Gilbert E.J. 2002. Issues in Environmental Science and Technology, 18, 73–102.

Thermoselect. 2003. Thermoselect SA, Locarno, Switzerland.

Verstraete W. and Vandevivere P. 1999. Critical Reviews in Environmental Science and Technology, 28, 151–173.

Warmer Bulletin 29. 1991. Compost, Warmer Fact Sheet. The World Resource Foundation, Tonbridge, Kent, UK.

Wheatley A. 1990. Anaerobic Digestion: A Waste Treatment Technology. Elsevier Applied Science, London.

Whiting K.J. 2001. Proceedings of the 3rd International Symposium on Waste Treatment Technologies, Brussels. Institution of Chemical Engineers Publications, Rugby, Vol. 1, p. 13.

Whiting K.J. 2003. Incineration of Municipal Solid Waste, University of Leeds Short Course. University of Leeds.

Williams P.T. 2001. The Chemical Engineer, 717, 24–27.

Williams P.T. 2003. Proceedings of the Fourth International Symposium on Waste Treatment Technologies, Sheffield. Institution of Chemical Engineers Publications, Rugby, Vol. 1, p. 12.

Williams P.T. and Bagri R. 2004. International Journal of Energy Research, 28, 31–44.

Williams P.T. and Besler S. 1992. Journal of the Institute of Energy, 65, 192–200.

Williams P.T. and Besler S. 1996. Renewable Energy, 7, 233–250.

Williams P.T. and Brindle A.J. 2002. Waste Management and Research, 20, 546–555.

Williams P.T. and Brindle A.J. 2003. Journal of Analytical and Applied Pyrolysis, 67, 143–164.

Williams P.T. and Horne P.A. 1994. Biomass and Bioenergy, 7, 223–236.

Williams P.T. and Reed A.R. 2003. Journal of Analytical and Applied Pyrolysis, 70, 563–577.

Williams P.T. and Reed A.R. 2004. Journal of Analytical and Applied Pyrolysis, 71, 971–986.

Williams E.A and Williams P.T. 1997. Journal of Chemical Technology and Biotechnology, 70, 9–20.

Williams P.T., Besler S., Taylor D.T. and Bottrill R.P. 1995. Journal of the Institute of Energy, 68, 11–21.

7

Integrated Waste Management

Summary

This concluding chapter discusses the integration of treatment and disposal options described in the previous chapters to introduce the concept of 'integrated waste management'. The different approaches to integrated waste management are described.

7.1 Integrated Waste Management

The treatment and disposal of waste has developed from its early beginnings of mere dumping to a sophisticated range of options including re-use, recycling, incineration with energy recovery, advanced landfill design and engineering and a range of alternative technologies, including pyrolysis, gasification, composting and anaerobic digestion. The further development of the industry is towards integration of the various options to produce an environmentally and economically sustainable waste management system.

Integrated waste management has been defined as the integration of waste streams, collection and treatment methods, environmental benefit, economic optimisation and

Waste Treatment and Disposal, Second Edition   Paul T. Williams
© 2005 John Wiley & Sons, Ltd   ISBNs: 0-470-84912-6 (HB); 0-470-84913-4 (PB)

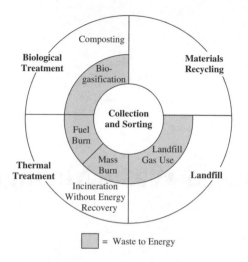

Figure 7.1   *Elements of an integrated waste management system. Source: Warmer Bulletin 49, 1996; McDougall et al 2001. Reproduced by permission of R.C. Strange.*

societal acceptability into a practical system for any region (Warmer Bulletin 49, 1996). Integrated waste management implies the use of a range of different treatment and disposal options, including the areas covered in this book, i.e., waste reduction, re-use and recycling, landfill, incineration and alternative options such as pyrolysis, gasification, composting and anaerobic digestion. However, integration also implies that no one option of treatment and disposal is better than another and each option has a role to play, but that the overall waste management system is the best environmentally and economically sustainable one for a particular region (Figure 7.1, Warmer Bulletin 49, 1996; McDougall et al 2001).

Environmental sustainability means that the options and integration of those options should produce a waste management system that reduces the overall environmental impacts of waste management, including energy consumption, pollution of land, air and water and loss of amenity (White et al 1995; Warmer Bulletin 49, 1996). Economic sustainability means that the overall costs of the waste management system should operate at a cost level acceptable to all areas of the community, including householders, businesses, institutions and government (White et al 1995; Warmer Bulletin 49, 1996). In assessing the most environmentally and economically sustainable system, the local existing waste management infrastructure, such as availability of landfill sites, existing incinerators, the types of waste to be managed, waste tonnages generated, etc., should all be considered.

Figure 7.1 (Warmer Bulletin 49, 1996; McDougall et al 2001) shows that at the centre of an integrated waste management system is the collection and sorting of the waste, since this influences the treatment and disposal options of the waste, for example, recycling, composting, use for energy recovery, etc. (White et al 1995). Materials recycling enables the useable materials of the waste to be removed, e.g., paper, glass, metals, etc., in a materials recycling facility. The residual waste may then be processed as refuse derived fuel or combusted in an incinerator to recover energy. The waste may be land-filled to produce landfill gas and energy recovered from the combustion of the derived gas. Biological treatment of the waste via anaerobic digestion to produce a combustible gas, or treatment to produce compost, may also be an option. In most cases the treatment options require landfill as a final disposal route for the residual product. An integrated waste management system would include one or all of the above options (White et al 1995).

Integrated waste management may also be interpreted as integration in terms of the management of wastes from different sources such, as commercial, household and indus-trial, or else in terms of different materials, such as metals, paper and putrescible wastes, or of waste from different product areas, such as packaging waste, white goods, etc. (Warmer Bulletin 49, 1996). In a truly integrated waste management system, wastes such as demolition products, sewage sludge, hazardous, agricultural, industrial and household wastes, would all be included in the waste management system. However, such diverse wastes are often covered by different authorities, are subject to different legislation and arise in different amounts, and are therefore more difficult to integrate than, for example, 'municipal solid waste'.

Tchobanoglous et al (1993) define integrated waste management in terms of the inte-gration of six functional elements.

1. Waste generation – Assessment of waste generation and evaluation of waste reduction.
2. Waste handling and separation, storage and processing at the source – Involves the activities associated with the management of wastes until they are placed in storage containers for collection. This may include source separation of household waste into recyclable and non-recyclable materials. Provision for suitable storage for the wastes, which may encompass a wide variety of different types, is also part of this element. Processing includes such processes as compaction or composting of putrescible materials.
3. Collection – This element of the waste management system covers the collection and transport of the waste to the location where the collection vehicle is emptied. This location may be for example, a materials recycling facility, waste transfer station or landfill disposal site.
4. Separation, processing and transformation of solid waste – The recovery of separated materials, the separation and processing of waste components and transformation of wastes are elements which occur primarily in locations away from the source of waste generation. This category includes waste treatment at materials recycling facilities, activities at waste transfer stations, anaerobic digestion, composting and incineration with energy recovery.
5. Transfer and transport – This element involves the transfer of wastes from the smaller collection vehicles to the larger transport equipment and the subsequent transport of the wastes, usually over long distances, to a processing or disposal site. The transfer usually takes place at a waste transfer station.
6. Disposal – Final disposal is usually landfill or landspreading, i.e., the disposal of waste directly from source to a landfill site, and the disposal of residual materials from

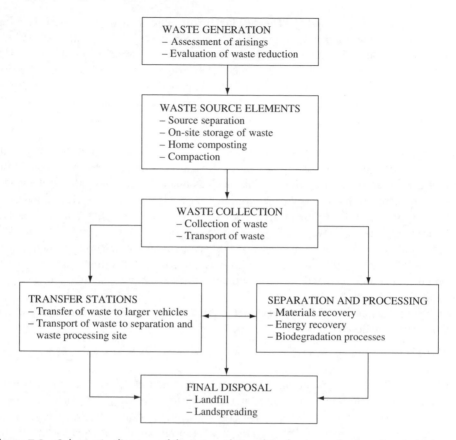

Figure 7.2   *Schematic diagram of the inter-relationships between the functional elements in a solid waste management. Source: Tchobanoglous et al 1993.*

materials recycling facilities, residue from waste incineration, residues from composting or anaerobic digestion, etc., to the final disposal in landfill.

The inter-relationships of the six functional elements of an integrated solid waste management system are shown in Figure 7.2 (Tchobanoglous et al 1993).

Integrated waste management as described by Tchobanoglous et al (1993) involves evaluation of the use of the functional elements and the effectiveness and economy of all the interfaces and connections, to produce an integrated waste management system. They define integrated waste management as the selection and application of suitable techniques, technologies and management programmes, to achieve specific waste management objectives and goals.

Examples of integrated waste management systems are described in Boxes 7.1–7.5 (Riley 1996; Collis 2002; ASSURE 1998).

Box 7.1
Integrated Waste Management in Hampshire, UK

Project INTEGRA is an integrated waste management plan for Hampshire, UK. The project involves waste minimisation, recycling, composting, anaerobic digestion, incineration with energy recovery and landfill with energy recovery from landfill gas. In terms of waste management Hampshire is divided into three regions, north, south east and south west. The total waste arisings from the population of about 1.6 million is 600 000 tonnes per year which is predicted to rise to 750 000 per year by the year 2010. The project aims to recycle 25% of the waste with a future target option of 40% recycling. Recycling will include materials recycling, composting, anaerobic digestion and energy from waste. Household waste recycling is currently carried out at three materials recycling facilities, which together recover approximately 13 000 tonnes/year of glass, plastics, paper and card, aluminium and steel. The collection system uses different recycling collection systems throughout the county including twin bin, split vehicles, bags, banks, trailers, boxes, etc. The material recycling facilities have the capability to handle all these types of input. Composting is to be carried out on a large scale of an approximate throughput of 50 000 tonnes/year. Anaerobic digestion of the biodegradable waste, segregated from the waste stream after the removal of dry recyclable materials, will be treated in large digesters and will generate methane gas for energy recovery. Waste incineration is also to be part of the waste management package, with three medium sized mass-burn incinerators located throughout the county, with throughputs of 90 000 tonnes in the north region, 165 000 in the South East region and 105 000 in the south west region. Landfill of waste with energy recovery by landfill gas utilisation is also an option, with several sites throughout the county. The projected throughputs of waste into the various sectors at a target of 25% waste recycling and the future option of a 40% target, is shown below, for an estimated waste generation of 750 000 tonnes/year by the year 2010.

Type of recycling	25% Recycling target		40% Recycling target	
	Tonnes/year (1000 s)	%	Tonnes/year (1000 s)	%
Dry Recycling	123	16	235	31
Composting	50	7	50	7
Anaerobic digestion	15	2	15	2
Energy from waste	360	48	360	48
Landfill	202	27	90	12
Total	750	100	750	100

Source: Riley 1996; Collis 2002.

Box 7.2

Integrated Waste Management in Copenhagen, Denmark

Approximately 870 000 tonnes per year of municipal solid waste is generated in Copenhagen by a population of about 555 000 inhabitants. Householders are required by law to source separate their waste. Commercial waste must also be source separated. Organic waste and residual waste are collected from households. The collected organic waste is composted in centralised composting facilities using the windrow method. Approximately 60 000 tonnes of organic waste are processed per year resulting in 25 000 tonnes of compost each year. There is also an extensive bring system with kerbside containers for glass, newspaper, magazines, cardboard and hazardous waste, located in local neighbourhoods. All the source segregated recyclable material is sent to material recycling facilities. The centralised recycling facility is a very large operation, processing more than 500 000 tonnes of material each year, including more than 300 000 tonnes of construction waste. The residue from the recycling facilities, together with the residual waste collected from households, is incinerated in two incinerators. Similarly, residual commercial and industrial waste is incinerated. The throughput of the incinerators is approximately 240 000 tonnes per year. The energy generated from the Copenhagen incinerators is used for district heating, but they are under development for combined heat and power. The increase in recycling has resulted in a reduction of the non-combustible fraction of municipal solid waste, increasing the calorific value of the waste. The incinerator ash is used for building material.

The integrated management of the 870 000 tonnes of municipal solid waste results in only 35 000 tonnes of inert material going to landfill, approximately 4% of the total.

Source: ASSURE 1998.

Box 7.3

Integrated Waste Management in Brescia, Italy

The population of Brescia in Northern Italy comprises about 192 000 inhabitants and 13 000 commercial businesses, together generating 113 000 tonnes of municipal solid waste each year. Approximately 60% of municipal solid waste comes from households and 40% from commerce. The composition of the municipal solid waste is 40% organics, 4% green waste, 10% glass, 18% cardboard, 4% textiles and wood, 4% metals, 8% plastics and rubber and 12% others. The waste collection in Brescia is arranged around kerbside collection points for both refuse and recyclable materials. Collections are very frequent at six times per week. Organic material from large commercial enterprises and from household collection is composted in a central facility, processing 9500 tonnes of organic waste to produce about 4000 tonnes of compost each year. A 260 000 tonne per year incinerator plant is under development, generating heat for the district heating system and also for power generation using a combined heat and power system. Recycling recovers glass, metal, paper, plastics, and textiles/wood, totalling approximately 12 500 tonnes of recyclable materials. The residual waste is landfilled with the production of landfill gas, which is used to generate electricity.

Source: ASSURE 1998.

Box 7.4
Integrated Waste Management in Malmo, Sweden

The Malmo region of Sweden includes Malmo and Lund and contains 500 000 inhabitants. The waste generated from households is about 230 000 tonnes per year and, together with commercial and industrial waste, the total municipal and industrial waste management amounts to 550 000 tonnes per year. Home composting of organic waste is very common, with estimates of 14% of all homes involved. The waste collection system is dependent on the different municipalities. In Lund, bi-weekly collection of refuse, glass and paper is carried out. The refuse is incinerated. In addition, recyclable waste collection centres are able to collect glass, paper, metal, plastics, batteries, etc. The inhabitants of Malmo use two central bring stations to recycle waste, including glass, plastic, metals, paper and board, and also to dispose of hazardous waste and large objects. The residual waste is collected on a weekly basis and sent directly for incineration. Recycling centres are located throughout the region which handle 45 000 tonnes of recyclable material each year. There is also a collection system for organic waste which is sent for central composting, handling 16 000 tonnes per year of organic waste and generating 5000 tonnes of compost. The waste incinerator in the Malmo region has a capacity of 220 000 tonnes per year, handling about 60% household waste and 40% commercial/industrial waste. The bottom ash is sorted to recover 6700 tonnes of metal and the residue, amounting to 28 000 tonnes per year, is used in the secondary aggregates industry. The energy generated is used in the district heating system for the Malmo region. A significant proportion of the municipal solid waste generated, 156 000 tonnes is landfilled, with recovery of landfill gas for energy utilisation.

Source: ASSURE 1998.

Box 7.5
Integrated Waste Management in Saarbrucken, Germany

The city of Saarbrucken in Germany has 187 000 inhabitants. Total waste generation from households, commerce and industry, handled by the municipality, comprises 102 000 tonnes per year. Collection from households is on a door-to-door basis, additionally 30% of households have a bi-weekly collection of organic and garden waste, although no meat or cooked product waste is allowed. Packaging materials are collected in a yellow-sack system twice per month from households and also paper, glass and textiles are collected from large containers at bring centres. The total packaging waste collected from households and centres amounts to 28 000 tonnes per year. The non-recyclable waste is collected from households and sent for incineration. The recyclable waste is sorted in a central materials recycling facility where the waste is sorted, mainly by hand, into paper, composite containers, mixed plastics, plastic film and aluminium and ferrous fractions. There is also a sorting plant for paper which separates the paper into mixed cardboard, carton, printed and newspaper fractions. The organic and garden waste collected from

Continued on page 374

Continued from page 373

households is composted in a central composting facility which handles approximately 8400 tonnes per year and generates about 3000 tonnes per year of compost. The compost process takes place in a covered hall and generates high quality compost. The waste incinerator for non-recyclable waste is of 110 000 tonne per year capacity, taking about 20 000 tonnes per year of Saarbrucken waste. The remaining capacity is taken up by other municipalities. Waste is also landfilled generating landfill gas, which is used for energy recovery to produce electricity.

Source: ASSURE 1998.

Bibliography

ASSURE. 1998. Towards integrated management of municipal solid waste. Association for Sustainable Use and Recovery of Resources in Europe (ASSURE), Brussels, Belgium.

Collis J. 2002. Incineration of municipal solid waste, Industrial Short Course. University of Leeds, UK.

McDougall F.R., White P.R., Franke M. and Hindle P. 2001. Integrated Solid Waste Management: A Lifecycle Inventory. Blackwell Science, London.

Riley K. 1996. Hampshire Waste Services. Otterbourne, Hampshire.

Tchobanoglous G., Theisen H. and Vigil S. 1993. Integrated Solid Waste Management: Engineering Principles and Management Issues. McGraw-Hill Inc., New York.

Warmer Bulletin 49. 1996. White P. So what is integrated waste management? Journal of the World Resource Foundation. Tonbridge, Kent.

White P., Franke M. and Hindle P. 1995. Integrated Solid Waste Management: A Lifecycle Inventory. Blackie Academic and Professional, London.

Index

Waste Treatment and Disposal, Second Edition   Paul T. Williams
© 2005 John Wiley & Sons, Ltd   ISBNs: 0-470-84912-6 (HB); 0-470-84913-4 (PB)